CELL ADHESION MOLECULES

MOLECULES

Cellular Recognition Mechanisms

A Continuation Order Plan is available for this series. A continuation order will bring delivery of each new volume immediately upon publication. Volumes are billed only upon actual shipment. For further information please contact the publisher.

CELL ADHESION MOLECULES

Cellular Recognition Mechanisms

Edited by

Martin E. Hemler

Dana–Farber Cancer Institute
Boston, Massachusetts

and

Enrico Mihich

Roswell Park Cancer Institute
Buffalo, New York

SPRINGER SCIENCE+BUSINESS MEDIA, LLC

Library of Congress Cataloging in Publication Data

Cell adhesion molecules: cellular recognition mechanisms / edited by Martin E. Hemler and Enrico Mihich.

    p.   cm.—(Pezcoller Foundation symposia; 4)

    "Proceedings of the Pezcoller Foundation Symposium on Adhesion Molecules: Cellular Recognition Mechanisms, held June 24–26, 1992, in Rovereto, Italy"—T.p. verso.

    Includes bibliographical references and index.

    ISBN 978-1-4613-6224-1

    1. Cell adhesion molecules—Congresses. I. Hemler, Martin E. II. Mihich, Enrico. III. Pezcoller Foundation Symposium on Adhesion Molecules: Cellular Recognition Mechanisms (1992: Rovereto, Trento, Italy) IV. Series.

QP552.C42C45   1993   93-18554

574.87—dc20   CIP

Proceedings of the Pezcoller Foundation Symposium on
Adhesion Molecules: Cellular Recognition Mechanisms,
held June 24–26, 1992, in Revereto, Italy

ISBN 978-1-4613-6224-1   ISBN 978-1-4615-2830-2 (eBook)
DOI 10.1007/ 978-1-4615-2830-2

© 1993 Springer Science+Business Media New York
Originally published by Plenum Press, New York in 1993
Softcover reprint of the hardcover 1st edition 1993

Professor Alessio Pezcoller
(Photo by Dino Panato, Trento, Italy)

THE PEZCOLLER FOUNDATION

The Pezcoller Foundation was created in 1979 by Professor Alessio Pezcoller (1896-1993), who was the chief surgeon of the S.Chiara Hospital in Trento from 1937 to 1966 and who gave a substantial portion of his estate to support its activities; the Foundation also benefits from the cooperation of the Saving Bank Cassa di Risparmio di Trento e Rovereto.

The main goal of this non-profit foundation is to provide and recognize scientific progress on life-threatening diseases, currently focusing on cancer. Towards this goal, the Pezcoller Foundation awards, every two years, the Pezcoller Prize, recognizing highly meritorious contributions to medical research; it also sponsors a series of annual symposia promoting interactions among scientists working at the cutting edge of basic oncological sciences.

The award selection process is managed by the European School of Oncology in Milan, Italy, with the aid of an international committee of experts chaired by Professor U. Veronesi.

The symposia are held in the Trentino Region of Northern Italy and their scientific focus is selected by Enrico Mihich with the collaboration of an international Standing Symposia Committee. A Program Committee determines the content of each symposium.

The first symposium focused on *Drug Resistance: Mechanisms and Reversal* (E. Mihich, Chairman, 1989); the second on *The Therapeutic Implications of the Molecular Biology of Breast Cancer* (M.E. Lippman and E. Mihich, Co-Chairmen, 1990); and the third on *Tumor Suppressor Genes* (D.M. Livingston and E. Mihich, Co-Chairmen, 1991). The fifth symposium (1993) will be focused on apoptosis (E. Mihich and R.T. Schimke, Co-Chairmen).

Starting from this, the fourth symposium, the proceedings are published by Plenum Press, New York.

PREFACE

The Fourth Annual Pezcoller Symposium entitled Adhesion Molecules: Cellular Recognition Mechanisms was held in Rovereto, Italy, June 24-26, 1992 and was focussed on the detailed mechanisms whereby cells utilize certain integral membrane proteins to perceive their surrounding environment and interact with it. With timely presentations and stimulating discussions this Symposium addressed the genetics and biochemistry of adhesion molecules, the regulation of their functions and their role in cancer and the immune system. Emphasis was given to adhesion proteins in the integrin family because of the widespread distribution of this group of molecules and its important role in essentially all eukaryotic biological systems.

The regulation of integrin genes and their expression are discussed in detail, as are specific aspects of the genetics of fibronectin. The molecular basis for the regulation of certain integrins, the function of these proteins in determining cell adhesion, and the consequences of this adhesion for the function of the cells involved are discussed. The role of certain integrins in stimulating signal transduction, the essential involvement of integrins in conditioning the function of T and NK cells function, the heterogeneity of integrins and its biological consequences, and the role of cell adhesion molecules in tumor cells invasion and metastases are all extensively analyzed. New information was presented on the role of CD44 and splice variants in normal differentiation and tumor progression. From the proceedings of this Symposium, it is clear that outstanding progress is being made in the field of cell adhesion research. With the increasing identification of relevant molecules, it has become possible to carry out precise molecular, genetic and biochemical studies such as described herein.

We wish to thank the participants in the Symposium for their substantial contributions and their participation in the spirited discussions which followed. We would also like to thank Drs. G. Della Porta, A. Levine, D. Livingston and G. Nicolson for their valuable input as members of the Program Committee, Dr. E. Kastelec (Rome) for her essential assistance in the organization of the Symposium, and Ms. A. Toscani (Roswell Park Cancer Institute) for her invaluable assistance in securing the prompt publication of this volume. The aid of the Bank Cassa di Risparmio di Trento and Rovereto, and the Municipal, Provincial and Regional Administrations in supporting this Symposium through the Pezcoller Foundation are also acknowledged with deep appreciation.

Finally, we wish to thank the staff of Plenum Publishing Corporation for their invaluable cooperation in the production of these Proceedings.

Martin Hemler
Enrico Mihich

CONTENTS

TRANSCRIPTIONAL REGULATION OF THE αIIβ INTEGRIN GENE

Gerard Marguerie and Georges Uzan

Laboratoire d'Hématolgoie
Départmement de Biologie Molécularire et Structurale
INSERM unité 217
CENG 85 X, 38041 Grenoble, Cedex, FRANCE

INTRODUCTION

The commitment of a totipotent hematopoietic stem cell and its terminal differentiation into the different hematopoietic lineages are certainly controlled by switching on and off a number of genes and by controlling the level of transcription of these genes. Despite the large amount of information that has been accumulated on the role of growth factors and cytokines in haematopoiesis, the role of transcriptional factors that are implicated in the developmental program of haematopoietic cells is less understood. Within this process, the megakaryocytic and erythroid lineage have multiple features in common. Both cells express receptors for erythropoietin, which is one of the major regulatory factors of erythrocyte production (Fraser et al., 1989). Erythroid specific transactivating factors are also produced in megakaryocytes and this includes GATA1 (Martin et al., 1990, Romeo ct al., 1990), and NFE6 (Colin et al., 1990). Finally, the existence of a bipotent erythro-megakaryocytic burst forming cell has been suggested by clonal assays, using chromosomal markers (McLeod et al., 1989). Thus, there is ample evidence to suggest that the erythroid and megakaryocyte lineage are genetically related.

Different strategies can be used to identify the mechanisms which control this development phenomenon. One way would consist in examining the role of specific integrin molecules. Adhesion to the stroma cells and their extra-cellular matrices is probably part of these mechanisms. Integrin mediated adhesion can either locate progenitor cells in proximity to high concentrations of specific growth factors or transduce a signal that ultimately will lead to gene activation (Guau et al., 1991). On the other hand lineage restricted factors implicated in the establishment of a phenotype can be activated through their binding to cis-acting elements of promoter regions of cell specific genes. In our studies of the megakaryocytic development leading to the production of circulating platelets, we use the gene coding for the αIIβ molecule as a specific marker.

The molecule is the chain of the arginine-glycine-aspartic acid (RGD) sensitive platelet integrin αIIβ3, which expresses receptor function for fibrinogen, fibronectin and von Willebrand factor (Plow et al., 1987), and is exclusive implicated in platelet aggregation. While the β3 subunit is expressed in a variety of cells, including megakaryocyte fibroblasts, endothelial cells and macrophages, the αIIb mRNA is only produced in megakaryocyte and at an early stage of megakaryocytopoiesis (Molla et al., submitted [1992]). For these reasons we have focused our attentions on the transcriptional regulation of this gene, which constitutes a good candidate to analyze megakaryocyte development.

```
                    GAATTCCCCGGATCAGAAAATAGAAATCAAAAGGAAAATGTGGCTATGGTTAC-1201
CCCTAGCGGACCTCTTAAATCTTCCTGAGAACCTGCTTTTTTGGGAAGGCATGAGTGCCAGTAAGACTTGGCACTCCTCC-1121
TCTTCCGCTTACCGAGAGAAAATGACTTTGCCTTTCTGCTCAAAACTCATCCCTTCACTTTGTCACCCTATGTTTGCATC-1041
TTCCATCCTTAGTGTGTGTTTCCATCCAGTCTTTCAGCAATACACGTACTACACATTGGACTCTTGGGTAGTCTCTAGGG -961
CTGTAGCAAGGAGCCTTGCTCCCAAGGGACTCATTTACACAATCCTGTGAACGGACCAAGAGTAAACAGTGTGCTCAATG -881
```
                        Bgl I
                          ▼
```
CTGTGCCTACGTGTGTTAGCCCACGCGGCCAGCCTGAGGAGTCAGGGAAGGCTCCCCTAGGCAAAGCCCCAACCAGAATC -801
AAGTCTTAATGGTTAAAGAGCTCCATCACCCAAAAAGGATTGAGGGCCTACCTTCAACTGAACAGCTAATGCATAATCTC -721
AGAAACTGTGAGTCAAAATTCCCTGGAATAACTCCACTTTATCCCCAATCTCCTTGCCACCTAGACCAAGGTCCATTCAC -641
```
                                                                        ets
```
CACCCTGTCCCCAGCACTGACTGCACTGCTGTGGCCACACTAAAGCTTGGCTCAAGACGGAGGAGGAGTGAGGAAGCTGC -561
```
                                                                        IR1
```
                                      ets
TGCACCAATATGGCTGGTTGAGG[CCGCCC]AAGGTCCTAGAAGGAGGAAGTGGGTAAATGCCATATCCAAAAAGATACAGA -481
→
```
                                                          PvuII
                                                            ▼
```
AGCCTCAGGTTTTATCGGGGGCAGCAGCTTCCTTCTCCTTCCCCGACCTGTGGCCAAGTCACAAAGCACCACAGCTGTAC -401
    IR1
AGCCAGATGGGGGAAGGGAGGAGATTAGAACTGTAGGCTAGAGTAGACAAGTATGGACCAGTTCACAATCACGCTATCCC -321
AAGCCAGAAAGTGATGGTGGCTTGGACTAGCACGGTGGTAGTAGAGATGGGGTAAAGATTCAAGAGACATCATTGATAGG -241
```
```
[CAGAACCAAT]AGGACATGGTAATAAATCTATTCTCAGGAAAGGGGAGGAGTCATGGCTTTCAGCCATGAGCATCCACCCT -161
                    IR2 →
```
                                                                Ddel
                                                                  ▼
```
CTGGGTGGCCTCACCCACTTCCTGGCAATTCTAGCCACCATGAGTCCAGGGGCTATAGCCCTTTGCTCTGCCCGTTGCTC  -81
← IR2
```
```
AGCAAGTTACTTGGGGTTCCAGTTTGATAAGAAAAGACTTCCTGTGGAGGAATCTGAAGGGAAGGAGGAGGAGCTGGCCC   -1
```
+1 →   ets   Ⓑ
```
ATTCCTGCCTGGGAGGTTGTGGAAGAAGGAAG ATG GCC AGA GCT TTG TGT CCA CTG CAA GCC CTC TGG   68
                                 Met Ala Arg Ala Leu Cys Pro Leu Gln Ala Leu Trp
```
```
CTT CTG GAG TGG GTG CTG CTG CTC TTG GGA CCT TGT GCT GCC CCT CCA GCC TGG GCC TTG  128
Leu Leu Glu Trp Val Leu Leu Leu Leu Gly Pro Cys Ala Ala Pro Pro Ala Trp Ala Leu
```
                                                      Bgl I
                                                        ▼
```
AAC CTG GAC CCA GTG CAG CTC ACC TTC TAT GCA GGC CCC AAT GGC AGC CAG TTT GGA TTT  188
Asn Leu Asp Pro Val Gln Leu Thr Phe Tyr Ala Gly Pro Asn Gly Ser Gln Phe Gly Phe
            Ⓐ                       NcoI
```
```
TCA CTG GAC TTC CAC AAG GAC AGC'CAT GGG AG GTGAGCCGTAAGGGAAGTTGGGGTATTGGGAGAGAGC  258
Ser Leu Asp Phe His Lys Asp Ser His Gly
```
```
AGGACCCCTCCCCATCACTGCTTCTGGGGGCTTCGAGTTTCCCATTTGCGATAGCAGTTGAGCAAGGTGACTTGTGGGGG  338
CCTATTCAGGTTGATTTCTTGTCAAGATGTGGGTCCCAGGGACTGGCTCAGGTGAAGGTATAAGGGCAGGGCACATGTGG  418
GCTGATGGGCACTGAAAACTACAGCAAGAACAAAGGGAAGACAATAGTTGATGCTTTATTTTTTCCCAAGGGTCAGTTGT  498
ATGAACCACTCCACCCTCAACACCTTGAAATGCAGAGAGGAGGCCGGGCGCGGTGCTCATGCCTGTAATCCCAGCACTTT  578
GGGAGGCCGAGGCGGGCAGATCACCTGAGGTCGAGAATT                                           616
```

Figure 1. DNA Sequence of the First Exon and 5' Flanking Region of the Human GPIIb Gene. The A designated as +1 is the transcription start site, the arrow below shows the direction of transcription. Potential regulatory elements are boxed. Horizontal arrows show inverted repeats. The 5' splice is indicated with a dotted line (from Prandini et al., 1988).

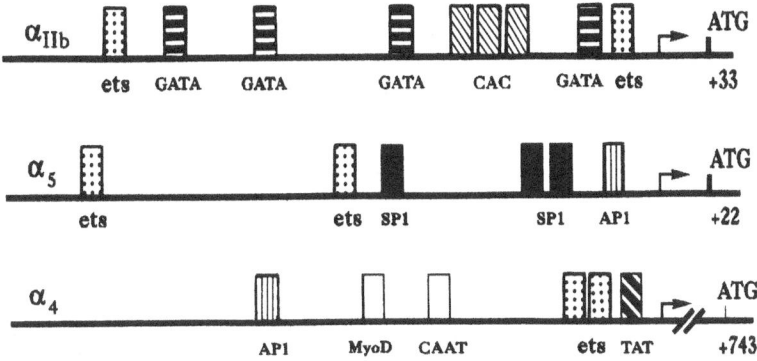

Figure 2. Comparison of the Structure of the αIIb Promoters of Other Integrin's α Subunits Does Not Reveal Obvious Similarities, Suggesting that They May Be Controlled by Different Mechanisms.

STRUCTURE OF THE αIIβ GENE

The gene for αIIβ spans approximately 17kb (Prandini et al., 1988, Heidenreich et al., 1990) and is located on the long arm of chromosome 17 at position q21.1-q21.2 (Bray et al., 1987; Sosnoski et al., 1988; Van Cong et al., 1988). Though functional domains of the αIIβ subunit have not been all identified, there is no obvious relationship between the position of the 30 exons of the gene and the functional structure of the protein (Heidenreich et al., 1990). In addition, the αIIβ gene contains seven intragenic complete AluI repeats, the function of which is unknown at the present time.

Since αIIβ expression is limited to megakaryocyte the structure of the 5' flanking region of the gene was carefully examined. The promoter contains a single transcriptional start site which has been identified by independent groups using different approaches including RNase mapping, primer extension and S1 nuclease experiments (Prandini et al., 1988, Heidenreich et al., 1990). The site corresponds to an A, located 33 base pair upstream from the only ATG in correct reading frame (Figure 1). The region upstream the transcriptional start site does not contain the canonical TATA and CAAT boxes which are normally present in proximity to the start site of polymerase II transcribed genes. This is not unusual, since a variety of TATA-less promoters have now been described. Some of these TATA-less promoters contain multiple GC boxes and are mainly involved in the transcription of house keeping genes. Other TATA-less promoter are GC-poor and mostly involved in transcription of genes that are implicated in cell proliferation or differentiation (Smale and Baltimore, 1989). The αII gene seems to fall into the latter category.

Comparison of the αIIβ promoter with the sequence of known promoters of other a integrins does not reveal obvious similarities (Figure 2). The 5' flanking domain of the gene encoding the α subunit of the fibronectin receptor α5β1 has recently been sequenced (Birkenmeier et al., 1991). This region also lacks the TATA box but contains three SP1 binding sites with the consensus sequence CCGCCC. The gene coding for the α4 subunit which associates with the β1 subunit to form the α4β1 complex which functions as a receptor for fibronectin and VCAM1 has also been characterized (Rosen et al., 1991). Expression of this gene is restricted to the lymphoid and the myeloid lineages (Hemler et al., 1990). In contrast to the αIIβ and α5 promoters, the α4 promoter contains TATA and CAAT boxes and is poor in GC sequences. Thus it is likely that cell specific expression of the different α subunits of the integrin family is dictated by different sets of transcriptional factors. Interestingly enough, the αIIβ, α5, and α4 promoters all contain multiple binding sites for the C-*ets* proteins, with the consensus motif GAAGGA. The translation products of the C-*ets* protooncogenes constitute a family of transcriptional factors involved in cell proliferation and acting in cooperation with AP1 (Watson et al., 1988; Wasylyk et al., 1990) which mediates transcriptional stimulation by phorbol esters or TGFβ (Angel et al, 1987; Kim et al., 1990). In that respect, it is worth noting that the α5 and α4 promoters, contain consensus binding site for the Ap1, complex. Since, α5β1 and α4β1 integrins are both regulated by these inducers, it is attractive to speculate that during growth stimulation, their expression is regulated by a combination of AP1 and C-*ets* protein. In contrast, the αIIβ promoter does not contain an apparent AP1 site. Again, this suggests that specific expression of αIIβ during megakaryocytopoiesis may be controlled by an entirely different mechanism.

Comparison of the nucleotide sequence of the αIIβ promoter with erythroid specific promoters reveals interesting features (Figure 3). The αIIβ promoter contains multiple binding sites for the erythroid factor GATA1, with the motif TGATAA. This transcriptional factor is a zinc finger protein which belongs to the GATA family of which three members, GATA1, GATA2 and GATA3, have been identified. GATA1 is expressed in erythrocytes, megakaryocytes and mast cells. This factor is critical for the development of the erythroid lineage as gene knock out experiments in transgenic mice, resulted in a profound anaemia and was lethal for most of the homozygote animals (Pevny et al., 1990). The presence of multiple binding sites for GATA1 in the sequence of the αIIβ promoter contains multiple CACCC motifs prsent at position -165, -155 and -148 and a CCAAT motif at position -228. The DNA binding proteins interacting with these sites have been shown to cooperate with GATA1 sites in erythroid genes (Frampton et al., 1990). When analyzed in DNasesI protection assays these sites are protected by nuclear factors indicating that they may be

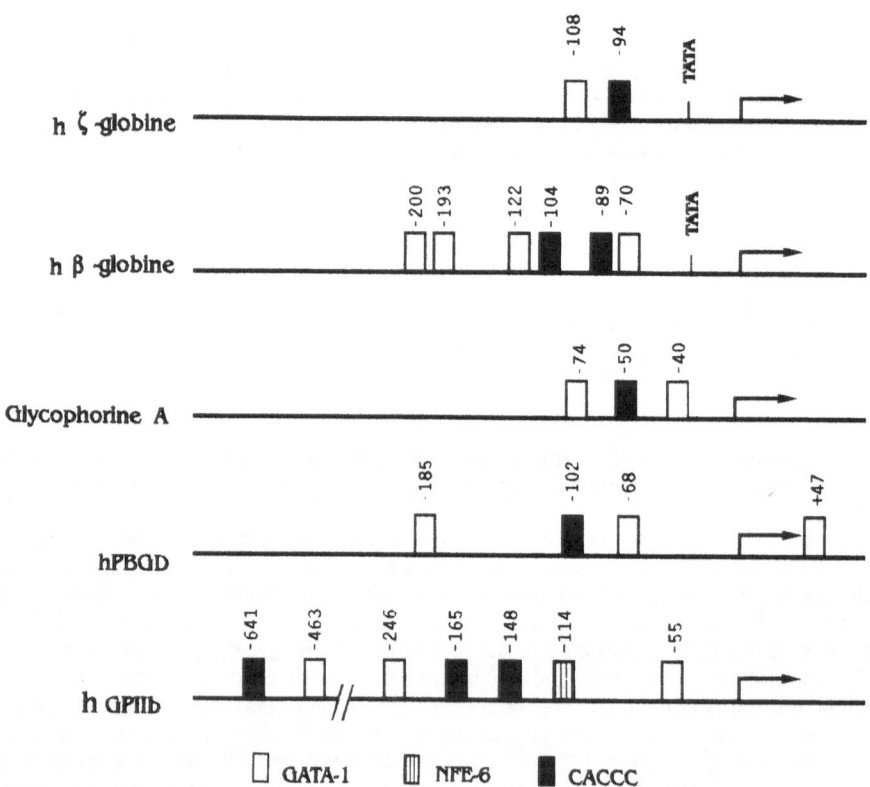

Figure 3. Comparison of the Structure of the Promoter of GPIIb Gene with Erythroid Promoter.

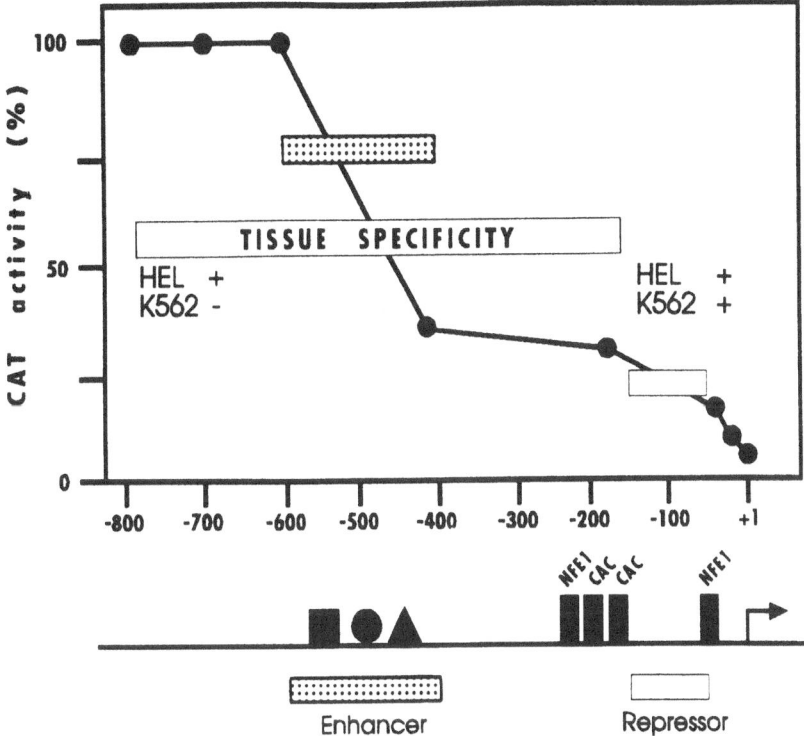

Figure 4. Deletion Analysis of αIIb Gene Promoter Indicates the Presence of Three Different Functional Domains. The nucleotide sequence located between 400 and 600 functions as a erythroid-megakaryocytic specific enhancer. The domain located between nucleotides 70 and 150 exhibits negative regulatory function and is probably responsible for the tissue-specific expression of the gene.

functional. In summary, one can conclude from sequence analysis, that the αIIβ promoter contains all the elements that are implicated in the expression of erythroid genes. Yet, αIIβ is not produced in these cells.

This does not seem to be unique to αIIβ. Other megakaryocyte genes have recently been examined, and contain elements of erythroid-like promoter, including the PF4 gene and the GPIb gene. It is therefore attractive to speculate that these erythroid-like promoters are not functional in erythrocyte and are turned off by some unknown mechanism.

TISSUE SPECIFIC ACTIVITY OF THE αIIβ PROMOTER

A series of experiments have been performed to verify the tissue-specific activity of the αIIβ promoter. Using transient CAT assays we have demonstrated that a 676 pb fragment of the promoter contains sufficient information for cell-specific expression. When fused to the CAT gene and transfected in cell lines this promoter region was able to drive the expression of the enzyme in HEL cells but not in K562 or Hela cells (Uzan et al., 1991). The activity of the αIIβ promoter in this case corresponded to 15% of that observed with the ubiquituous RSV promoter taken as reference. From these results, it was concluded that the 5' flanking region of the αIIβ gene contains responsive DNA sequences that control its transcription in a tissue specific manner and that these DNA elements are located within the first 600 bp of the promoter domain.

THE αIIβ PROMOTER CONTAINS A SPECIFIC ERYTHROMEGAKARYO-CYTIC ENHANCER

Nucleotide sequences that are implicated in the transcriptional control of the αIIβ gene were then examined using 5' deletion mutants of the promoter, fused with the CAT reporter gene, and transfection experiments in HEL cells (Uzan et al. 1991; Prandini et al., 1992; Uzan et al., 1992). The 5' flanking region contains at least four distinct functional domains (Figure 4).

The region located in between nucleotides -554 and -414 functions as a positive regulatory domain. Removal of this domain produces a 70% decrease of the promoter activity. Further deletion, to nucleotide -113, had no detectable effect, while deletion of the sequence between nucleotide -113 and nucleotide -29 produced an additional 20% decrease of the activity. Finally, the sequence, adjacent to the transcriptional start site, from -13 to +33, exhibited a significant residual activity of about 6% when compared to a promoter-less construct, suggesting that this TATA-less domain contains specific sequences that are critical for the positioning of the transcriptional complex.

The promoter sequence between -514 and -414 functions as an erythromegakaryocytic specific enhancer. This domain alone, produces a 5- to 6-fold increase of the activity of a ubiquitous enhancer-less viral promoter SV40. This enhancer activity is orientation- and position-independent and was observed in HEL cells, a megakaryocytic cell line which expresses αIIβ, and in K562 cells, an erythroid cell line which does not express αIIβ, but not in the fibroblastic Hela cells. DNA fast printing and mobility shift assays, have shown that this enhancer domain contains multiple binding sites for nuclear factors that are present in K562 and HEL cells but not in Hela cells. Mutagenesis of the contact site of these nuclear proteins abolished the enhancer activity, thus establishing the direct implication of these factors in the activity of the αIIβ promoter.

Transcriptional enhancers have been discovered in eukaryotic and prokaryotic organisms. They are binding sites for proteins and may function either as positive or negative regulatory domains. It has been proposed that proteins interacting with enhancers may control the rate limiting step in initiation of transcription by interfering with the TFIID transcriptional complex. It is worth noting that the αIIβ enhancer contains binding sites that are homologous to the binding site for the erythroid factor GATA1 and for the c-*ets* protein. A cooperation of these two factors in the transcriptional activity of the αIIβ promoter cannot be excluded. It remains, however, to know how this enhancer communicates with other functional domains of the promoter.

NEGATIVE REGULATION OF THE αIIβ GENE IN NON-MEGAKARYOCYTIC CELLS

The presence of a tissue-specific enhancer within the structure of the αIIβ promoter does not explain why the transcription of the gene is restricted to the megakaryocyte. As indicated by the above summarized results this enhancer interacts with erythroid and megakaryocytic factors and is active in both lineages. Since the promoter contains information to direct a megakaryocyte-specific expression of the gene, DNA responsive elements that are responsible for this lineage-specificity, must be present within the structure of this promoter. To delineate these elements, 5' progressive deletion mutants were generated and analyzed for their capacity to express the CAT reporter gene in megakaryocytic- and non-megakaryocytic cells. These experiments allowed the identification of a DNA core sequence from nucleotide -113 to nucleotide -75 which functions as a potential negative regulatory element in non-megakaryocytic cells (Figure 4). Mutation of the corresponding sequence resulted in a complete loss of the cell specific activity of the promoter, but did not affect promoter activity in megakaryocytic cells.

The presence of negative responsive elements with silencing activity has been identified in a variety of genes; some of these elements are promoter specific, others are ubiquituous. We still do not know whether the negative regulatory sequence present within the structure of the αIIβ promoter is αIIβ-specific. A comparison of the sequence with other known sequences for protein-binding sites did not reveal obvious similarities. It is tempting to speculate that the silencer is not functional in megakaryocyte because these cells do not provide functional nuclear factors with binding capacity to this care sequence. This negative element is located precisely within the erythroid-like promoter. Thus, one can also hypothesize that this element turns off the erythroid-like promoter in non-megakaryocytic cells. Further investigations are required to understand the communication between this negative domain and other functional domains of the promoter and its precise implication in the lineage-restricted transcription of the αIIβ gene.

CONCLUSION

Megakaryocytes play a central role in the production of circulating platelets and consequently in the human hemostatic response. These cells differ from other hematopoietic cells by their capacity to undergo non-mitotic development prior to the production of platelet. They are directly or indirectly involved in a number of physiological disorders including inflammation, atherogenesis, thrombosis and immunological diseases. Therefore, understanding the mechanisms that are implicated in the establishment or the maintaining of the megakaryocyte phenotype is of real interest. In our efforts to understand these mechanisms we use the αIIβ gene whose expression is limited to this lineage. Domains of the promoter of this gene, that are necessary for a tissue-specific activity have been identified. It remains to verify if these elements are also functional *in vivo*. Transgenesis should be the method of choice to address this question. Production of transgenic mice using megakaryocyte specific promoters will indeed allow targeting of genes in the megakaryocytic lineage and in platelet. By over- or down-expressing these genes, the role of platelet in vascular biology *in vivo* will be better understood.

REFERENCES

Angel, P., Imagawa, M., Chiu, R., Imbra, R.J., Rahmsdorf, M.J., Jonat, C., Herrlich, P., and Karin, M., 1987, Phorbol ester-inducible genes contain a common *cis* element recognized by a TPA-modulated trans-acting factor. *Cell* 49:729-739.

Birkenmeir, T.M., Quillan, J.J., Boedeker, E.D., Argraves, W.S., Ruoslahti, E., and Dean, D.G., 1991, The α5β1 fibronectin receptor characterization of the α5 gene promoter. *J. Biol. Chem.* 266:20544-20549.

Bray, P.F., Rosa, J.P., Johnston, G., Shin, D.T., Cook, R.G., Lan, C., Kan, Y.W., McEver, R.P., and Shuman, M.A., 1987, Platelet glycoprotein IIβ. Chromosomal localization and tissue expression. *J. Clin. Invest.* 80:1812-1817.

Colin, Y., Joulin, V., Levan Kim, C., Romeo, P.H., and Carton, J.P., 1990, Characterization of a new erythroid megakaryocyte specific nuclear factor that binds the parameter of the housekeeping human glycophorin C gene. *J. Biol. Chem.* 265:16729-16732.

Frampton, J., Walker, M., Plumb, M., and Harrison, P.H., 1990, Synergy between the NFE1 erythroid specific transcription factor and the CACCC factor in the erythroid specific promoter of the human prophobilinogen deaminase gene. *Mol. Cell. Biol.* 10:3838-3842.

Fraser, J.K., Tan, A.S., Lin, F., and Berrindge, M., 1989, Expression of specific high affinity binding rites for erythropoietin on rat and mouse megakaryocytes. *Exp. Heamtol.* 17:10-16.

Guau, J.P., Trevithick, J.E., and Hynes, R.O., 1991, Fibronectin/integrin interaction induces tyrosine phosphorylation of a 120 kDa protein. *Cell Regulation* 2:951-964.

Heidenreich, R., Eisman, R., Surrey, S., Delgrosso, K., Bennett, J.S., Schwartz, E., and Poncz, M.,1990, Organization of the gene for platelet glycoprotein IIβ. *Biochem.* 29:1232-1244.

Hemler, M.E., Elices, M.J., Parker, C., and Takada Y., 1990, Structure of the integrin VLA-4 and its cell-cell and cell-matrix adhesion functions. *Immunol. Rev.* 114:45-65.

Kim, S.J., Angel, P., Lafyatis, R., Hattori, K., Kim, K-Y., Sporn, M.B., Karin, M., and Roberts, A.B., 1990, Autoinduction of transforming growth factor β1 is mediated by the AP-1 complex. *Mol. Cell. Biol.* 10:1492-1497.

Martin, D., Zon, L.I., Mutter, G., and Orkin, S.H., 1990, Expression of an erythroid transcription factor in megakaryocytic and mast cell lineages. *Nature* 344:444-447.

Pevuy, L., Simon, M.C., Robertson, E., Klein, W.H., Tsai, S.F., D'Agati, V., Orkin, S.M., and Constantini, F., 1991, Erythroid differentiation in chimeric mice blocked by a targeted mutation in the gene for transcription factor GATA-1. *Nature* 349:257-260.

Plow, E.F., Marguerie, G., and Ginsberg, M.H., 1987, in: "Perspectives in Inflammation, Neoplasias and Vascular Cell Biology," Alan R. Liss, Inc., New York, NY, pp 267-275.

Prandini, M.H., Denarier, E., Frachet, P., Uzan, G., and Marguerie, G., 1988, Isolation of the human glycoproteins IIβ gene and characterization of the 5' flanking region. *Biochem. Biophys. Res. Comm.* 156:595-601.

Romeo, P.H., Prandini, M.H., Moulin, V., Mignotte,V., Prenant, M., Vainchencker, W., Marguerie, G., and Uzan, G., 1990, Megakaryocytic and erythrocytic lineages share specific transcription factor. *Nature* 344:447-449.

Rosen, G.D., Birkenmeier, T.M., and Dean, D.G., 1991, Characterization of the α4 integrin gene promoter. *Proc. Natl. Acad. Sci. USA* 88:4094-4098.

Smale, S.T. and Baltimore, D., 1989, The initiator as a transcription control element. *Cell* 57:103-113.

Sosnoski, D.M., Emanuel, B.S., Hawkins, A.L., Van Tuinen, P., Ledbetter, D.H., Nussbaum, R.L., Laos, F-T., Schwartz, E., Phillips, D., Bennett, J., Fitzgerald, L.A., and Poncz, M., 1988, Chromosomal localization of the genes for the vitronectin and fibronectin receptor α subunits and for platelet glycoproteins IIβ and IIIα. *J. Clin. Invest.* 81:1993-1998.

Van Cong, N., Uzan, G., Gross, M.S., Jegon-Faubert, C., Frachet, P., Boucheix, C., Marguerie, G., and Frezal, J., 1988, Assignment of human platelet GPIIβ gene to chromosome 17, region q21.1-q21.2. *Human Genet.* 80:389-392.

Watson, D.K., McWilliams, M.J., Lapis, P., Lantenberger, J.A., Schweinfast, C.W., and Pagas, T.S., 1988, Mammalian *ets* 1 and *ets* 2 encode highly conserved proteins. *Proc. Natl. Acad. Sci. USA* 85:7862-7866.

Wasylyk, B., Wasylyk, C., Flores, P., Begue, A., Leprince, D., and Stehelin, D , 1990, The c-*ets* protooncogenes encode transcription factors that cooperate with c-*fos* and c-*jun* for transcriptional activity. *Nature* 346:191-193.

DISCUSSION

R. JULIANO

Is there a clearly defined enhancer?

G. MARGUERIE

Yes, the positive element is the enhancer domain, but that enhancer domain is not a regulatory domain since it is also functional in erythroid cells, so it is only an enhancer domain which increases the activity of the promoter, in fact we are trying now to multimerize this enhancer domain to improve the activity of the promoter into the cells.

R. JULIANO

Do you think any of the transcription elements are cell-type specific?

G. MARGUERIE

No, we have been looking for that two years and I was eager to find a megakaryocyte specific transcriptional factor. So far the factors that we have found are also expressed in erythrocytes. I think that it is going to be mostly a combination of tissue-specific, probably tissue-specific, but what we mean by tissue-specific is hematopoietic factors.

G. NICOLSON

I think the surprising finding is not that the promoter sequences are so different, it is that they are very similar and the factors that they use are very similar. This suggests that the conformation of the sequence or perhaps other nucleoproteins are really very important in determining whether you get a positive or negative signal.

G. MARGUERIE

This is a good question. In fact several possibilities can be hypothesized. The erythroid like promoter is functional in megakaryocyte but not in erythrocyte because there is a specific megakaryocyte sequence which interacts with a negative regulatory protein in all the cells and turn off the erythroid promoter of the megakaryocytic genes in all the cells. The other possibility is that the promoter exhibits a specific conformation in such a way that the system is working in megakaryocytes but not in other cells. In the megakaryocyte either there is a functional erythroid-like promoter or a megakaryocytic-specific promoter that we have not identified as yet.

G. NICOLSON

But in fact it could be the conformation?

G. MARGUERIE

It could be the conformation. All these experiments, all the results that I have shown were obtained by transfection into mammalian cells. We are now trying to verify this in transgenic mice. We have to verify all this *in vivo* and see how inclusion of the DNA construct into the chromosome or into the chromatid will affect the expression of the construct.

G. NICOLSON

What if they are different?

G. MARGUERIE

We will see. These are the obvious experiments to be done in the future.

A. ANICHINI

If I understood correctly your experiments, you have shown that by culturing progenitor cells in the presence of erythropoietin you can switch these cells towards the erythroid lineage. When doing that can you show or did you try to show the actual synthesis of the negative regulatory factor?

G. MARGUERIE

No, we have not tried that, in fact, we cannot do that with blast cells. It is a very complicated experiment, you need a lot of human bone marrow and it is not easy to isolate sufficient blast cells to look at transcription factors. We just wanted to check whether you had lineage-specific expression. We cannot use this system to purify transcriptional factors. What we are going to do in fact is to look at permanent cell lines, K562 or HEL cells, and see if we can induce the expression or the activity of these proteins. That is the only way to do it. Alternatively, one can mutagenize the binding sites in the promoter and look into the transgenic mice if you have expression into the erythrocyte. One of the first questions that we are going to ask with the mice is in fact whether you have a bi-potent progenitor for the erythrocytes and megakaryocytes.

S. GOODMAN

That was a very beautiful talk. I am always fascinated by the complexity of the promoters. How conserved is this negative regulatory element? That is, have you looked at the lineages from the point of view of evolution of the promoters? Are they found in other places?

G. MARGUERIE

We have looked at the data bank for the sequence. The only consensus sequence that we found was a binding sites for the *ets* protein. So we thought the *ets* protein was probably responsible for that and we mutagenized the *ets* binding site but it did not work.

S. GOODMAN

But that negative regulatory region of the protein itself is not conserved anywhere else?

G. MARGUERIE

We did not find a lot of conservation and obvious homologies. We only find these *ets* binding sites. That is all.

S. GOODMAN

You made this remark about TATA box-less and GC-less promoters. Could you mention other proteins where these have been found?

G. MARGUERIE

This is from the literature. Usually the GC-rich sequences interacts with SP1 and these are mostly found in housekeeping gene. Many of the TATA-less promoters have been reported in proteins that are in differentiation.

M. HEMLER

Do you know of any examples of this in basthenic patients that have altered promoter regions?

G. MARGUERIE

We have not looked. Patients with no production of α_{IIb} will be good candidates.

V. QUARANTA

I would like to ask a more general question. If I understand you correctly, most of or all the regulatory elements of the α_{IIb} gene are very close to the transcription unit. Could this be a general property of α-chain genes?

G. MARGUERIE

I do not see any reason why it should be like that for all these promoters. The guess is that they are controlled by totally different mechanisms and I do not see why they should have a general mechanism that will be the same for all the integrins.

V. QUARANTA

Are there any results concerning the regulation of expression of β-chain genes? Any similarities with α-chain gene regulatory elements?

G. MARGUERIE

We cloned the β_3 gene but we never found the initiation start site.

G. TARONE

First of all, relative to the other promoters α_5, hemopoietic cells express the fibronectin receptor but during maturation and differention they start to lose it. So I would expect that the α_5 promoter will be regulated negatively during the differentiation exactly in the opposite way as the α_{IIb} promoter you have shown for the megakaryocytes. So the difference in the sequence of the α_5 promoter and the α_{IIb} may well relate to a completely different regulation during hemopoietic differentiation.

G. MARGUERIE

This is a good point.

G. TARONE

The other point that I would like to ask is about the β_1 promoter. In the laboratory we are looking at the β_1 promoter which in humans does not have the TATA box. It is very rich in GC boxes so this is a different situation. But we are still at the beginning of the characterization.

J. McCARTHY

Is there anything known about how all of these binding interactions take place between all these factors and the various sequences from the standpoint of the proteins? How does the specificity actually occur? For example, are there basic residues clustered?

G. MARGUERIE

Yes, there are, we are not doing these types of experiments, but for instance the GATA-1 has been cloned, the GATA-3 has been cloned too, and people are trying to express mutants and look at the binding conditions. There are also some mechanisms by which the cell controls the activity of these proteins. Some of the cells can produce for instance a messenger coding for the whole domain of these transcription factors, and post-transcriptional mechanisms can also splice messenger and produce inhibitors of the protein itself which will bind but will not have the transcriptional activity.

R. JULIANO

I wonder if I can ask a sort of general question, namely, is the expression of IIb controlled purely at the transcriptional level or are there other controls in terms of message stability and other phenomena.

G. MARGUERIE

This is a good question. I do not have experimental proof for this, but we know that all cells producing the protein have the messenger and cells that do not produce the protein do not express the messenger.

INTEGRIN EXPRESSION AND EPITHELIAL CELL DIFFERENTIATION

Vito Quaranta

Department of Cell Biology, NX-7
The Scripps Research Institute
10666 North Torrey Pines Road
La Jolla, California 92037 U.S.A.

INTRODUCTION

The purpose of this paper is to discuss the contribution of a type of cell adhesion receptors, the integrins, to the biology of epithelial cells. While our knowledge on this subject is far from satisfactory, the available evidence suggests that integrins play a critical role in determining many attributes of epithelial cells.

Epithelial cells are a heterogenous group. However, they share certain common traits that, together, define the "epithelial phenotype" (Rodriguez-Boulan and Nelson, 1989). Among these, pertinent ones include: i) the polarized state, such that apical, lateral and basal domains of the epithelial cell plasma membrane can be distinguished at a structural and functional level; ii) a capacity to form morphogenetically meaningful structures (sheets, tubes, furrows, etc.); iii) a potential to migrate as a sheet (epiboly) or individually, depending upon environmental clues; iv) a capacity to undergo mitosis for the purpose of epithelial regeneration or renewal, respecting basement membrane boundaries and architecture of the epithelium.

It is self-evident that the regulation of adhesion must play a key role in implementing such epithelial cell traits. Recent work supports this view, by showing direct involvement of various types of adhesion molecules in epithelial morphogenesis (Rodriguez-Boulan and Nelson, 1989; Gumbiner, 1992). Several laboratories, including ours, have focused on the adhesion receptors of the integrin family which, because of their versatility, are likely to explain many adhesive properties of differentiated epithelia.

OVERVIEW OF THE INTEGRIN FAMILY

In this section, the main structural and functional principles of integrins will be recalled, to better visualize the potential connections between these molecules and the epithelial phenotype.

The typical integrin receptor (Hynes, 1992) is comprised of two glycoprotein subunits, α and β, which span the plasma membrane and are noncovalently associated. Both the α and the β subunit are encoded by gene families. At least fourteen distinct α gene products and eight β gene products have been recognized in man. It appears that a similar degree of diversity exists in other vertebrates, while in invertebrates the integrin family may be less numerous. A high degree of structural conservation among all species, however, has been observed at the amino acid sequence level, suggesting an important evolutionary role for these adhesion molecules.

Cell Adhesion Molecules, Edited by M.E. Hemler
and E. Mihich, Plenum Press, New York, 1993

Ligands for integrins (Languino et al., 1989; Sonnenberg et al., 1988; Pytela et al., 1985b; Pytela et al., 1985a; Argraves et al., 1987; Suzuki et al., 1987; Takada et al., 1987; Wayner and Carter, 1987; Wayner et al., 1989) are components: i) of the extracellular matrix, such as fibronectin and laminin; ii) of plasma, such as fibrinogen and vonWillebrand factor; or iii) of the plasma membrane, such as the ICAM or VCAM counter-receptors. Therefore, integrins have the capability of becoming involved in both cell-cell and cell-matrix interactions. Both situations occurr in epithelial sheets.

A fundamental feature of integrin subunits is that their expression is controlled in a cell type-specific manner. Thus, unique migratory and adhesive properties may be acquired by cells by changing their integrin repertoire, possibly in a dynamic relationship with other differentiation steps. The rules governing cell-type specific integrin expression are poorly understood. A possible connection with pathways controlling cell proliferation and activation was suggested by the observations that TGF-β regulates integrin expression at several levels (Ignotz et al., 1989; Heino et al., 1989). A direct link was also observed between chemical (Dedhar and Saulnier, 1990) or oncogene-induced (Giancotti and Ruoslahti, 1990) carcinogenesis and modulation of integrin expression. Some integrins are tightly controlled like, for instance, the $\beta2$ integrins that are exclusively found in leukocytes (Springer, 1990), or the $\alpha IIb\beta3$ integrin which is platelet-specific (Wayner et al., 1989). Other integrins are less tightly controlled in terms of cell lineage, but still display a degree of specific expression, e.g., dependence on differentiation.

In general, an integrin α chain tends to associate with a preferred β chain partner. Several exceptions to this rule, however, have been described, consisting of α chains that may choose their partner among two or more β chains. For example, the integrin subunit $\alpha6$ may associate with either $\beta1$ or $\beta4$. Interestingly, in epithelial cells, $\alpha6$ is mostly found in a complex with $\beta4$, in spite of the presence of $\beta1$, suggesting cell type specific regulation of integrin heterodimer formation.

Initially, integrins were described as transmembrane links between extracellular matrix and cytoskeleton. On the extracellular side these receptors would engage with ligands of the matrix, such as collagen, fibronectin or laminin. On the intracellular side, they would interact with the actin microfilament network through some link proteins (e.g., talin, vinculin, α-actinin), causing its organization. Adhesion complexes, recognizable by immunofluorescence and referred to as adhesion plaques or focal adhesions (Burridge et al., 1988), would then form. Overall, such paradigm is still valid, but several modifications and additions have become necessary. For example, it is likely that cytoskeletal components other than microfilaments interface integrins, as is the case for the integrin $\alpha6\beta4$ which was shown to interface keratin-containing intermediate filaments (Jones et al., 1991; Quaranta and Jones, 1991). It has also become clear that integrins are not simply "cell attachment" receptors. Rather, they are capable of transmitting signals to the cell interior which may directly affect gene expression (Werb et al., 1989) and intracellular pH or Ca^{++} (Schwartz et al., 1990). Such effects are not a generic consequence of cell adhesion or spreading, but are specifically transduced by the integrin receptor, through pathways that are beginning to be characterized and include the tyrosine phosphorylation of specific proteins (Hynes, 1992).

An important point to be made is that several integrins exist in two or more states of affinity for ligand. Integrin "activation", with subsequent changes in cell adhesive properties, may occur upon cell differentiation or engagement of other cellular receptor systems. Perhaps a most striking example is T lymphocytes, where a higher affinity state (activation) of LFA-1 or laminin receptor ($\alpha6\beta1$) integrins can be induced, with fast kinetics, by cross-linking of the antigen receptor, of the CD3 complex, or by protein kinase activators (Shimizu et al., 1990). Thus, cross-talking of integrins with other receptor systems occurs, most likely via intracellular pathways. The cytoplasmic domains of both the α (O'Toole et al., 1991) and the β (Hibbs et al., 1991) integrin subunits have been involved in this process.

INTEGRINS AND EPITHELIAL PHENOTYPE

From the properties highlighted above, it seems logical that integrins are in a position to contribute to several aspects of the epithelial phenotype (Simons and Fuller, 1985). For instance, they could be involved in establishing and/or maintaining epithelial polarity, in distinguishing lateral (cell-cell) from basal (cell-substratum) adhesive interactions, or in supporting epiboly or short-distance migration. Some or all of these phenomena may exploit the ability of integrins to interact with ligands in the matrix or on the cell surface, to modulate cytoskeletal organization, to transduce signals that may influence proliferation, or to change their own state of affinity for ligand by cross-talking with other types of cellular receptors.

In general, in those epithelial cell types investigated for their integrin repertoire, many distinct integrin heterodimers were detected. This is true both of normal and transformed phenotypes. A summary of "epithelial" integrins can be found in Table 1. The significance of such abundance of integrins in epithelial cells is not entirely clear. Some possible explanations, however, are beginning to emerge. For instance, the integrin α6β4 appears to be a receptor for the hemidesmosome adhesion devices that anchor epithelial cells to the basement membrane, and is perhaps exclusively involved in mediating cell-matrix interactions. On the contrary, the α2 and α3 integrins appear to be involved in lateral cell-cell interactions, based on their topography. These findings begin to suggest distinct roles for individual integrin types in the shaping of epithelium architecture.

Table 1. Integrins Expressed in Epithelial Cells

β Chain	α Chain	Ligands	Expression
	α2	Ln, Cn	epithelium, endothelium, others, *carcinoma*
	α3	Ln, Fn, Cn, Eg	epithelium, endothelium, others, *carcinoma*
β1	α5	Fn	epithelium, fibroblasts, others, *carcinoma*
	α6	Ln	epithelium, others, *carcinoma*
	α8	?	epithelium, ?
β4	α6	Ln?	epithelium, Schwann cells, *carcinoma*
β5	αV	Vn	epithelium, others
β6	αV	Fn	epithelium, ?

Ln = laminin; Fn = fibronectin; Vn = vitronectin; Cn = collagen; Eg = epiligrin

Direct links need to be established between a particular epithelial function, and a particular integrin. This task will be laborious, because of the complex repertoire of integrins in epithelial cells (DeLuca et al., 1990; Quaranta, 1991; Larjava et al., 1990; Carter et al., 1990b) (Table 1). Furthermore, such repertoire may vary according to lineage and differentiation and, in general, includes five or more distinct integrins. Note that both transformed and normal epithelial cells, e.g., human keratinocytes, bear out this complexity (Quaranta, 1991).

Within a given cell lineage, changes in levels or types of integrins may mark various stages of differentiation. This is particularly clear in the case of stratifying keratinocytes. In this epithelium, those cells in direct contact with the basement membrane express a variety of integrin types, while cells undergoing further differentiation in the upper layer of the epithelial sheet almost abruptly shut down integrin expression.

Changes in integrin expression may also be associated with benign or malignant transformation of epithelial cells. In some cases, such changes have been proposed as potential markers for staging transformation, e.g., in breast cancer or head and neck tumors (Koukoulis et al., 1991; Wolf et al., 1990).

An interesting question is whether there exist integrins whose expression is epithelial specific, since such specificity would imply functions directly related to the epithelial phenotype. Among α or β integrin subunits two, β4 and β6, appear to be preferentially expressed in epithelial cells (Tamura et al., 1990; Sheppard et al., 1990). In the case of β4, expression in few other cell types has been reported, namely sympathetic neurons and

Schwann cells.   These latter cell types share with epithelial cells the property of being polarized, i.e., they possess defined plasma membrane domains that are equivalent to a basal or an apical surface.   Since in epithelial cells the β4 integrins specialize in adhesive interactions between the cell basal surface and basement membranes, it is possible that analogous functions are carried out by these integrins in the basal plasma membrane domain equivalents of neurons and Schwann cells.   In the case of β6, its association with αV (Sheppard et al., 1990) and its RGD-dependent recognition of fibronectin (Busk et al., 1992) makes it possible that it may represent "the" fibronectin receptor of epithelial cells, since other types of fibronectin receptors are often not found in these cells (Cheresh et al., 1989).

An important finding concerns the topography of integrin expression within the spatially defined domains of the epithelial plasma membrane.   In keratinocytes, α2β1 and α3β1 are almost exclusively found on lateral surfaces, in areas of intercellular contacts, while α6β4 is on the basal surface, at the cell-basal lamina interface.   Such exquisite spatial regulation may be lost upon malignant transformation (Quaranta and Jones, 1991; Gumbiner, 1992), suggesting that it may play a key role in maintaining the integrity of epithelial sheets.

THE INTEGRIN α6β4

The remainder of this review will be on the integrin, α6β4, which, as mentioned, appears to play a special role in epithelial cell interactions with the basement membrane.

Recently, we (Jones et al., 1991; Quaranta and Jones, 1991; Kurpakus et al., 1991) and others (Stepp et al., 1990; Carter et al., 1990a) showed that the integrin α6β4 co-localizes with epithelial structures, called hemidesmosomes, which are part of the basement membrane zone and control adhesive interactions with the basement membrane.   We also found that α6β4 can influence hemidesmosome assembly (Jones et al., 1991; Kurpakus et al., 1991), supporting the importance of this integrin in regulating molecular interactions occurring in the basement membrane zone.

The hemidesmosome (Staehelin, 1974) is a specialized adhesion device that in many epithelia supports the anchoring of cells to the underlying connective tissue.   Like focal adhesions (Geiger, 1981; Burridge et al., 1987), hemidesmosomes act as plasma membrane attachment sites for cytoskeletal components.   A major difference, however, is that focal adhesions are the site of interaction for actin microfilaments, while hemidesmosomes are a site of interaction for the intermediate filament network (Gipson et al., 1989; Jones and Green, 1991).   The molecular components that make up both the intracellular and the extracellular part of the hemidesmosome are just beginning to be understood.   Some of these have been recognized because they are the targets of autoantibodies associated with certain skin blistering diseases, e.g., bullous pemphigoid (Quaranta and Jones, 1991).

The finding that α6β4 co-localizes with hemidesmosomal structures has opened new venues of structural and functional investigations.   Even though such co-localization is thus far based only on morphological criteria (light and electron immuno-microscopy), it is sufficient to warrant further molecular studies.

Moreover, we showed that addition of antibodies to β4 prevents the assembly of hemidesmosomes in cultured epithelial cell systems (Kurpakus et al., 1991), suggesting a direct role for α6β4 integrins in promoting assembly and, perhaps, disassembly of these adhesive structures.

STRUCTURE OF β4

The β4 protein contains an intriguing structural feature, unique among integrins (Hogervorst et al., 1990; Suzuki and Naitoh, 1990; Tamura et al., 1990):   its cytoplasmic domain extends for more than 1,000 amino acids, about 20 times the length of other βchains' cytoplasmic domains.   Three type III fibronectin repeats are located within this region of β4, but no other significant homologies to known proteins are otherwise found (Quaranta, 1991; Suzuki and Naitoh, 1990).   Potential functions for the β4 cytoplasmic domain include interaction with cytoskeletal components, and organization of hemidesmosomal plaques.   However, a likely possibility is that some catalytic function is carried out by such a conspicuous receptor domain on the inner side of the plasma membrane.   In this regard, it may be relevant that, when added to live cells, antibodies to β4 can prevent formation of hemidesmosomal structures (Jones et al., 1991; Kurpakus et al., 1991). This result suggests that proper orientation of this integrin in the plasma membrane is critical to initiate assembly of hemidesmosomes. It is possible that clusters of $\beta 4$ domains on the inner plasma membrane, produced by receptor-ligand interaction, are the initiating factor of such assembly.

It is important to note that $\alpha 6 \beta 4$ is also expressed by cell types that do not assemble hemidesmosomes as judged by transmission electron-microscopy, e.g., Schwann cells or gut epithelial cells. It remains to be seen whether adhesion structures equivalent to hemidesmosomes exist in these cell types. A criterion for assimilating adhesion structures to hemidesmosomes rather than, e.g., to focal adhesions, is that they be connected to intermediate filaments, as opposed to actin microfilaments. Alternatively, there may exist cell-type variability in the structural components of hemidesmosomes, such that in certain cells, e.g., intestinal epithelial cells, the classical hemidesmosomal plaque may no longer be recognizable in transmission electron microscopy images.

Structural heterogeneity of $\beta 4$ may also produce different types of adhesion structures in distinct cell types. For example, we described variant forms of $\beta 4$ cytoplasmic domains, distinguished by the presence of extra amino acid stretches presumably derived from alternative mRNA splicing. Three splice variants of $\beta 4$ may exist, according to whether one of two possible inserts of 70 or 53 residues, or neither of them, are included in the cytoplasmic domain (Tamura et al., 1990). The insertions are localized in the area of the type III fibronectin repeats and, by PCR amplification with diagnostic primers, we found that they occur in a cell-type specific fashion.

ALTERNATIVE $\alpha 6$ CYTOPLASMIC DOMAINS

Antisera to the cytoplasmic domain of $\alpha 6$, whose sequence was recently determined (Tamura et al., 1990), were capable of immunoprecipitating $\alpha 6 \beta 4$ or $\alpha 6 \beta 1$ heterodimers from carcinoma cells or platelet detergent lysates, respectively (Tamura et al., 1990). In contrast, these same antisera were negative with radiolabeled lysates from F9 teratocarcinomas and embryonic stem (ES) cells. Since these latter cell lines were shown to be $\alpha 6$ positive by immunoprecipitation with an anti-$\alpha 6$ monoclonal antibody, this result suggested structural variability in the $\alpha 6$ cytoplasmic domain. By the use of PCR (Mullis and Faloona, 1987), the cytoplasmic region of $\alpha 6$ mRNA was cloned and sequenced from several cell lines. Alignment of the nucleotide sequences showed that $\alpha 6$ mRNA from ES or F9 cells (heretofore referred to as $\alpha 6B$) lacks a stretch of 130 nucleotides in the open reading frame, corresponding exactly to the region encoding the cytoplasmic domain. Alternative mRNA splicing is likely responsible for the deletion of this 130 bp segment, which causes 53 downstream codons to become in-frame and encode the cytoplasmic tail of $\alpha 6B$. The existence of the $\alpha 6B$ splice variants was then proven unequivocally by the fact that antisera to its cytoplasmic peptide sequence immunoprecipitated $\alpha 6B$ from a variety of cell lines (Cooper et al., 1991; Tamura et al., 1991), including ES and F9 cells.

The first five residues of both the $\alpha 6A$ and the $\alpha 6B$ cytoplasmic domains are GFFKR, representing a sequence found in all integrin α chains (Tamura et al., 1990). The remainder of the A and B cytoplasmic sequences bear no resemblance to each other, and they exhibit no antigenic cross-reactivity (Cooper et al., 1991; Tamura et al., 1991). Both $\alpha 6A$ and $\alpha 6B$ can associate to either $\beta 1$ or $\beta 4$, as indicated by sequential immunodepletions (Tamura et al., 1991).

Most cultured transformed cell lines express both forms of $\alpha 6$. The relative ratio of the two mRNAs varied according to the cell line. However, it was consistent in multiple independent determinations, suggesting it is an intrinsic property of a given cell line. We found two cell lines (A431 and 804G) that are exclusively $\alpha 6A$, and two (ES and F9) that only express $\alpha 6B$. The exclusive presence of $\alpha 6A$ in 804G cells may have some bearing on their ability to form hemidesmosomes (Quaranta and Jones, 1991).

Each of three possible patterns was also found in normal tissues. Kidney, brain, and ovary were $\alpha 6A-$, $\alpha 6B+$. Liver, lung, cervix and spleen had the opposite pattern. Other tissues were positive for both forms, with varying ratios of A to B. Because tissues are inevitably a mixture of cell types, these patterns need to be refined. Recent experiments suggest that $\alpha 6B$ may be expressed by totipotent cells, while $\alpha 6A$ appears to be found in differentiated cells. Thus, when ES cells were grown for eight days in the absence of differentiation inhibitory factors (DIA/LIF) (Rathjen et al., 1990), they switched from α6A-, α6B+ to α6A+, α6B+. Likewise, when F9 cells were induced to differentiate by exposure to retinoic acid, they also became α6A+, α6B+.

It is also intriguing that cultured normal epidermal keratinocytes as well as 804G, a unique carcinoma cell line that assembles hemidesmosomes *in vitro* (Jones et al., 1991), presumably a highly differentiated function, are α6A+, α6B-.

The functional significance of the alternative α6 tail is a fascinating question, because of its potential connection with cell differentiation. It is possible that distinct cytoskeletal connections are established by the two α6 chains. Another option is that these domains may have a regulatory role. For instance, they may impart a different character to the heterodimers they form, e.g., activatable vs. constitutively activated.

CONCLUSION

In summary, the study of integrins in epithelial cells will explain many properties of these cell types. Integrins with a degree of epithelial specific expression, such as α6β4, are particularly promising in this regard. In a more general sense, understanding integrin functions in epithelial cells will give us a measure of how profoundly this family of adhesion receptors impacts on cell differentiation.

REFERENCES

Argraves, W.S., Suzuki, S., Arai, H., Thompson, K., Pierschbacher, M.D., and Ruoslahti, E., 1987, Amino acid sequence of the human fibronectin receptor. *J. Cell Biol.* 105:1183-1190.

Burridge, K., Molony, L., and Kelly, T., 1987, Adhesion plaques: sites of transmembrane interaction between the extracellular matrix and the actin cytoskeleton. *J. Cell Sci. Suppl.* 8:211-229.

Burridge, K., Fath, K., Kelly, T., Nuckolls, G., and Turner, C., 1988, Focal adhesions: Transmembrane junctions between the extracellular matrix and the cytoskeleton. *Ann. Rev. Cell. Biol.* 4:487-525.

Busk, M., Pytela, R., and Sheppard, D., 1992, Characterization of the integrin αVβ6 as a fibronectin-binding protein. *J. Biol. Chem.* 267:5790-5796.

Carter, W.G., Kaur, P., Gil, S.G., Gahr, P.J., and Wayner, E.A., 1990a, Distinct functions for integrins alpha-6-beta-4-bullous pemphigoid antigen in a new stable anchoring contact (SAC) of keratinocytes: Relation to hemidesmosomes. *J. Cell Biol.* 111:3141-3154.

Carter, W.G., Wayner, E.A., Bouchard, T.S., and Kaur, P., 1990b, The role of integrins alpha 2 beta 1 and alpha 3 beta 1 in cell-cell and cell-substrate adhesion of human epidermal cells. *J. Cell Biol.* 110:1387-1404.

Cheresh, D.A., Smith, J.W., Cooper, H.M., and Quaranta, V., 1989, A novel vitronectin receptor integrin ($\alpha_V\beta_X$) is responsible for distinctive adhesive properties of carcinoma cells. *Cell* 57:59-69.

Cooper, H.M., Tamura, R.N., and Quaranta, V., 1991, The major laminin receptor of mouse embryonic stem cells is a novel isoform of the α6β1 integrin. *J. Cell Biol.* 115:843-850.

Dedhar, S. and Saulnier, R., 1990, Alterations in integrin receptor expression in chemically transformed human cells: Specific enhancement of laminin and collagen receptor complexes. *J. Cell Biol.* 110:481-489.

DeLuca, M., Tamura, R.N., Kajiji, S., Bondanza, S., Rossino, P., Cancedda, R., Marchisio, P.C., and Quaranta, V., 1990, Polarized integrin mediates keratinocyte adhesion to basal lamina. *Proc. Natl. Acad. Sci. USA* 87:6888-6892.

Geiger, B., 1981, Involvement of vinculin in contact-induced cytoskeletal interactions. *Cold Spring Harbor Quant. Biol.* XLVI:671-682 (*abstract*).

Giancotti, F.G., and Ruoslahti, E., 1990, Elevated levels on the $\alpha_5\beta_1$ fibronectin receptor suppress the transformed phenotype of Chinese hamster ovary cells. *Cell* 60:849-859.

Gipson, I.K., Spurr-Michaud, S., Tisdale, A., and Keough, M., 1989, Reassembly of the anchoring structures of the corneal epithelium during wound repair in the rabbit. *Invest. Ophthal. Vis. Sci.* 30:425-434.(*abstract*)

Gumbiner, B., 1992, Epithelial morphogenesis. *Cell* 69:385-387.

Heino, J., Ignotz, R.A., Hemler, M.E., Crouse, C., and Massague, J., 1989, Regulation of cell adhesion receptors by transforming growth factor-β. Concomitant regulation of integrins that share a common β_1 subunit. *J. Biol. Chem.* 264:380-388.

Hibbs, M.L., Xu, H., Stacker, S.A., and Springer, T.A., 1991, Regulation of adhesion of ICAM-1 by the cytoplasmin domain of LFA-1 integrin β subunit. *Science* 251:1611-1613.

Hogervorst, F., Kuikman, I., vondemBorne, A.E.G.Kr., and Sonnenberg, A., 1990, Cloning and sequence analysis of β_4 cDNA: An integrin subunit that contains a unique 118kDa cytoplasmic domain. *EMBO J.* 9:765-770.

Hynes, R.O., 1992, Integrins: Versatility, modulation and signaling in cell adhesion. *Cell* 69:11-25.

Ignotz, R.A., Heino, J., and Massague, J., 1989, Regulation of cell adhesion receptors by transforming growth factor-β. Regulation of vitronectin receptor and LFA-1. *J. Biol. Chem.* 264:389-392.

Jones, J.C.R., and Green, K.J., 1991, Intermediate filament-plasma membrane interactions. *Curr. Opin. Cell Biol.* 3:127-132.

Jones, J.C.R., Kurpakus, M.A., Cooper, H.M., and Quaranta, V., 1991, A function for the integrin $\alpha_6\beta_4$ in the hemidesmosome. *Cell Regul.* 2:427-438.

Koukoulis, G.K., Virtanen, I., Korhonen, M., Laitinen, L., Quaranta, V., and Gould, V.E., 1991, Immunohistochemical localization of integrins in the normal, hyperplastic and neoplastic breast: Correlation with their functions as receptors and cell adhesion molecules. *Am. J. Pathol.* 139:787-799.

Kurpakus, M.A., Quaranta, V., and Jones, J.C.R., 1991, Surface relocation of α6β4 integrins and assembly of hemidesmosomes in an *in vitro* model of wound healing. *J. Cell Biol.* 115:1737-1750.

Languino, L.R., Gehlsen, K.R., Wayner, E., Carter, W.G., Engvall, E., and Ruoslahti, E., 1989, Endothelial cells use $\alpha_2\beta_1$ integrin as a laminin receptor. *J. Cell. Biol.* 109:2455-2462.

Larjava, H., Peltonen, J., Akiyama, S.K., Yamada, S.S., Gralnick, H.R., Uitto, J., and Yamada, K.M., 1990, Novel function for beta 1 integrins in keratinocyte cell-cell interactions. *J. Cell Biol.* 110:803-815.

Mullis, K.B. and Faloona, F.A., 1987, Specific synthesis of DNA *in vitro* via a polymerase-catalyzed reaction. *Methods. Enzymol.* 155:335-350.

O'Toole, T.E., Mandelman, D., Forsyth, J., Shattil, S.J., Plow, E.F., and Ginsberg, M.H., 1991, Modulation of the affinity of (integrin αIIbβ3 (GPIIb-IIIa) by the cytoplasmic domain of αIIb. *Science (in press)*.

Pytela, R., Pierschbacher, M.D., and Ruoslahti, E., 1985a, Identification and isolation of a 140 kd cell surface glycoprotein with properties expected of a fibronectin receptor. *Cell* 40:191-198.

Pytela, R., Pierschbacher, M.D., and Ruoslahti, E., 1985b, A 125/115-kDa cell surface receptor specific for vitronectin interacts with the arginine-glycine-aspartic acid adhesion sequence derived from fibonectin. *Proc. Natl. Acad. Sci. USA* 82:5766-5770.

Quaranta, V. and Jones, J.C.R., 1991, The internal affairs of an integrin. *Trends Cell Biol.* 1:2-4.

Quaranta, V., 1991, Epithelial integrins. *Cell Differ. Develop.* 32:361-366.

Rathjen, P.D., Toth, S., Willis, A., Heath, J.K., and Smith, A.G., 1990, Differentiation inhibiting activity is produced in matrix-associated and diffusible forms that are generated by alternate promoter usage. *Cell* 62:1105-1114.

Rodriguez-Boulan, E. and Nelson, W.J., 1989, Morphogenesis of the polarized epithelial cell phenotype. *Science* 245:718-725.

Schwartz, M.A., Lechene, C.P., and Ingber, D.E., 1990, Activation of a cytoplasmic signal by integrin α5β1. *J. Cell Biol.* 111:263a.

Sheppard, D., Rozzo, C., Starr, L., Quaranta, V., Erle, D.J., and Pytela, R., 1990, Complete amino acid sequence of a novel integrin β subunit (β_6) identified in epithelial cells using the polymerase chain reaction. *J. Biol. Chem.* 265:11502-11507.

Shimizu, Y., Van Seventer, G.A., Horgan, K.J., and Shaw, S., 1990, Regulated expression and binding of three VLA (beta 1) integrin receptors on T cells. *Nature* 345:250-253.

Simons, K. and Fuller, S.D., 1985, Cell surface polarity in epithelia. *Ann. Rev. Cell. Biol.* 1:243-288.

Sonnenberg, A., Modderman, P.W., and Hogervorst, F., 1988, Laminin receptor on platelets is the integrin VLA-6. *Nature* 336:487-489.

Springer, T.A., 1990, Adhesion receptors of the immune system. *Nature* 346:425-434.

Staehelin, L.A., 1974, Structure and function of intercellular junctions. *Int. Rev. Cytol.* 39:191-278.

Stepp, M.A., Spurrmichaud, S., Tisdale, A., Elwell, J., and Gipson, I.K., 1990, Alpha-6-beta-4 integrin heterodimer is a component of hemidesmosomes. *Proc. Natl. Acad. Sci. USA* 87:8970-8974.

Suzuki, S., Argraves, W.S., Arai, H., Languino, L.R., Pierschbacher, M.D., and Ruoslahti, E., 1987, Amino acid sequence of the vitronectin receptor alpha subunit and comparative expression of adhesion receptor mRNAs. *J. Biol. Chem.*

Suzuki, S. and Naitoh, Y., 1990, Amino acid sequence of a novel integrin β_4 subunit and primary expressio of the mRNA in epithelial cells. *EMBO J.* 9:757-763.

Takada, Y., Huang, C., and Hemler, M.E., 1987, Fibronectin receptor structures in the VLA family of heterodimers. *Nature* 326:607-609.

Tamura, R.N., Rozzo, C., Starr, L., Chambers, J., Reichardt, L.F., Cooper, H.M., and Quaranta, V., 1990, Epithelial integrin $\alpha_6\beta_4$: Complete primary structure of α_6 and variant forms of β_4. *J. Cell Biol.* 111:1593-1604.

Tamura, R.N., Cooper, H.M., Collo, G., and Quaranta, V., 1991, Cell-type specific integrin variants with alternative α chain cytoplasmic domains. *Proc. Natl. Acad. Sci. USA* 88:10183-10187.

Wayner, E.A. and Carter, W.G., 1987, Identification of multiple cell adhesion receptors for collagen and fibronectin in human fibrosarcoma cells possessing unique α and common β subunits. *J. Cell. Biol.* 105:1873-1884.

Wayner, E.A., Carter, W.G., Piotrowicz, R.S., and Kunicki, T.J., 1989, The function of multiple extracellular matrix receptors in mediating cell adhesion to extracellular matrix receptors: Preparation of monoclonal antibodies to the fibronectin receptor that specifically inhibit cell adhesion to fibronectin and react with platelet glycoproteins Ic-IIa. *J. Cell. Biol.* 107:1881-1891.

Werb, Z., Tremble, P.M., Behrendtsen, O., Crowley, E., and Damsky, C.H., 1989, Signal transduction through the fibronectin receptor induces collagenase and stromelysin gene expression. *J. Cell Biol.* 109:877-889.

Wolf, G.T., Carey, T.E., Schmaltz, S.P., McClatchey, K.D., Poore, J., Glaser, L., Hayashida, D.J.S., and Hsu, S., 1990, Altered antigen expression predicts outcome in squamous cell carcinoma of the head and neck. *J. Natl. Cancer Inst.* 82:1566-1572.

DISCUSSION

R. JULIANO

Are focal contacts also formed?

V. QUARANTA

In vitro, yes. At least in the 804G cell line it looks like there must be focal contacts. We have not looked at that in detail as yet but as you could see in that experiment I showed you, the cells were in fact attached even though the hemidesmosomes have not formed. There are also other laboratories, for instance, Carter's laboratory, that described the presence of α_3-dependent adhesion plaques in cells that make $\alpha_6\beta_4$ and put it on the basal surface. I do not know what the situation is *in vivo*.

T. CAREY

First question, when you did the separation of the corneal epithelium with anti-α_6 antibody was that a monoclonal antibody?

V. QUARANTA

Yes, that is GOH3. And that also works with a different monoclonal antibody to α_6.

T. CAREY

Most of the experiments I have seen published with respect to detachment of cells with anti-β_4 reagents has been with the 5710 antiserum. What about other anti-β_4 antibodies, do you have any monoclonal antibodies that do the same thing?

V. QUARANTA

No, we have tried I think three or four different anti-β_4 monoclonal antibodies, with no effect. I should say also we were quite surprised by that result with GOH3.

T. CAREY

Can you explain why you were surprised?

V. QUARANTA

Because we expected it was not going to do anything. No, we were surprised because we felt that all the adhesion specificity was going to be in the β_4 subunit. It is possible, however, that the α_6-chain contains a sort of "molecular switch" that can be operated by GOH3 and causes cell detachment.

T. CAREY

My second question had to do with the A and B form expression in the keratinocytes.

V. QUARANTA

What we have recently done is to use both PCR and Western blotting to detect α_6A and -B expression at the mRNA and protein levels. We obtained human keratinocytes primary cultures from Clonetics (San Diego, CA), and grew them in serum-free, defined media (KGM). Both by PCR and Western, we only detect α_6A, not B, expression in normal human keratinocytes.

T. CAREY

We repeated the experiment with primers for the α_6A and α_6B forms using PCR and we found amplification of both forms. I like the idea of there being only one form expressed though, because it helps to explain a lot of things and so I was just curious about whether we might have a different culture environment than the one you use. We have not been able to show differences based on culture conditions ourselves. We found that both forms were present even after we induced differentiation with phorbol ester and after we grew the cells with calcium. I worried about our primers at first but they appear to be correct. Then I thought that there may be some other explanation why we are seeing two forms and you see only one.

V. QUARANTA

I would be curious to see whether your culture conditions influence B expression. This is potentially a very important point.

A. SONNENBERG

Do you suggest that cells which express $\alpha_6\beta_4$ may only recognize the ligand for $\alpha_6\beta_4$, when the integrin is localized in a hemidesmosome, or also when it is outside the hemidesmosome. Your data with keratinocytes suggest the former, because you only see inhibition of adhesion by antibodies against $\alpha_6\beta_4$ after 12 hours of culture, when hemidesmosomes have been formed.

V. QUARANTA

Well yes, what that suggested is that there are cells that do not assemble hemidesmosomes and yet express $\alpha_6\beta_4$ otherwise, I do not see any reason the Schwann cells, for example, would express $\alpha_6\beta_4$.

A. SONNENBERG

But in the keratinocytes, if it is not assembled in the hemidesmosomes is it functional?

V. QUARANTA

What I suggested is that in the keratinocytes, the $\alpha_6\beta_4$ may have a role in initiating the corneal hemidesmosomes. When you look at keratinocytes by immunostaining, one of the first molecules to appear, *in vitro*, where the hemidesmosomes will be is $\alpha_6\beta_4$.

A. SONNENBERG

However, again, I would like to point out that from your results it appears that $\alpha_6\beta$ does not become functional in keratinocytes until after 12 hours of culture, simultaneously with its incorporation in hemidesmosomes.

V. QUARANTA

The data we have published concern epiboly, a situation when epithelial cells migrate altogether, as a sheet advancing on one, or several, fronts. With that system, when you look at what happens to $\alpha_6\beta_4$ as cells are migrating, the migrating cells do express $\alpha_6\beta_4$ but, at least in our hands, it is diffused in the cytoplasm. As soon as the epithelial cells have reached a position in which they are going to stop, then $\alpha_6\beta_4$ becomes localized in the basal surface of the cell and then hemidesmosomes are assembled. Now of course while the cells are migrating they are also attaching. So, they have adhesion receptors, most likely the α_V or β_1 kind, that are mediating this attachment. If I understand your question correctly, the answer would be that, yes, epithelial cells or keratinocytes can be adherent cells in the absence of $\alpha_6 \beta_4$. In my mind, the way keratinocytes work is that there are initial stages of adhesion or migration that require integrins such as α_V. Then there is a second state of adhesion, once the epithelial sheet has formed, in which the $\alpha_6\beta_4$ becomes the major anchoring device and so the hemidesmosome then becomes the main means of attachment. What that suggests is also that, at one point, the hemidesmosome has to be disassembled perhaps in situations where the cell enter mitosis, or during self-renewal, or perhaps when migration is required. I was speculating that $\alpha_6\beta_4$ can be regulatory in such a way that it may cause hemidesmosomes to disassemble.

P. MARCHISIO

I have a comment and a general question. The comment is that the adhesive role of $\alpha_6 \beta_4$ in cultured keratinocytes appears when the cells have been kept for longer time in culture. This may suggest the possibility that these cells should assemble or organize molecules which are involved in the formation of a provisional basement membrane. This indicates that the adhesive role of $\alpha_6\beta_4$ requires the assembly of molecules which are formed by the epithelial cells themselves *in vitro*. This is the reason why we can see an adhesive role of $\alpha_6\beta_4$ when we carry out the adhesion tests on cells which have been kept for longer time in culture versus cell which have been just seeded in the culture. This is the general comment. The general question I would like to ask the rest of the audience and you, Vito, is, all the people who are involved in understanding the role of $\alpha_6\beta_4$ claim that the long cytoplasmic domain of $\alpha_6\beta_4$ is connected to the cytoskeleton epithelial cells. But this is just based on morphological evidence people have obtained. But is there any biochemical evidence that there may be a connection of the cytoplasmic domain of β_4 and the keratinocyte cytoskeleton. This is a very general topic we just address to everybody.

S. GOODMAN

Going back to this bovine epithelial-stripping experiment, I am even more surprised about that result, as the implication is that there is just the one integrin involved in mediating the attachment on the basement membrane. I mean, there are quite a few molecules present in the basement membrane. You are implying from your data that there is just the one adhesive structure, the hemidesmosome, which is not always the morphologically distinct structure which you can see. And all of the other integrins, surface molecules and ligands are not doing anything. Or, am I wrong? What is your interpretation of that result?

V. QUARANTA

I am glad you are bringing this up again. As I said, we were surprised. There are a number of explanations for this. My personal bias is that $\alpha_6\beta_4$, and hemidesmosomes do not explain everything. There are other things happening. For example, Lou Reichardt has shown that there is another integrin, $\alpha_8\beta_1$ localized, if I understand correctly, where $\alpha_6\beta_4$ also is. So, I do not know what that means. It could also be that α_6 is associated with other β-subunits, or that GOH3 cross-reacts with other α-chains. Another possibility is the existence of a "switch", located on the $\alpha6$ molecule, that is flipped by the GOH3 antibody, and that transmits a "detachment" signal to the cell interior. These are, of course, speculations.

N. HOGG

My colleague, Fiona Watt, who has a keratinocyte differentiation model, has shown that β_1 associated with α_2, α_3, and particularly α_5, have quite definite effects on the adherence of keratinocytes to matrix and determine whether they leave the basal layer or not, so it seems that there is evidence for β_1-type integrin association with basement membrane.

V. QUARANTA

Yes, I know that data well. I tend to agree that there must be something else happening.

G. NICOLSON

You passed over very quickly some of the changes that occur after transformation. For example, you mentioned that certain epithelial tissues show an increase in the expression of $\alpha_6\beta_4$, whereas in others there is a decrease of $\alpha_6\beta_4$. As you know, this oversimplification does not really give us much information about the tumors. For example, what is the relationship with tumor grade? The morphology from tumor to tumor is really quite different, the differentiation state is really quite different, and so on. What is the role of various integrins in maintaining tissue architecture? This could be very important. Could you give us more information on integrin expression related to cell differentiation, grade, and invasiveness?

V. QUARANTA

I am not a pathologist and I am not qualified to answer this but the way I understand it is that β_4 is almost always expressed in human carcinomas. There are also many indications of a possible correlation between β_4 polarization and tumor grade. For example, in tumors of the urinary tract, there seems to be some association between the way β_4 is expressed in topographical terms and the degree of malignancy. The idea is that if the polarization of β_4 is maintained, the tumor forms nests that express β_4 on the outside, in contact with the basement membrane laid down by the tumor or by the fibroblast. In this case, if I understand correctly, the degree of malignancy is lower. When the polarization of β_4 is lost, that seems to correlate with a more advanced degree of malignancy. A similar situation occurs in head and neck tumors.

G. NICOLSON

Well, it is obviously very complicated.

E. MIHICH

Is there any work trying to correlate the degree of "malignancy" in epithelium systems in culture as depicted by various keratin markers with the expression of the integrins?

V. QUARANTA

Like in skin cancer?

E. MIHICH

Yes.

V. QUARANTA

The data I know of say that basal cell carcinomas stop expressing β_4 while squamous carcinomas of the skin do express β_4 -- that is, the malignant cell does the opposite of the normal cell type it supposedly derived from in terms of β_4 expression.

P. MARCHISIO

We have recently compared the integrin distribution in basal cell carcinomas versus squamous cell carcinomas obtained from surgical samples and individually checked by pathologists. We have been able to find that the poorly malignant, non-metastasizing, basal cell carcinomas lose the expression of $\alpha_6\beta_4$ as well as the expression of one potential ligand of this integrin, the basement membrane protein nicein/kalinin. On the contrary, the highly malignant and metastasizing squamous cell carcinomas do express $\alpha_6\beta_4$ and nicein/kalinin. However, the cells of squamous cell carcinomas have lost the polarization of $\alpha_6\beta_4$ to the basal surface and $\alpha_6\beta_4$ is also found laterally. We may thus suggest that the metastasizing ability may depend on the capacity of recognizing the basement membrane and on the loss of polarized distribution of integrins, at least in these epidermal cancers.

V. QUARANTA

So, in summary, the normal basal cells express $\alpha_6\beta_4$, but when they transform they do not. In contrast, the spinous cells do not make $\alpha_6\beta_4$ but when they transform they do.

P. MARCHISIO

So the way is that normal basal cells make $\alpha_6\beta_4$ but, when they become transformed, they do not make it any longer. Instead, spinous cells, which do not make $\alpha_6\beta_4$ under normal conditions, express $\alpha_6\beta_4$ upon transformation.

D. LIVINGSTON

I wonder whether knocking out a potentially important integrin is a useful way to probe its biological function.

V. QUARANTA

I will give you an innocent answer. I think that the knock-out experiment is possible. I do not know if it is smart to do it in a cell line. I think another type of experiment would be the mirror image one, i.e., take a cell line that, for instance, does not express β_4 and make it express it and then look at the possible biological effects.

D. LIVINGSTON

I wondered also whether you might be able to ask whether the switch from the A to the B form in the α type integrin in the subunit of which we spoke might be also commanding in the development of certain organs?

V. QUARANTA

Yes.

D. LIVINGSTON

Could that also be done?

V. QUARANTA

Yes. Again, in that type of approach, the other experiment you could do is force the inappropriate expression of α_6A, for example. In fact those are, I think, critical experiments. Those are the experiments that are going to give us the answers that we need as to the relative importance of α_6A, α_6B, or β_4 in organ development. That is, whether or not, as you say, they take a commanding role during certain stages of development and differentiation.

G. MARGUERIE

Going back to the down-regulation of the expression, did you have a chance to look at samples from patients with skin disease at the expression of α_6 and β_4?

V. QUARANTA

Yes, but the data is so far dissappointing. There are some congenital or acquired skin diseases, in which blisters form. We are collaborating with a laboratory that has a large collection of tissue specimens from such epidermolysis bullosa cases. Thus far, we have nothing spectacular to report. I am aware of results from other laboratories, concerning a decrease in β_4 expression in skin-blistering diseases. We have seen nothing of the sort.

E. DEJANA

I am pretty confused. Is β_4 a unique requirement to go to hemidesmosome? I mean do you have any evidence that other integrins can also be organized or recruited like β_1, for instance, $\alpha_6\beta_1$?

V. QUARANTA

We have not been able to co-stain hemidesmosomes with anti-β_1 antibodies. That does not mean that other integrins may not be there. In fact, I suspect that something like that may be going on. Arnaud, do you have any data on that?

A. SONNENBERG

Together with David Garrod, we have done the following experiment. We have treated skin with NaCl which causes separation of the epidermis from the dermis. Subsequently, we have incubated the separated epidermis with antibodies against α_6, β_4, or β_1. Only antibodies against α_6 and β_4 stain the hemisdesmosome. The antigene which are stained by anti-β_1 is localized in between hemidesmosomes. Therefore, we have no evidence that β_1 integrins are localized in hemidesmosomes.

G. TARONE

May I ask one short question too, you showed that there are epithelial cell types and in particular the monostratified epithilium which do not form hemidesmosomes but still they express $\alpha_6\beta_4$. You have also identified two isoforms, A and B, of the α_6 subunit. Can you speculate whether perhaps the monostratified epithelia and keratinocytes express two different α_6 isoforms. In other words, is it possible that the information to organize hemidesmosomes reside in the specific $\alpha6$ cytoplasmic domain?

V. QUARANTA

Yes, we have looked in gut epithelium in the small intestine. There is an interesting situation there because the cells express $\alpha_6\beta_4$, on their basal surface, but this epithelium is continuously migrating from the bottom of the crypts, where probably the stem cells are. Continuous migration should be incompatible with the formation of hemidesmosomes -- and, in fact, no conventional hemidesmosomes are seen by electron microscopy in the small intestinal epithelium -- so, here you have a situation in which $\alpha_6\beta_4$ is contributing to substratum adhesion of cells that are mobile. I should mention, though, that we do not know what the precise structure of β_4 is in those cells. It is also possible that the $\alpha6$ cytoplasmic domain, A vs. B, may determine whether $\alpha_6\beta_4$ is a "migratory" receptor, e.g., in small intestine, or is an "anchoring" receptor, e.g., in skin.

G. TARONE

So you are implying that maybe β_4 is the one that plays the game in that case?

V. QUARANTA

That is one possibility in need of investigation.

A. SONNENBERG

Our results are different. We find $\alpha_6B\,\beta_4$, but not $\alpha_6A\,\beta_4$ in the intestine.

V. QUARANTA

But which part of the intestine?

A. SONNENBERG

It is the small intestine.

P. MARCHISIO

One possible model to study this problem could be the squamous-columnar junction in the uterine cervix, whether there is a very short transition between a columnar epithelium and a squamous epithelium and in that case you could see a sharp transition between α_6A and α_6B.

G. LENAZ

I have a more general question about the topographical localization in the cell. For instance is β_4 localized specifically after synthesis to the basal membrane or is it localized to any part of the membrane and then trapped where it finds its ligands? In the case of adhesive molecules the latter explanation could seem quite simple but there are molecules, as for example, membrane enzymes, which are also polarized but they probably do not have any ligands.

V. QUARANTA

That is a very interesting question. We find that there are situations in which a cell expresses $\alpha_6\beta_4$ in a non-polarized manner. The question is whether basal polarization of β_4 depends upon interaction with ligand or β_4 is redistributed as soon as the cell partitions the plasmamembrane in microdomains, i.e., in a ligand-independent way -- until the ligand for $\alpha_6\beta_4$ is well in hand. It will be cumbersome to obtain conclusive answers on this issue, which bears directly on the importance of $\alpha_6\beta_4$ to the establishment of the epithelial phenotype.

G. DOUGHERTY

I was wondering if there is any evidence that the crosslinking of $\alpha_6\beta_4$ or $\alpha_6\beta_1$ on carcinoma cells induces the release of metalloproteinases or other enzymes involved in the metastatic process?

V. QUARANTA

Not that I know of, but it is an interesting possibility. It brings up a view of $\alpha_6\beta_4$ as more than simply an adhesion receptor, capable of interfering with, or regulating other cellular pathways, such as proteolytic enzyme secretion.

FIBRONECTIN MUTATIONS IN MICE

Elisabeth N. Georges*, Elizabeth L. George, Helen Rayburn, and Richard O. Hynes

Center for Cancer Research, and
Howard Hughes Medical Institute
Department of Biology
Massachusetts Institute of Technology
Cambridge, MA 02139, U.S.A.

*Present address: L.G.M.E/C.N.R.S
U184 I.N.S.E.R.M
Faculté de Médecine
67000 Strasbourg, FRANCE

ABBREVIATIONS USED

FN = Fibronectin; **pFN** = Plasma Fibronectin; **cFN** = Cellular Fibronectin; **neo^R gene** = Neomycine Phosphotransferase Gene, conferring resistance to G418; **pgk-neo^R** = Neo gene under the control of the mouse phosphoglycerate kinase promoter; **pMC1neo^R** = Neo gene under the control of Herpes Simplex virus thymidine kinase promoter, and an enhancer from a mutant polyoma virus strain; **TK** = Thymidine Kinase Gene; **pgk-TK** = Thymidine Kinase Gene under the control of the mouse phosphoglycerate kinase promoter; **ES cells** = Embryonic Stem Cells

INTRODUCTION

Alternative Splicing Generates Isoforms of Fibronectin

Fibronectins (FNs) are a set of extracellular proteins found in the extracellular matrix and in blood plasma, which are involved in multiple functions, including cell adhesion and migration during development, hemostasis and thrombosis, or wound healing (see Hynes, 1990 for recent review). FNs comprise several isoforms, generated by alternative splicing of a single RNA transcript. Three sites of alternative splicing have been described: EIIIA, EIIIB, and V segments (sometimes called EDA, EDB, and IIICS). EIIIA and EIIIB can be completely included or excluded, while the V segment can be partially or totally included, depending on the splice site used (Figure 1).

The cell-type specificity of the splicing pattern has been demonstrated by several groups; for example, hepatocytes, which synthesize plasma FN (pFN) exclude EIIIA and EIIIB, while fibroblasts synthesizing cellular forms of FN largely include those two exons (Kornblihtt et al. 1985; Paul et al., 1986; Schwarzbauer et al., 1985; Schwarzbauer et al., 1987).

Specific Functions of FN Alternative Exons

The precise functions of FN isoforms are unclear. Schwarzbauer et al. (1989) have shown that homodimers of partial FNs lacking the V segment are not secreted, and that all pFN dimers contain at least one copy of the V segment. Recently, additional cell-binding sites for certain cell types have been found in the V segment (Humphries et al., 1986, 1987). In addition, in the study of recombinant FN isoforms, only the V form was found to be active in promoting the spreading of WEHI231 lymphoid cells (Guan and Hynes, 1990) The cell-binding site, contained in the V25 (or CS1) segment, appears to bind cells expressing a specific integrin receptor (α4-β1) (Wayner et al., 1989; Guan and Hynes, 1990; Mould et al., 1990). Thus, alternative splicing could be one mechanism to create FN molecules with different adhesive properties and binding only to certain cell types.

In the case of EIIIA and EIIIB, no such specific binding property has been found to date. However, a specific function for these exons is suggested by their tissue- and cell-specific expression, in situations such as embryogenesis, wound healing, or tumor formation. EIIIA and EIIIB are abundant in embryos but excluded in many adult tissues. Using specific antibodies, EIIIA and EIIIB expression have been demonstrated in fetal tissues and transformed cells or tumors (Castellani et al., 1986; Borsi et al., 1987; Vartio et al., 1987; Carnemolla et al., 1989). Ribonuclease protection experiments have shown that EIIIA and EIIIB segments were prevalent in mRNAs from early chicken embryos (Norton and Hynes, 1987; ffrench-Constant and Hynes, 1988, 1989) and in rat fetal liver and kidney (Pagani et al., 1991). At later stages of chick embryogenesis (E16), while EIIIA mRNAs were still predominant, EIIIB$^+$ mRNAs were less abundant (ffrench-Constant and Hynes, 1989).

When tissues are examined in more detail by *in situ* hybridization, EIIIA and EIIIB exons appear to be expressed in places and times when extensive cell migrations occur (Figure 2), such as the pathways of migrating neural crest cells, or in the cardiac jelly of the developing heart, in which endocardial cells migrate (ffrench-Constant and Hynes, 1988). Complex patterns of splicing with independent regulation of EIIIA and EIIIB inclusion or exclusion can be seen and some of the results are schematized in Figure 2. For example, in the liver of E16 chicken embryo, the vessel walls of large vessels contain A$^+$B$^-$V$^+$ FN mRNAs, while in adjacent smaller vessels, the FN mRNAs are A$^+$B$^+$V$^+$. As expected,

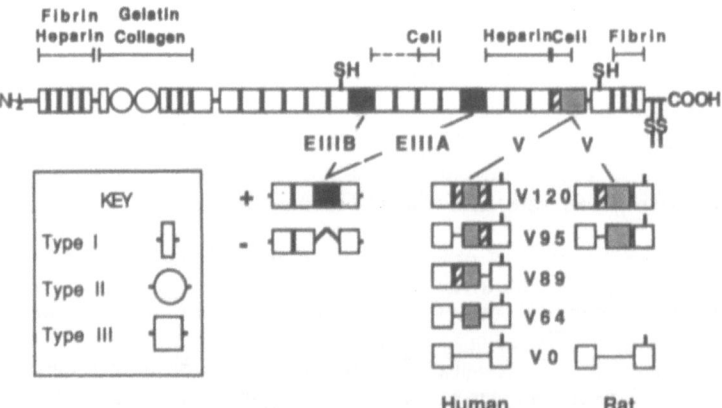

Figure 1. Structure of Fibronectin and Its Variants. Structure of an FN Subunit. Each box represents a repeating unit. Repeats can be of three types (Types I, II, III; see Hynes, 1990). Known binding sites, as well as positions of disulfide bonds or free sulfhydryl groups, are marked. The three alternative exons (EIIIA, EIIIB and V) are shown. EIIIA and EIIIB can either be included or excluded. The pattern of splicing of the V segment is complex, leading to several variants (V120, V95, V89, V64, V0 in human, V120, V95, V0 in rat) depending on splice sites. Modified after Hynes (1990).

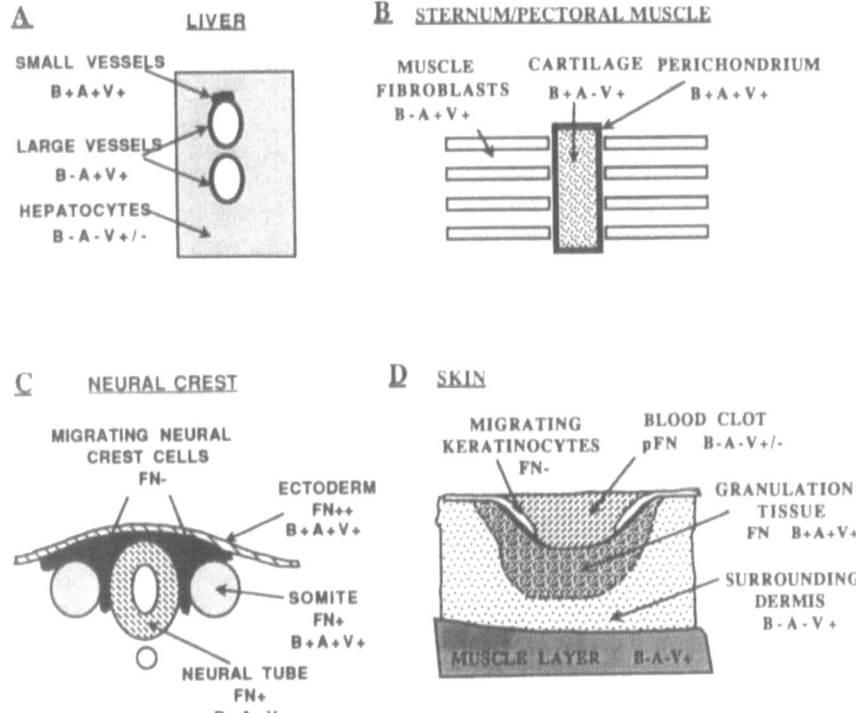

Figure 2. Examples of Alternative Splicing of Fibronectin. This figure is based on results of *in situ* hybridization experiments (ffrench-Constant and Hynes, 1988, and 1989; ffrench-Constant et al., 1989), with EIIIA, EIIIB and V segments as probes. These three exons are prevalent during early development in situations where cells are active in migration and proliferation (*e.g., neural crest as shown in C but also heart and area vasculosa*). In the embryonic liver (**A**) and pectoral muscle and sternum (**B**), EIIIA and EIIIB are included or excluded independently from each other. In rat adult skin after injury (**D**), there is an increase in FN synthesis in all layers, and the re-expression of EIIIA and EIIIB exons, normally absent from skin.

hepatocytes contain $A^-B^-V^+$ mRNA. Another example of splicing regulation of EIIIA and EIIIB is provided by the sternum and pectoral muscle (Figure 2). The perichondrium surrounding the developing cartilage contains a high level of A+B+V+ FN mRNA, while chondrocytes within the cartilage contain $A^-B^+V^+$ mRNA (ffrench-Constant and Hynes, 1989; Bennett et al.,1991). Instead, cells within the muscle were shown to contain mostly $A^+B^-V^+$ mRNA (Figure 2, ffrench-Constant and Hynes, 1989; Bennett et al., 1991).

Additional evidence for the role of EIIIA and EIIIB exons in cell migration and/or proliferation comes from the study of wound healing in adult rat skin. While those two exons are normally excluded in adult skin, they were shown to be reexpressed in cells at the base of a wound during the healing process (Figure 2, ffrench-Constant et al., 1989).

All these results point to possible roles of the EIIIA and EIIIB exons in the migration, proliferation or differentiation of specific cell types. However, such descriptive analyses can only suggest hypotheses for functions of FN isoforms, not test them. We were interested in conducting further analysis of the structure/function relationships of FN isoforms, and have recently taken a genetic approach to this question by generating mouse strains with an inactive FN gene, or alleles of the FN gene mutated in one alternative exon (EIIIB).

RESULTS AND DISCUSSION

The production of mutant mouse strains involves several steps. First, a targeting vector carrying the mutation of choice (knock-out or subtle mutation) is introduced into mouse embryonic stem (ES) cells. In a small fraction of cells, the vector will integrate into one of the two chromosomal loci by homologous recombination, yielding a heterozygous mutant cell (Thomas and Capecchi, 1987). The mutant cells are then injected into mouse blastocysts to generate chimaeric mice. Upon breeding, chimaeric mice in which injected ES cells have contributed to the germ-line will produce heterozygous progeny carrying the mutation, if the mutant allele is transmitted normally (i.e. is not a dominant negative mutation).

We have used two experimental strategies: the first was to generate a null allele of the FN gene, the second was to attempt to generate alleles carrying mutations in the alternative exon EIIIB.

Production of a FN-null Allele

Targeting Vector. Genomic clones of mouse FN gene were isolated from a lambda phage recombinant library of mouse genomic DNA using rat FN cDNA clones as probes (E. George et al., manuscript in preparation). DNA clones containing the promoter, 5' untranslated sequences and the first few coding exons of FN were used to construct the targeting vector. After removal of 0.8 kb including the basic promoter elements and the first ATG, a neo^R expression cassette (pgk-neo^R; neomycin phosphotransferase gene under control of the mouse phosphoglycerate kinase promoter region, conferring resistance to G418, Adra et al. 1987) was fused to the remainder of the first coding exon. At one end, the thymidine kinase gene of herpes simplex virus 1 (pgk-TK), conferring sensitivity to the drug Gancyclovir, was added to provide a selection against cells in which the TK gene is retained after random integration of the construct (Mansour et al. 1988). The resulting targeting vector is shown in Figure 3A.

Homologous Recombination in ES Cells. The targeting vector was electroporated into the D3 line of ES cells (Doetschman et al., 1985), which were then selected in culture medium containing G418 and Gancyclovir for 7-8 days. After electroporation of 1.7×10^8 cells, G418+GANC-resistant colonies were screened for correct integration by PCR in pools of 12. Clones from positive pools were then analyzed individually, and colonies were found that had integrated the construct by homologous recombination at the FN locus. Genomic DNA was isolated from those cells, and analyzed by Southern blot hybridization, using a FN probe not present in the vector. A restriction fragment of a different size was detected in addition to the endogenous fragment, confirming the presence of targeted alleles in the ES cell clones.

Production of FN-EIIIB$^-$ and FN-EIIIB$^+$ Mutant Alleles

Targeting Vectors. A FN genomic fragment corresponding to the EIIIB region was isolated from a lambda phage library as already mentioned. Two types of targeting vectors were constructed (Figure 3B). The first vector was designed to target an EIIIB-minus mutation in the FN gene. A fragment of 0.7 kb containing EIIIB sequences was deleted and replaced by a pMC1neo^R cassette (Thomas and Capecchi, 1987) in a transcriptional orientation opposite to that of FN. The second vector was designed to target a mutation so that the EIIIB exon should always be retained (EIIIB$^+$ mutation). To that end, a cDNA segment was synthesized, starting from 3T3 total RNAs, and amplified by PCR using primers in the III7b and III8a exons flanking EIIIB. This cDNA was substituted for the corresponding region containing EIIIB in the genomic clone. A pMC1neo^R cassette was inserted into the intron downstream of III8a to serve as a selection marker in transfected ES cells. These two targeting vectors also contain one (EIIIB$^+$) or two (EIIIB$^-$) copies of the pgkTK gene. The total length of FN sequences were 7.7 kb for the EIIIB$^-$ vector, and 6.6 kb for the EIIIB$^+$ vector.

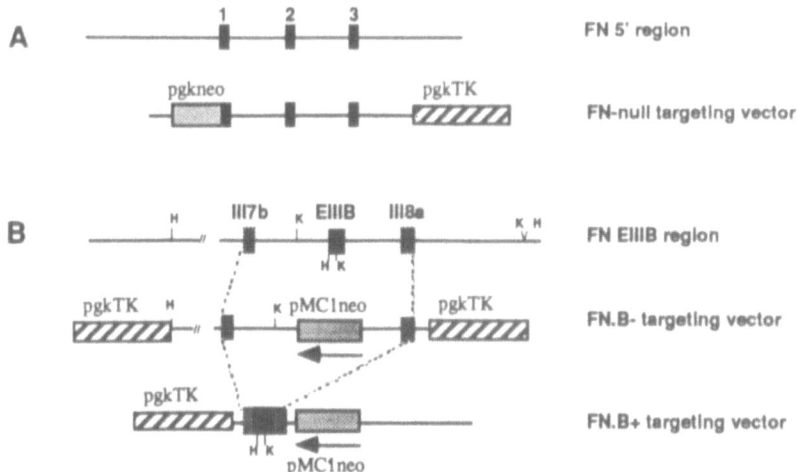

Figure 3. Targeting Vectors for Homologous Recombination.

A) For targeting a null mutation, a deletion was made in a genomic clone of the FN 5' region, and a pgk-neo^R cassette with its own promoter and polyadenylation signals (*shaded box*) was inserted. At the 3' end of the construct, a pgk-TK cassette (*cross-hatched box*) was added to allow negative selection against random integrants. Dark boxes represent exons 1, 2, and 3 of the fibronectin gene.

B) EIIIB Mutations. Top: the region of the fibronectin gene containing EIIIB (*black box*) and the two flanking exons, III7b and III8a (*shaded boxes*). For the FN.B- vector, a 0.7 kb deletion was made, encompassing the EIIIB exon and flanking intron sequences and a pMC1-neo^R cassette (*shaded box*) was inserted. For the FN.B+ vector, a cDNA segment (III7b/EIIIB/III8a, see text) was substituted for the genomic fragment containing those three exons. A pMC1-neo^R cassette was inserted as a selection marker. These two vectors both also contained pgk-TK cassette(s) for negative selection. The positions of restriction sites used for diagnosis of targeting events are marked (H:Hind III; K:KpnI).

Homologous Recombination in ES Cells. The EIIIB⁻ targeting vector was electroporated into 3.8 x 10^7 D3 ES cells, and cells were selected for 7 days in culture medium containing G418 and Gancyclovir. Double-resistant colonies were isolated and expanded and their genomic DNAs analyzed by Southern blot with a FN probe outside the construct. For one clone (4A1) a KpnI fragment of the expected size for a targeted allele was detected, in addition to the fragment from the wild-type allele. For the EIIIB+ mutation, 6x10^7 cells were electroporated, and double-resistant colonies were isolated after 6-7 days in selective medium, and analyzed by Southern blot hybridization, with a probe outside the vector. Two clones (6-10, 7-16) were identified as homologous recombinants, by the presence of additional restriction fragments after Hind III digestion.

FN Synthesis in FN.B⁻ and FN.B⁺ Targeted ES Cells

To test whether or not the FN.B⁻ and FN.B⁺ alleles lead to abnormal FN expression in the targeted ES cells, ES 4A1 (B⁻), 7-16 (B⁺), and D3 as a control, were analyzed by Northern blotting. Only the expected sized mRNAs for the FN and neo^R genes were detected. There were no RNAs containing both sequences and no accumulation of unspliced precursor. These experiments cannot rule out instability of the RNA transcribed from the mutant allele. To analyze synthesis of FN protein, the cells were pulse-labeled for 10 minutes with ^{35}S-Methionine and ^{35}S-Cysteine, and chased for varying times. FNs in the cell lysates were then immunoprecipitated using anti-FN antibodies and analyzed on SDS-polyacrylamide gels. No additional FN fragment was observed in 4A1 or 7-16 cell lines, when compared to the D3 control. This suggests that the mutated FN alleles that we introduced into those cells do not lead to the synthesis of abnormal or truncated FN molecules, although the presence of unstable abnormal FN molecules cannot be ruled out. No clear differences in the level of FN were observed among the different cell lines but, since clones 4A1 and 7-16 carry also a wild-type FN allele, no clear conclusion can be drawn concerning the *level* of expression from the mutated alleles.

Generation of Chimaeric Mice and Germ-Line Transmission of Targeted Alleles

To generate chimaeric mice, ES cells were injected into blastocysts from C57Bl6 females. The extent of chimaerism was judged by the amount of agouti hair in the coat. Males showing extensive coat-color chimaerism were then bred to C57Bl6 females. Contribution to the germ line, i.e. the presence of agouti progeny, varied depending on the cell line. In the case of the FN-null allele, three independent clones contributed to the germ line of male chimaeras (up to 57%). As for the EIIIB mutations, the EIIIB-minus clone and one of the EIIIB-plus clones were able to contribute to the germ line at a high frequency (up to 100%).

Agouti mice were then assessed for the presence of the different targeted alleles, by Southern blot analysis of tail DNAs. The results are shown in Table 1. Out of 55 agouti mice of the F1 generation, 34 were shown to carry one copy of the FN-null allele and 21 were homozygous wildtype. For the EIIIB⁻ mutation, 71 mice were wild-type, and 60 heterozygous and for the EIIIB⁺ mutation, 46 mice were wild-type, and 37 heterozygous. Thus, in all cases, the targeted allele was detected in numbers close to half of the agouti progeny, suggesting no dominant lethal effect by any of these three alleles.

Table 1. Transmission of Mutations in F1 Agouti Progeny

	+/+	+/-
FN-null	21	34
FN.B-	71	60
FN.B+	46	37

Analysis of Homozygous Progeny: FN-null, FN.B⁻, and FN.B⁺ Are Recessive Embryonic Lethal Mutations

FN-null Mutation. Heterozygous mice carrying the FN-null allele were interbred, and their progeny analyzed by tail-blot. Out of 139 mice of the F2 generation, 50 were wild-type homozygotes and 89 were heterozygotes. No homozygous mutant mice were found (Table 2). Litters from heterozygous intercrosses then were analyzed at different times of gestation. At the blastocyst stage (E3.5), FN-null homozygous embryos were found in the expected ratio, and appeared normal. In contrast, no homozygous embryos were present at E14, indicative of embryonic lethality after implantation. At E8.5, out of 31 embryos analyzed, 7 were FN-null homozygotes; those embryos appeared delayed compared to their littermates. This suggests that embryonic defects leading to the death of FN-null homozygotes begin around E8.5.

FN.B⁻ and FN.B⁺ Mutations. Breeding of FN.B⁻ heterozygotes also yielded wild-type and heterozygous mice in a ratio close to 1:2 (Table 2). No FN.B⁻ homozygous mutant live progeny were found, indicating that FN.B⁻ is a recessive embryonic lethal mutation, as is the FN-null mutation. Similarly, for the FN.B⁺ mutation, no homozygous mutant mice were found for the small number of animals examined to date (Table 2).

Litters from heterozygous intercrosses were analyzed at different stages of development. For the FN.B⁻ mutation, the presence of resorbed implantation sites was observed as early as E10.5. At E9.5, two sites contained only extra-embryonic membranes, which were shown by Southern blotting to be FN.B⁻/FN.B⁻. At E8.5, two homozygous mutant embryos were found and appeared delayed compared to their littermates. This suggests that embryonic defects of FN.B⁻ homozygotes begin around or before E8.5.

Table 2. Breeding of Heterozygotes

	+/+	+/-	-/-
FN-null	50	89	-
FN.B-	16	38	-
FN.B+	5	8	-

Similar results were obtained when FN.B$^+$ heterozygotes were intercrossed. No normal, homozygous, mutant embryos were obtained at E9.5-17.5. Instead, some resorbed implantation sites were observed, some of which contained a placenta. Although the number of embryos analyzed is small (24), these results suggest that embryonic defects leading to death of homozygous mutant embryos begin around E8.5.

CONCLUSION

We have generated three mouse strains carrying FN alleles altered by homologous recombination. Mice heterozygous for the FN-null allele, as well as for the FN.EIIIB mutant alleles are viable and do not show any obvious defects. It will be interesting to look at their responses to situations such as wound healing or tumor formation, where changes in FN expression have been documented. As for the defects leading to embryonic death of homozygous mutants, studies are in progress to define the nature and time of the defects, and to compare carefully the three mutant strains. The fact that defects begin to appear at or around 8.5 days suggests that gastrulation and the early steps of morphogenesis may be affected. This would be consistent with the postulated role of FNs in cell migration. It will be particularly important to determine whether the defects in the FN.B$^+$ and FN.B$^-$ strains are similar or different from those in the FN-null strain since such results may shed light on the functions of the EIIIB segment. However, it remains to be tested whether or not the mutations introduced have produced the desired biochemical defects or have had some other unanticipated effects on the function of the gene. Only by analyzing FN mRNA and protein expression in homozygous mutant cells or embryos will we be able to know whether the mutated alleles are expressed at normal levels and spliced in the way we expect.

Embryonic lethality confirms the importance of the FN gene for embryonic development. Further analyses involving transgenic mice expressing different alternatively spliced forms of FN interbred with the mutant strains described here should provide insights into the role of the different alternatively spliced segments.

ACKNOWLEGEMENTS

This work was supported by grants from the National Heart, Lung and Blood Institue and from the Howard Hughes Medical Institute. E.N.G. was supported in part by NATO and Fondation Philippe.

REFERENCES

Adra, C.N., Boer P.H., and McBurney, M.W., 1987, Cloning and expression of the mouse *pgk*-1 gene and the nucleotide sequence of its promoter. *Gene* 60:65-74.

Bennett, V.D., Pallante, K.M., and Adams, S.L., 1991, The splicing pattern of fibronectin mRNA changes during chondrogenesis resulting in an unusual form of the mRNA in cartilage. *J. Biol. Chem.* 266:5918-5924.

Borsi, L., Carnemolla, B., Castellani, P., Rosellini, C., Vecchio, D., Allemanni, G., Chang, S.E., Taylor-Papadimitriou, J., Pande, H., and Zardi, L., 1987, Monoclonal antibodies in the analysis of fibronectin isoforms generated by alternative splicing of mRNA precursors in normal and transformed human cells. *J. Cell Biol.* 104:595-600.

Carnemolla, B., Balza, E., Siri, A., Zardi, L., Nicotra, M.R., Bigotti, A., and Natali, P.G., 1989, A tumor-associated fibronectin isoform generated by alternative splicing of messenger RNA precursors. *J. Cell Biol.* 108:1139-1148.

Castellani, P., Siri, A., Rosellini, C., Infusini, E., Borsi, L., and Zardi, L., 1986, Transformed human cells release different fibronectin variants than do normal cells. *J. Cell Biol.* 103:1671-1677.

Doetschman, T. C., Eistetter, H., Katz, M., Schmidt, W. and Kemler, R., 1985, The *in vitro* development of blastocyst-derived embryonic stem cell lines: Formation of visceral yolk sac, blood islands and myocardium. *J. Embryol. Exp. Morphol.* 87, 27-45.

ffrench-Constant, C. and Hynes, R.O., 1988, Patterns of fibronectin gene expression and splicing during cell migration in chicken embryos. *Development* 104:369-382.

ffrench-Constant, C. and Hynes, R.O., 1989, Alternative splicing of fibronectin is temporally and spatially regulated in the chicken embryo. *Development* 106:375-388.

ffrench-Constant, C., Van De Water, L., Dvorak, H.F., and Hynes, R.O., 1989, Reappearance of an embryonic pattern of fibronectin splicing during wound healing in the adult rat. *J. Cell Biol.* 109:903-914.

Guan, J-L. and Hynes, R.O., 1990, Lymphoid cells recognize an alternatively spliced segment of fibronectin via the integrin receptor $\alpha 4 \beta 1$. *Cell* 60:51-63.

Humphries, M.J., Akiyama, S.K., Komoriya, A., Olden, K., and Yamada, K.M., 1986, Identification of an alternatively spliced site in human plasma fibronectin that mediates cell type-specific adhesion. *J. Cell Biol.* 103:2637-2647.

Humphries, M.J., Akiyama, S.K., Komoriya, A., Olden, K., and Yamada, K.M., 1988, Neurite extension of chicken peripheral nervous system neurons on fibronectin: Relative importance of specific adhesion sites in the central cell-binding domain and the alternatively spliced type III connecting segment. *J. Cell Biol.* 106:1289-1297.

Hynes, R.O., 1990, "Fibronectins," Springer-Verlag, New York.

Kornblihtt, A.R., Umezawa, K., Vibe-Pedersen, K., and Baralle, F.E., 1985, Primary structure of human fibronectin: Differential splicing may generate at least 10 polypeptides from a single gene. *EMBO J.* 4:1755-1759.

Mould, A.P., Wheldon, L.A., Komoriya, A., Wayner, E.A., Yamada, K.M., and Humphries, M.J., 1990, Affinity chromatographic isolation of the melanoma adhesion receptor for the IIICS region of fibronectin and its identification as the integrin a4b1. *J. Biol. Chem.* 265:4020-4024.

Mansour, S. L., Thomas, K. R., and Capecchi, M. R., 1988, Disruption of the proto-oncogene int-2 in mouse embryo-derived stem cells: A general strategy for targeting mutations to non-selectable genes. *Nature* 336:348-352.

Norton, P.A. and Hynes, R.O., 1987, Alternative splicing of chicken fibronectin in embryos and in normal and transformed cells. *Mol. Cell Biol.* 7:4297-4307.

Pagani, F., Zagato, L., Vergani, C., Casari, G., Sidoli, A., and Baralle, F.E., 1991, Tissue-specific splicing pattern of fibronectin messenger RNA precursor during development and aging in rat. *J. Cell Biol.* 113:1223-1229.

Paul, J.I., Schwarzbauer, J.E., Tamkun, J.W., and Hynes, R.O., 1986, Cell-type-specific fibronectin subunits generated by alternative splicing. *J. Biol. Chem.* 261:12258-12265.

Schwarzbauer, J.E., Paul, J.I., and Hynes, R.O., 1985, On the origin of species of fibronectin. *Proc. Natl. Acad. Sci. USA* 82:1424-1428.

Schwarzbauer, J.E., Patel, R.S., Fonda, D., and Hynes, R.O., 1987, Multiple sites of alternative splicing of the rat fibronectin gene transcript. *EMBO J.* 6:2573-2580.

Schwarzbauer, J.E., Spencer, C.S., and Wilson C.L., 1989, Selective secretion of alternatively spliced fibronectin variants. *J. Cell Biol.* 109:3445-3453.

Thomas, K.R. and Capecchi, M.R., 1987, Site-directed mutagenesis by gene targeting in mouse embryo-derived stem cells. *Cell* 51:503-512.

Vartio, T., Laitinen, L., Narvanen, O., Cutolo, M., Thornell, L.E., Zardi, L., and Virtanen, I., 1987, Differential expression of the ED sequence-containing form of cellular fibronectin in embryonic and adult human tissues. *J. Cell Sci.* 88:419-430.

Wayner, E.A., Garcia-Pardo, A., Humphries, M.J., McDonald, J.A., and Carter, W.G., 1989, Identification and characterization of the T lymphocyte adhesion receptor for an alternative cell attachment domain (CS-1) in plasma fibronectin. *J. Cell Biol.* 109:1321-1330.

DISCUSSION

D. LIVINGSTON

Can you isolate homozygous mutant cells and inject them into normal embryos?

E. GEORGES

Yes, I could do it two ways. The first would be to target ES cells twice, one could also isolate ES cells from homozygous blastocysts.

D. LIVINGSTON

And let the chimera go to term, you might expect to find B⁻ cells in the appropriate organs. Have you tried that experiment?

E. GEORGES

No I have not tried that experiment. I have not isolated homozygous mutant ES cells yet. That would be interesting to try. Fibronectin is secreted, so we do not know if mutant cells would not be rescued by a normal environment. It seems possible to me that some organs would contain a mixture of matrix made by normal cells and by mutant cells, maybe in some places you would see a defect.

D. LIVINGSTON

You may have a normal matrix loaded with B⁻, but also with wild-type protein. As long as it is not something mysterious, while a normal cell is loading the matrix with wild-type fibronectin, the neighbor cell can then load it with mutant fibronectin. Then, in theory, you ought to get chimeras, and, therefore, mosaic organs.

E. GEORGES

I think it was shown in the work by Guan et al. with recombinant fibronectin that, *in vitro*, B⁻ fibronectin can incorporate into preexisting matrices.

G. TARONE

You have shown that the B⁻ mutants synthesize fibronectin. Do you know if they assemble it into the matrix?

E. GEORGES

We are studying this now.

G. TARONE

This may be interesting because the B region fibronectin is close to the RGD side that interact with the cell surface receptor and there are some indications, from the work of K. Yamada, showing that this region close to the RGD, which includes the B region, regulate the affinity of fibronectin for its receptor.

E. GEORGES

We do not know for sure that the mutant embryos are secreting fibronectin. For now we have just run a gel with culture medium from embryo cultures. We are doing the immunoprecipitation now, and in parallel we are looking at the cultured embryos by immunofluorescence after staining with an anti-fibronectin, to address the question of matrix assembly. I do not know the answers yet.

G. TARONE

So is it possible that removing the B region you decrease the total affinity of the fibronectin for the receptor and that is why fibronectin may not go into the matrix?

E. GEORGES

This is quite possible, we do not know.

J. McCARTHY

Have you examined the ability of these mutants to produce other extracellular matrix components?

E. GEORGES

I have not looked at that yet, we need markers to show that.

G. MARGUERIE

Can B⁻ be suicidal in heterozygous animals?

E. GEORGES

If you are talking about a dominant negative effect, we have not observed any obvious defect in the heterozygous animals.

V. QUARANTA

Did you do any staining with antibodies to the B region of fibronectin?

E. GEORGES

We have not been able to do it yet.

V. QUARANTA

What are your methods for embryo studies?

E. GEORGES

I dissect out litters from heterozygotes intercrosses at day 7.5 to 8 of gestation and then dissociate the embryos, and culture the pieces for a few days. Then we analyze the culture medium for the presence of fibronectin.

M. HEMLER

Would it be feasible to try to immortalize any of these cells at an early stage to make cell lines? Since fibronectin sometimes has a tumor-suppressive effect, would you expect that the absence of fibronectin would make any of these mice more sensitive to carcinogens?

E. GEORGES

Maybe, we have not done anything yet about tumorigenicity, but we are monitoring heterozygous animals for tumor incidence.

G. TARONE

In adult tissue, the B^+ form of fibronectin is switched in tumors or during angiogenesis, so it will be interesting to look at this heterozygous which can only express low levels of B form.

S. GOODMAN

Does your B^+ transgenic mean that there is no B^- expressed in the animal?

E. GEORGES

Yes, that is what it means.

V. QUARANTA

Did you try to generate aggregation chimeras between homozygous B^- embryos and normal embryos?

E. GEORGES

We need to do it.

GENERAL DISCUSSION -- SESSION I

G. TARONE

There was a lot of discussion on the keratinocytes integrins and some questions have been left open.

M. HEMLER

I would like to bring up a somewhat controversial issue. Is $\alpha_6\beta_4$ a receptor for laminin? There have been multiple papers suggesting that it is a laminin receptor, and even more papers suggesting that it is not.

T. CAREY

We began to wonder about the issue of what $\alpha_6\beta_4$ actually does bind to, and I think the evidence reported from Mercurio's laboratory is that in the colon carcinoma $\alpha_6\beta_4$ serves as a receptor for laminin and I think it was shown fairly clearly that there is no β_1 there. Vito Quaranta also said that α_6 associates predominantly with β_4, and β_4 is there. But I suspect that there is another ligand for $\alpha_6\beta_4$ and perhaps laminin is a receptor for migration and there is another receptor for anchorage. Perhaps Dr. Sonnenberg might want to comment on this, since he has said several times that $\alpha_6\beta_1$ is a laminin receptor and $\alpha_6\beta_4$ is not.

A. SONNENBERG

In none of the cell lines that we have used, did we found evidence that $\alpha_6\beta_4$ binds to laminin. Binding of these cells to laminin could always be completely inhibited by both anti-α_6 and anti-β_1 antibodies. Also, by laminin-affinity chromatography of a human mammary gland cell line, which expresses both $\alpha_6\beta_1$ and $\alpha_6\beta_4$, we could only isolate $\alpha_6\beta_1$. Maybe the affinity of $\alpha_6\beta_4$ for laminin depends on activation.

V. QUARANTA

We had a similar experience. We repeatedly tried passing radiolabeled lysates from pancreatic or lung carcinoma cell lines over affinity columns made with either human or mouse laminin. We could never recover bound $\alpha_6\beta_4$, while we could easily recover $\alpha_3\beta_1$. We insisted on these trials by using freshly purified mouse laminin, again with no luck. I am also aware of analogous experiments in Dr. Eva Engvall's laboratory, with a colon carcinoma cell line that expresses high levels of $\alpha_6\beta_4$. As far as I am aware, they also could not show binding of $\alpha_6\beta_4$ to laminin, although I do not know the details. Again, as Arnaud pointed out, it is possible that we are all using the wrong cell lines. We obtained the clone A cells from Art Mercurio, who published that $\alpha_6\beta_4$ is a ligand for laminin, but have yet performed chromatography. It is possible that $\alpha_6\beta_4$ is "activated" in some cell lines and not in others. Subtle structural variations cannot also be dismissed. Concerning the point Martin Hemler made on the inhibition of laminin adhesion by the anti-$\alpha6$ antibody GOH3 in cells that are positive for $\alpha_6\beta_4$ but apparently lack $\alpha_6\beta_1$, I would like to mention another puzzling observation we made. Together with J. Jones group, we found that GOH3 causes detachment of corneal keratinocytes from their natural basement membrane, to which they presumably adhere through $\alpha_6\beta_4$-based hemidesmosomes. In contrast, anti-β_4 polyclonal antibodies caused hemidesmosome disassembly, but no cell detachment (Kurpakus et al., *J. Cell Biol. 115:1737, 1991*).

P. MARCHISIO

In normal human keratinocytes in culture, which is our favorite model, we find that there is a deposition, within the matrix which is assembled by the keratinocytes, of laminin, collagen type 4, the bullous pemphigold antigen and kalinin/nicein or as you want to call it. These molecular structures which are either intracellular or extracellular, do colocalize with $\alpha_6\beta_4$ and when we have studied the time course of their exposure and their deposition within the culture, we have found that there is a perfect coherence between the production of $\alpha_6\beta_4$ and its exposure and extracellular organization of different molecules indicating that $\alpha_6\beta_4$ may have a role in organizing the extracellular matrix spatially and topographically.

G. TARONE

It is surprising that $\alpha_6\beta_4$ is one of the few integrins for which a ligand is not yet been identified. Is it possible that $\alpha_6\beta_4$ is usually inactive and it requires some sort of activation to bind its ligand?

P. MARCHISIO

May I stress again the concept that $\alpha_6\beta_4$ may work in organizing the basement membrane and the topography of the basement membrane components as $\alpha_5\beta_1$ does for organizing fibronectin fibers outside the cells.

G. NICOLSON

The fact that we have these disparate experiments suggests that it might be something other than fibronectin involved, such as the carbohydrate on laminin. Is there any evidence that this particular integrin might bind to carbohydrate and that might explain its lack of retention in an affinity column unless the right carbohydrate is used?

A. SONNENBERG

We used two sources of laminin, murine EHS laminin and human placenta laminin, the latter being a mixture of different laminin isoforms.

T. CAREY

We found that from squamous carcinoma cells we can detect $\alpha_6\beta_4$ but we can precipitate no β_1 with anti-α_6 antibodies. We have plenty of β_1 on the cells, the cells express α_2 in abundance, α_3 in abundance, β_1 in abundance, no α_1, they make a small amount of α_5 and lots of α_6 and lots of β_4. These cells over express $\alpha_6\beta_4$, that means they do not polarize it and they make a huge amount of it. They also stick to human laminin very well and to mouse laminin less well. We can inhibit laminin binding with anti-α_6, α_2, and α_3 so, based on the immunoprecipitation data, we think that it binds to laminin at least in part through the $\alpha_6\beta_4$ receptor but we can't prove it unless we can show that $\alpha_6\beta_4$ binds to a laminin column. Anti-β_1 blocks binding to laminin in these cells, so $\alpha_2\beta_1$ and $\alpha_3\beta_1$ serve as laminin receptors as well. Alternatively, since the anti-β_1 antibody that everybody uses to inhibit was selected by screening or blocking adhesion, maybe there's something special about that anti-β_1 antibody because I do not think that is true of all anti-β_1 antibodies.

M. HEMLER

I will just briefly comment that for every $\alpha_6\beta_4$-positive cell line that we have looked at, adhesion to laminin has been nearly 100% blocked by antibodies to either α_6 or to β_1. Therefore, although α_6 may appear to be associated with β_4 by immunoprecipitation, perhaps in some cellular compartment α_6 may actually be associated with β_1. That would possibly explain the mysterious anti-β_1-blocking result that we have not been able to understand for several years.

T. CAREY

Have you done colocalization studies on those experiments to see if when the cells are stuck to laminin they colocalize the integrin with the laminin? It could be that like in the chicken laminin receptor, there may be trimer formation. With tissues co-localization of β_1 integrins with basal lamina does not seem to be the case. With tissues it seems to be that $\alpha_6\beta_4$ is in the hemidesmosome and β_1-containing integrins are not there. Does anyone think that multimer formation by these integrins is important for attachment?

S. GOODMAN

Can we summarize by saying that the $\alpha_6\beta_4$ ligand is not known at the moment? Do you agree with that? Then the cytoplasmic side of β_4 is very interesting. Hemidesmosomes form so beautifully in these cells. So there is a difference between the literature which I was involved with for a number of years about hemidesmosomes and desmosomes. Several structural components found in the desmosomes are also found in the hemidesmosomes, so there is this question of what is keeping $\alpha_6\beta_4$ out of the desmosome.

A. SONNENBERG

The term hemidesmosome is used because under the electron microscope, they look as if they are a half desmosome. Furthermore, both hemidesmosomes and desmosomes are sites for attachment of the keratin-filaments to the plasma membrane. However, at the molecular level, as far as I know, none of the components that have been identified in desmosomes have also been found in hemidesmosomes and vice versa.

REGULATION OF β1 INTEGRIN-MEDIATED ADHESIVE FUNCTIONS

Francisco Sánchez-Madrid, Alicia G. Arroyo, Miguel R. Campanero, and Paloma Sánchez-Mateos

Servicio de Inmunología
Hospital de la Princesa
Universidad Autónoma de Madrid
c/ Diego de León, n° 62 - 28006
Madrid, SPAIN

ABBREVIATIONS USED

COL = Collagen; **ECM** = Extracellular Matrix; **FN** = Fibronectin; **ICAM-1** = Intercellular Adhesion Molecule-1; **LFA-1** = Leukocyte Function-associated Antigen-1; **LN** = Laminin; **RGD** = Arginine-Glycine-Aspartic Acid Sequence; **VCAM-1** = Vascular Cell Adhesion Molecule-1; **VLA** = Very Late Activation Antigen

INTRODUCTION

The integrin family includes receptors for extracellular matrix (ECM) components as well as receptors involved in cell-cell adhesive interactions (Hynes, R.O, 1992; Hemler, M.E, 1990; Springer, T.A, 1990; Ruoslahti, E, 1991). The VLA subfamily of integrins is comprised of at least eight heterodimers, each with a common β subunit (the β1 chain) non-covalently associated with one of eight different α subunits (Hynes, R.O, 1992; Hemler, M.E, 1990). In addition to α:β1 heterodimers, novel associations of α and alternative β subunits have been found which expand the molecular and functional repertoire of integrins (Hynes, R.O, 1992; Hemler, M.E, 1990). The involvement of VLA integrins in cell interactions either with extracellular matrix (ECM) components or with cellular ligands, appears to be essential for lymphocyte homing (Springer, T.A, 1990; Holzmann, B. and Weissman, I.L., 1989), embryogenesis and histogenesis (Darribière, T. et al., 1990), leukocyte activation (Davis, L.S. et al., 1990; Matsuyama, T. et al., 1989; Nojima, Y. et al., 1990; Shimizu, Y. et al., 1990), cytotoxic T lymphocyte activity (Clayberger, C. et al., 1987; Takada, Y. et al., 1989), and lympho-hemopoiesis (Miyake, K. et al., 1991).

All the members of the VLA integrin subfamily function as extracellular matrix receptors, as demonstrated by affinity chromatography, cell binding assays, and monoclonal antibody blocking experiments. The functional promiscuity of this recognition is illustrated by the fact that most β1 integrins recognize more than one extracellular ligand. Furthermore, the binding properties of individual VLA heterodimers depend on the cell line where it is expressed. Thus, VLA-2 may function either as a laminin (LN) or collagen (COL) receptor on different cell types (Elices, M.J. and Hemler, M.E., 1989; Kirchhofer, D. et al., 1990). Additional functional overlapping is obtained through the recognition of a single extracellular matrix component by several β1 heterodimers. Accordingly, laminin is recognized by VLA-1, VLA-2, VLA-3, VLA-6 and VLA-7 (Hynes, R.O, 1992; Hemler; M.E, 1990; Kramer, R.H. et al., 1989). Similarly, VLA-1, VLA-2, and VLA-3 recognize collagen, and cell attachment to fibronectin (FN) can be mediated through VLA-3, VLA-4, VLA-5, and the heterodimer αVβ1 (Hynes, R.O, 1992; Hemler, M.E, 1990). On the other hand, there are distinct cell-binding sites on fibronectin for VLA-4 (Wayner, E.A. et al., 1990; Mould, A.P. et al., 1991) and VLA-5 (Ruoslahti, E. and Pierschbacher, M.D., 1987), as well as different cell-binding sites on laminin for VLA-1 and VLA-6 (Hall, D.E. et al., 1990; Sonnenberg, A. et al., 1990).

In addition to their ECM-binding ability, some members of the VLA integrins mediate intercellular adhesion. Thus, VLA-4 is involved in lymphocyte, monocyte, eosinophil, and basophil attachment to cytokine-activated endothelium (Elices, M.J. et al., 1990; Dobrina, A. et al., 1991, Walsh, G.M. et al., 1991; Weller, P.F. et al., 1991; Schleimer, R.P. et al., 1992), through recognition of VCAM-1, an endothelial cell surface glycoprotein which belongs to the immunoglobulin gene superfamily (Osborn, L. et al., 1989). Additional cell surface counter-receptors for VLA-4 may exist that explain the VLA-4-induced homotypic aggregation of leukocytes (Campanero, M.R. et al., 1990; Bednarczyk, J.L. and McIntyre, B.W., 1990; Pulido, R. et al., 1991), the VLA-4 dependent/VCAM-1-independent adhesion of certain leukocytes to stimulated endothelial cells (Vonderheide, R.H. and Springer, T.A., 1992), and the VLA-4-dependent CTL-target interaction (Clayberger, C. et al., 1987; Takada, Y. et al., 1989).

The ability of integrins to undergo cycles of avidity regulation may be critical for cell migration on the surface of other cells or through the extracellular matrix. A gradient of integrin avidity along the cell membrane may drive migration (Dustin, M.L. and Springer, T.A., 1991). The adhesion-promoting activity of several members of the integrin family has been found to be rapidly and reversibly modulated. There is evidence for changes from low to high affinity ligand-binding states in different cellular systems. Thus, the active form of the integrin αIIbβ3 (platelet GPIIb-IIIa) is induced by peptides containing the arginine-glycine-aspartic acid (RGD) sequence (Du, X. et al., 1991), and the integrin CR3 (αMβ2) is regulated by a lipid (Hermanowski-Vosatka, A. et al., 1992). On resting T lymphocytes, the three integrins VLA-4, VLA-5 and VLA-6 have low constitutive avidity for FN and LN, respectively, which can be increased upon cell activation (Shimizu, Y. et al., 1990). Similarly, avidity of LFA-1 for its counterreceptor ICAM-1 is also regulated by cell activation (Dustin, M.L. and Springer, T.A., 1989; Van Kooyk, Y. et al., 1989). Activated B cells display an enhanced VLA-4-mediated binding to FN as well as to VCAM-1 on follicular dendritic cells as compared to resting B cells (Postigo, A. et al., 1991; Freedman, A.S. et al., 1990; Koopman, G. et al., 1991). This regulation also occurs "*in vivo*" as demonstrated by studies on the expression and function of VLA-4 and VLA-5 integrins on synovial T cells from rheumatoid arthritis patients. An upregulated function of these adhesion receptors was observed on "*in vivo*" activated synovial T cells suggesting that VLA-4 and VLA-5 receptors could play a role in the pathogenesis of certain inflammatory processes (Laffón, A. et al., 1991; Postigo, A.A. et al., 1992; García-Vicuña, R. et al., 1992).

The involvement of β-chains in regulating the affinity of different integrins for their corresponding ligands has been postulated. Evidence indicating that the ligand affinities of integrins can be modulated through the β1, β2, and β3 chains has been recently reported in different cellular systems (Figdor, C.G. et al., 1990; Neugebauer, K.M. and Reichardt. L.F.,1991; O'Toole, T.E. et al., 1990). Also, the cytoplasmic domain of the LFA-1 integrin β2 subunit is required for proper regulation of adhesion to ICAM-1 (Hibbs, M.L. et al., 1991):

We have investigated the possible role of the common β1 subunit of VLA integrins in the regulation of adhesive functions mediated by different heterodimers. The participation of different VLA integrins in leukocyte spreading has also been studied. In addition, we have examined protein tyrosine phosphorylation during cell attachment and spreading in different integrin-ligand interactions. Herein, we summarize our recent findings on the regulatory role of β1-chain in VLA integrin-mediated cell aggregation, cell attachment and spreading, and signal transmission.

Table 1. β1-mediated Induction of Cell Aggregation on Different Cell Lines

Cell Line/Type	Induction of Cell Aggregation	
	Induction by Anti-β 1 mAb	Induction by Anti-α4 mAb
Resting T lymphocytes	+	+
Resting B lymphocytes	ND	+
Neutrophils	-	-
T leukemic		
Jurkat	+	+
Peer	+	+
JM	+	+
B lymphoblastoid		
Ramos	+	+
JY	-	+
Erythroleukemic		
K562	-	-
α2-transfected K562	+	-
α4-transfected K562	+	+
Promyelocytic		
U937	+	+
HL60	+	+

Cell aggregation was induced on the different leukocyte cell lines/types with either the anti-β1 Lia 1/2 or the anti-α4 HP1/7 mAb. + and - indicate the ability or the unability to trigger cell aggregation, respectively. ND indicates not determined.

RESULTS AND DISCUSSION

Regulation of Cell Aggregation through the VLA β1 Chain

Whereas all of the integrins in the VLA protein subfamily are involved in cell-extracellular matrix interactions, only VLA-4 (through the α4 subunit) has been implicated in the triggering of intercellular adhesion. We have previously described the triggering of homotypic leukocyte aggregation by mAb specific for certain epitopes on VLA-α4 (Campanero, M.R. et al., 1990). We have found that the VLA β1 subunit (CD29) is also involved in the induction of homotypic cell aggregation.

We have obtained three novel anti-β1 mAb (Lia 1/2, Lia 1/5, and Alex 1/4) with the ability to induce cell aggregation on different leukocyte cell types (Table 1). These mAb recognize an antigenic site on the common β1 chain of VLA proteins which is topographically and/or functionally distinct from other epitopes previously defined by several prototype anti-β1 mAb. Induction of cell aggregation by anti-β1 mAb is epitope specific, isotype and Fc-independent, and displays kinetics similar to α4-mediated aggregation. This cell aggregation requires an intact cellular metabolism, the presence of divalent cations in the extracellular medium, and the integrity of cytoskeleton. We also have found that the Na+/H+ antiporter may be essential for this process. For Ramos cells, which bear only the VLA α4β1 heterodimer, intercellular adhesion induced through the β1-chain could be selectively inhibited by other anti-β1 mAb as well as by anti-α4 mAb (Campanero, M.R. et al., 1992).

Other components of the VLA family, such as the VLA-2 and VLA-3 integrins, which can bind to collagen, laminin, or fibronectin (Hemler, M.E, 1990), have also been suggested to be involved in cell-cell interactions. For example, anti-VLA-α3 mAb were able to block cell-cell interactions and anti-α2 mAb specifically bound to keratinocyte intercellular contact sites (Carter, W.G. et al., 1990). However, their involvement in cell-cell interactions has not been definitively proven. Interestingly, anti-β1 mAb which induced strong aggregation of VLA-α2 or VLA-α4 transfected K562 cells, had minimal effect on the α2⁻ α4⁻ α5⁺ K562 cell line (Table 1). Furthermore, the β1-mediated induction of cell aggregation on α2-K562 and α4-K562 transfected cells was blocked by preincubation with either anti-α2 or anti-α4 mAb, respectively, as well as by other anti-β1 mAb. These results provide strong evidence supporting the involvement of α2β1 and α4β1 heterodimers in intercellular interactions and underline the pivotal role of the common β1-chain of VLA proteins in the integrin-mediated induction of cell aggregation.

Altogether, these observations suggest the possibility that different members of the VLA subfamily of integrins could mediate cell-cell interactions on different cellular systems. Thus, VLA-4 could represent the VLA integrin used by leukocytes to interact either with other leukocytes or with endothelial cells, and VLA-2 could be also playing an important role in homotypic intercellular interactions among other cell types, such as epidermal cells. Further studies to elucidate the role of VLA-α2 in the triggering of cell aggregation are required. The availability of the aggregation-inducer anti-β1 mAb in conjunction with cDNA encoding other VLA-α subunits may be very useful to analyze the role of these α chains in intercellular interactions.

Regulation of Cellular Adhesion through the β1 Chain of VLA Integrins

We have investigated the functional regulatory role of the common β1 subunit of VLA integrins in cell binding to different extracellular matrix (ECM) components. As shown in Figure 1, the treatment of purified peripheral blood T lymphocytes with anti-β1 TS2/16 mAb promoted a strong enhancement of T cell interactions with different ECM components such as two proteolytic fragments of FN of 38 and 80 Kd, containing binding sites for VLA-4 and VLA-5, or to LN. Similarly, a remarkable increase in T cell attachment to a recombinant soluble form of VCAM-1 was observed after treatment with anti-β1 TS2/16 mAb. The adhesion of either α2- or α4-transfected K562 cells to COL or to FN38 was also strongly enhanced by anti-β1 mAb (Figure 2). These enhancing effects were comparable in magnitude to those triggered by activating agents such as phorbol esters or anti-CD3 mAb (Shimizu, Y. et al., 1990; Arroyo, A.G. et al., 1992). They were specifically exerted on VLA integrin-ligand interactions since no effect was observed on lymphocyte binding to an unrelated endothelial cell adhesion ligand, the selectin ELAM-1 (Figure 1), and they could be virtually abrogated by specific anti-VLAα mAb.

The anti-β1-enhanced cell binding is isotype and Fc independent as can be triggered by both divalent and monovalent Fab fragments. The β1 regulatory effect is promoted in a dose dependence of ligand and does not require *de novo* protein synthesis since both it follows a very rapid kinetics and could be observed in the presence of inhibitors of protein and RNA synthesis. Moreover, the β1-mediated-enhanced adhesion does not involve any associated change in cell surface expression of the different VLA heterodimers (Arroyo, A.G. et al., 1992).

The enhancement of affinity promoted by anti-β1 mAb is not restricted to a unique member of VLA integrins nor to a single substrate. The β1-regulatory effects have been observed with at least VLA-2, VLA-4, VLA-5, and VLA-6 heterodimers and with both ECM proteins (FN, COL, and LN) and cellular ligands (VCAM-1). These effects have been shown to be substratum specific because they are only promoted on substrates known to be ligands of VLA integrins. This is similar to the regulatory effects previously described with either antibodies specific for some integrin α subunits or specific ligands (Bednarczyk, J.L. and McIntyre, B.W., 1990; Campanero, M.R. et al., 1990; Du, X. et al., 1991; Gulino, D. et

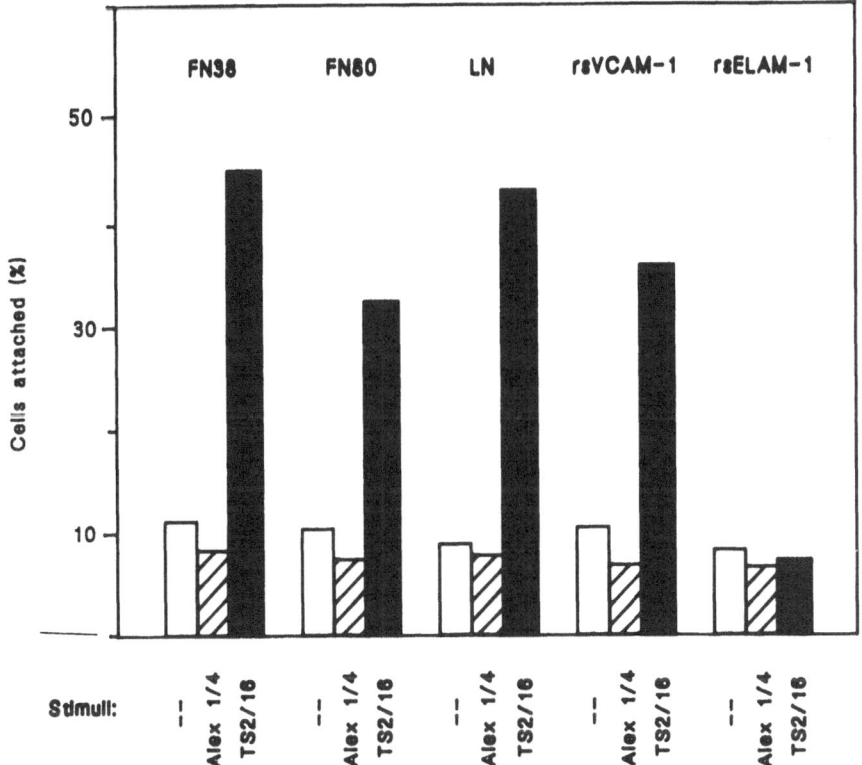

Figure 1. Upregulation of the Attachment of T lymphocytes to Different Ligands by the Anti-β1 TS2/16 mAb. Purified peripheral blood T cells were preincubated with either anti-β1 TS2/16 or Alex 1/4 mAb. Cells then were assayed for binding to plates coated with FN38, FN80, LN, rsVCAM-1, or rsELAM-1, as previously described (Arroyo et al., 1992). Control without mAb is also included. The percentage of cells attached is shown. Note the enhanced attachment of T cells to specific integrin ligands induced by the anti-β1 TS2/16 mAb.

al., 1990; Keizer, G.D. et al., 1988; van-Kooyk, Y. et al., 1991). Very recently, similar effects have been described with antibodies specific for β-chains (Neugebauer, K.M. and Reichardt. L.F.,1991; Arroyo, A.G. et al., 1992; Kovach, N.L. et al., 1992; van de Wiel-van Kemenade, E. et al., 1992; Robinson, M.K. et al., 1992). Taken together, these results suggest that anti-β1 mAb could function by increasing the affinity of different VLA αβ1 heterodimers for their respective ligands.

Leukocyte Spreading during β1-regulated Integrin Receptor Ligand Interactions

Adhesion and subsequent spreading of cells to the endothelial vessel wall are the initial steps in migration of circulating cells towards the surrounding tissues. We have investigated the role of β1 integrins in leukocyte spreading by using the anti-β1 TS2/16 mAb which increases the avidity of different β1-integrins for their ligands. This fact allowed us to study integrin-mediated cell interactions which have been amplified by the action of this particular anti-β1 mAb.

The anti-β1 TS2/16 mAb not only increased the number of cells attached to ECM proteins (FN, COL or LN) but most interestingly, it induced a dramatic increase in cellular spreading. Thus, endothelial cells treated with anti-β1 mAb displayed an extended morphology showing an increased number of cytoplasmic projections (Arroyo, A.G. et al., 1992).

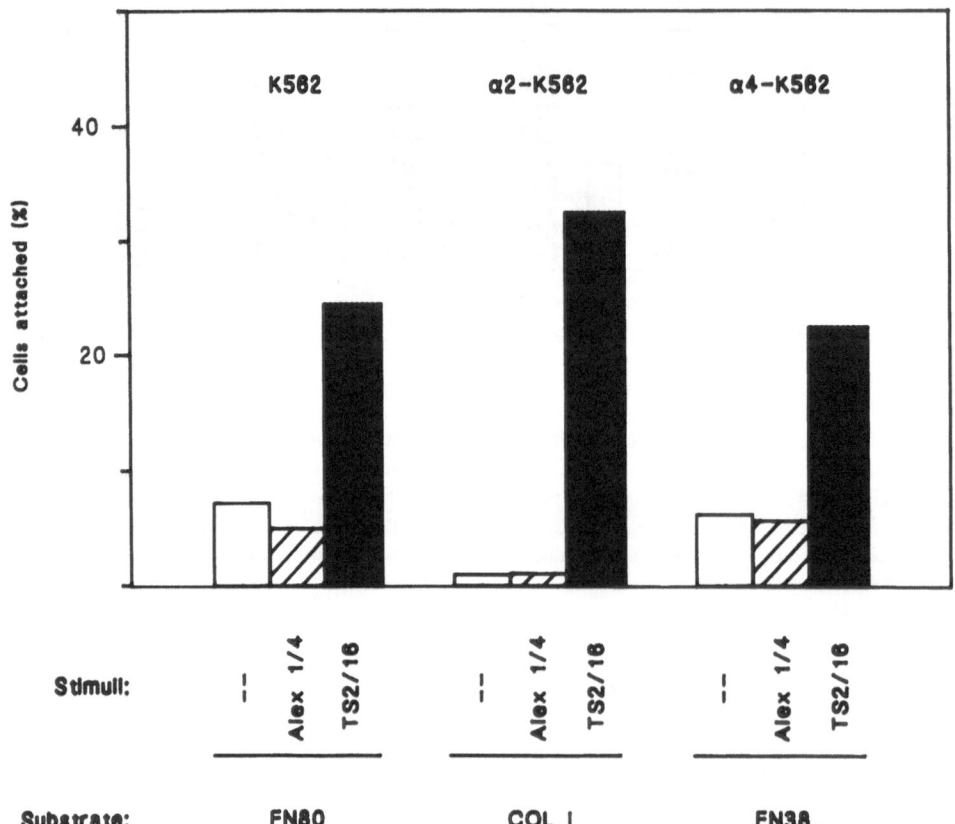

Figure 2. Enhanced Attachment of K562 and K562-transfected Cells to Specific Substrates Induced by the Anti-β1 mAb TS2/16. K562 cells (wild type and α2 or α4 transfected cells) were preincubated with anti-β1 TS2/16 or Alex 1/4 mAb. Cells then were assayed for binding to plates coated with FN80, COL I, or FN38, as previously described (Arroyo et al., 1992). Control without mAb also is included. The percentage of cells attached is shown.

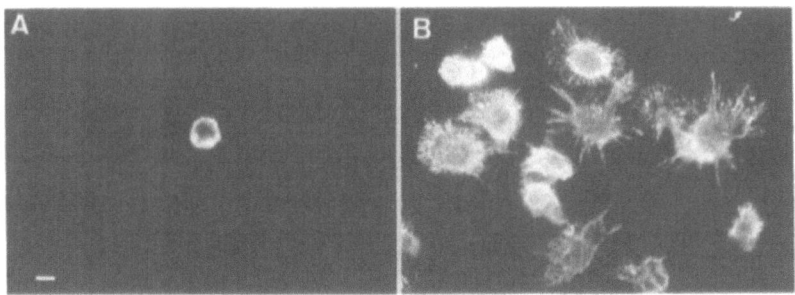

Figure 3. Anti-β1 TS2/16 mAb Induces U-937 Cell Spreading on FN. U-937 cells were incubated with either anti-β1 Alex 1/4 (**A**) or TS2/16 (**B**) mAb and allowed to spread on FN coated coverslips for 30 min at 37° C. Cells then were fixed with formaldehyde in PBS and processed for immunofluorescence. β1 integrin was stained with a FITC-labeled anti-mouse antibody. Bar, 10 μm.

We also analyzed morphological changes induced in U-937 by anti-β1 TS2/16 mAb. A dramatic change in U-937 cellular morphology was observed after specific treatment with the anti-β1 TS2/16 mAb, but not with the anti-β1 Alex 1/4 mAb (Figure 3). Cells which are normally round, extend many long thin protrusions known as microspikes or filopodia. These morphological changes were time-dependent, correlating with the kinetics of the enhancing effects triggered by anti-β1 TS2/16 mAb on U-937 cell adhesion to FN. The appearance of microspikes could be observed as early as 5 min after anti-β1 TS2/16 mAb addition.

Next, we investigated the participation of different VLA heterodimers in leukocyte spreading. Both VLA-4 and VLA-5 FN as well as VLA-2 COL receptors mediate the cell spreading and morphological changes induced through their common β1-chain. The spreading of U-937 and α4-transfected K-562 cells was also induced on VCAM-1, the endothelial counter-receptor of VLA-4, indicating that the morphological changes may be induced during integrin mediated cell-cell as well as cell-ECM interactions. It has been recently reported that the VLA-5 FN receptor, in addition to mediate cell adhesion and motility on FN, is required for cell motility even when other integrins such as the vitronectin receptor mediate the adhesion of the cell to the substratum (Bauer, J.S. et al., 1992). However, cell-spreading on VCAM-1 and COL appears to take place independently of the VLA-5 FN receptor function (Sánchez-Mateos, P. et al., *submitted* [1992]).

The VLA-5 fibronectin receptor also plays a role in the regulation of cell spreading and migration, matrix assembly, and cytoskeletal organization (Akiyama, S.K. et al., 1989), likely through the interaction of the cytoplasmic tail of β1 chain with the cytoskeletal proteins talin and α-actinin (Mueller, S.C. et al., 1989; Burn, P. et al., 1988; Otey, C. et al., 1990; Horwitz, A. et al., 1986). We have found that both the β1 integrin and the cytoskeletal protein talin colocalized in the TS2/16-induced microspikes (Figure 4), suggesting that anti-β1 regulatory effect on cell adhesion could be associated with organization of cytoskeleton.

Signal Transduction during β1-regulated Integrin Receptor-ligand Interactions

Adhesion molecules on T cells appear to have multiple functions. In addition to mediate adhesion, integrins have been implicated in transducing signals to T cells upon binding of mAb or interaction with the ligand. The implication of β1 integrins in the proliferative response of T lymphocytes has also been documented. The interaction of VLA-4 and VLA-5 with fibronectin, VLA-6 with laminin, and VLA-3 with collagen, as well as their engagement with specific anti-VLAα and β1 antibodies induced CD3-directed proliferation and lymphokine secretion by CD4+ T cells (Davis, L.S. et al., 1990; Matsuyama, T. et al., 1989; Nojima, Y. et al., 1990; Shimizu, Y. et al., 1990; Dang, N.H. et al., 1990; Yamada, A. et al., 1991). Moreover, VCAM-1, the intercellular ligand of VLA-4, also induced T-cell antigen receptor-dependent activation of CD4+ cells (Damle, N.K. and Aruffo, A., 1991; Burkly, L.C. et al., 1991; van Seventer, G.A. et al., 1991). Taken together, these observations indicate that the synergistic collaboration between integrins and

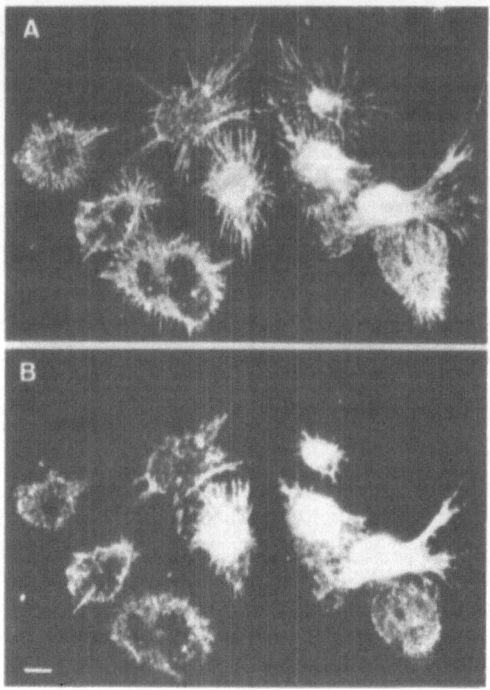

Figure 4. Co-localization of the β1 Integrin and the Cytoskeletal Protein Talin along the Microspikes. U-937 cell spreading was induced with anti-β1 TS2/16 mAb on FN-coated coverlips for 30 min at 37° C. After fixation and permeabilization, the coverslips were double immunostained for β1 integrin: (**A**) with RITC-secondary antibody; and for talin (**B**) with a rabbit anti-talin serum, followed by FITC-anti-rabbit secondary antibody. Bar, 10 μm.

CD3-T cell antigen receptor plays an important role in the regulation of specific immune responses.

The cellular adherence to fibronectin or the ligation of fibronectin receptor by specific antibodies also induce gene expression in multiple cellular types. The interaction VLA-5/fibronectin on CD4+ T cells induces the AP-1 transcription factor (Yamada, A. et al., 1991). Furthermore, the VLA-5-mediated induction of collagenase and stromelysin genes on fibroblasts, and the isolation and characterization of a number of immediate-early genes induced by monocyte adherence have been described (Werb, Z. et al., 1989; Haskill, S. et al., 1991).

The interaction of integrins with their protein ligands or specific antibodies also influences the levels of intracellular second messengers. In this regard, the binding of β1 mAb induces the accumulation of cAMP in activated T cells (Groux, H. et al., 1989), and insoluble fibronectin as well as anti-β1 antibodies activate the Na+/H+ antiporter (Schwartz, M.A. et al., 1991). We also studied possible intracellular signals involved in anti-β1 TS2/16 mAb-enhancing effect. No changes in intracellular levels of cAMP and Ca2+ were observed after treatment of T lymphocytes with anti-β1 TS2/16 mAb (data not shown). Only minor alterations of intracellular pH (an increase 0.03-0.06 pH after treatment of T lymphocytes with TS2/16 mAb) could be detected, in agreement with results reported in a recent study using the anti-β1 TS2/16 mAb (Schwartz, M.A. et al., 1991).

Integrins span the plasma membrane and provide a link between the extracellular matrix and the cytoskeleton, clustering into structures known as focal adhesions. The identification of potential regulatory proteins, including viral tyrosine kinases within focal adhesions, supports the idea that these regions may be involved in transmembrane signalling (Burridge,

K. et al., 1988). Elevated levels of phosphotyrosine have been detected in the focal adhesions of nontransformed fibroblasts (Maher, P.A. et al., 1985). More recently, tyrosine phosphorylation of proteins of 105-130 Kd has been described to be induced after cross-linking of β1-integrins with mAb (Kornberg, L.J. et al., 1991; Nojima, Y. et al., 1992). The attachment of fibroblasts to FN also induced tyrosine phosphorylation of a polypeptide of 120 Kd (Guan, J.L. et al., 1991).

We have also investigated triggering of tyrosine kinase activation during β1-stimulated integrin ligand interaction. We have found the colocalization of β1 integrins and tyrosine phosphorylated proteins during U-937, as well as α2- and α4-transfected K-562 cells spread on both ECM (FN and COL) and cellular (VCAM-1) ligands, thus implicating tyrosine kinase activation during integrin transmembrane signalling. The tyrosine phosphorylation of polypeptides of 130-140 Kd can be triggered in U-937 cells by the interaction of β1-stimulated VLA-5 and VLA-4 integrins with FN80 and VCAM-1, respectively. In addition, a tyrosine phosphorylated polypeptide of 77 Kd was specifically induced during the VLA-5/FN80 interaction, but was absent on the VLA-4/VCAM-1 mediated (Sánchez-Mateos, P. et al., *submitted* [1992]). Therefore, tyrosine kinase activation after VLA integrin-ligand interaction may be a common event found in a variety of cell types. The nature of the pp130-140 protein complex is presently unknown. In this regard, it has been recently cloned a cDNA encoding a new protein-tyrosine kinase, termed pp125^{FAK}, which is localized to focal adhesions (Schaller, M.D. et al., 1992). Since pp130-140 is induced through both VLA-4 and VLA-5, it may be distinct from either pp105 or pp130 which have been described to occur only with anti-α4 or anti-α3 mAb, respectively. It is noteworthy our finding of a pp77 Kd that was selectively induced by cell adhesion to FN80, which may indicate VLA-5 specificity. Our findings indicate that β1 chain is playing a pivotal role in the modulation of these intracellular signals. However, it still remains undetermined the contribution of each integrin α chain in such functions.

CONCLUSION

VLA integrins are, commonly, in a low avidity state unable to mediate either cell adhesion or cell aggregation but, following cellular activation, VLA heterodimers become activated and display enhanced avidity for their ligands. Herewith, we have shown that the different VLA-mediated adhesive functions can also be modulated from the outside of the cell with specific antibodies directed against different functional epitopes on the common β1 chain of VLA polypeptides. Taking into account all the results herein presented, we can propose a model for β1-mediated regulation of cell adhesive functions (Figure 5).

The anti-β1 TS2/16 and Lia 1/2 mAb recognize two different functional epitopes on the common VLA-β1 chain. Whereas TS2/16 induces cell adhesion and prevents cell aggregation, Lia 1/2 displays just the opposite functional effects (Arroyo, A.G. et al., 1992; Campanero, M.R. et al., 1992). These effects can be attributed to differential conformational changes induced by the interaction of each mAb with the β1 polypeptide. Binding of either pro-adhesive or pro-aggregatory anti-β1 mAb would induce the exposure of the binding site for either adhesion or aggregation-related ligands, respectively, thus allowing cellular interaction with ECM components or with cellular counter-receptors. The newly exposed ligand-binding site could be defined by the contribution of both the α- and the β1-chain of the VLA heterodimers since the β1-mediated enhancing effects on cell adhesive functions can be blocked by mAb-binding either to the β1 or to the α subunits of VLA integrins. Antigen cross-linking on the cell surface could, in turn, take place and mediate the triggering of intracellular signals, such as tyrosine kinase activation, intracellular alkalinization, and induction of gene expression. Since induction of these intracellular second messengers is known to be crucial in many cellular processes such as cellular proliferation or cell differentiation, conformational changes induced on the VLA integrins by interaction with specific mAb, and likely by physiologic ligands, as it occurs with the β3 integrins, could be the responsible of the regulation of important cellular functions. Further studies to elucidate each of the steps of the proposed model will help to better understand the mechanisms of regulation of the VLA integrin-mediated cellular activities.

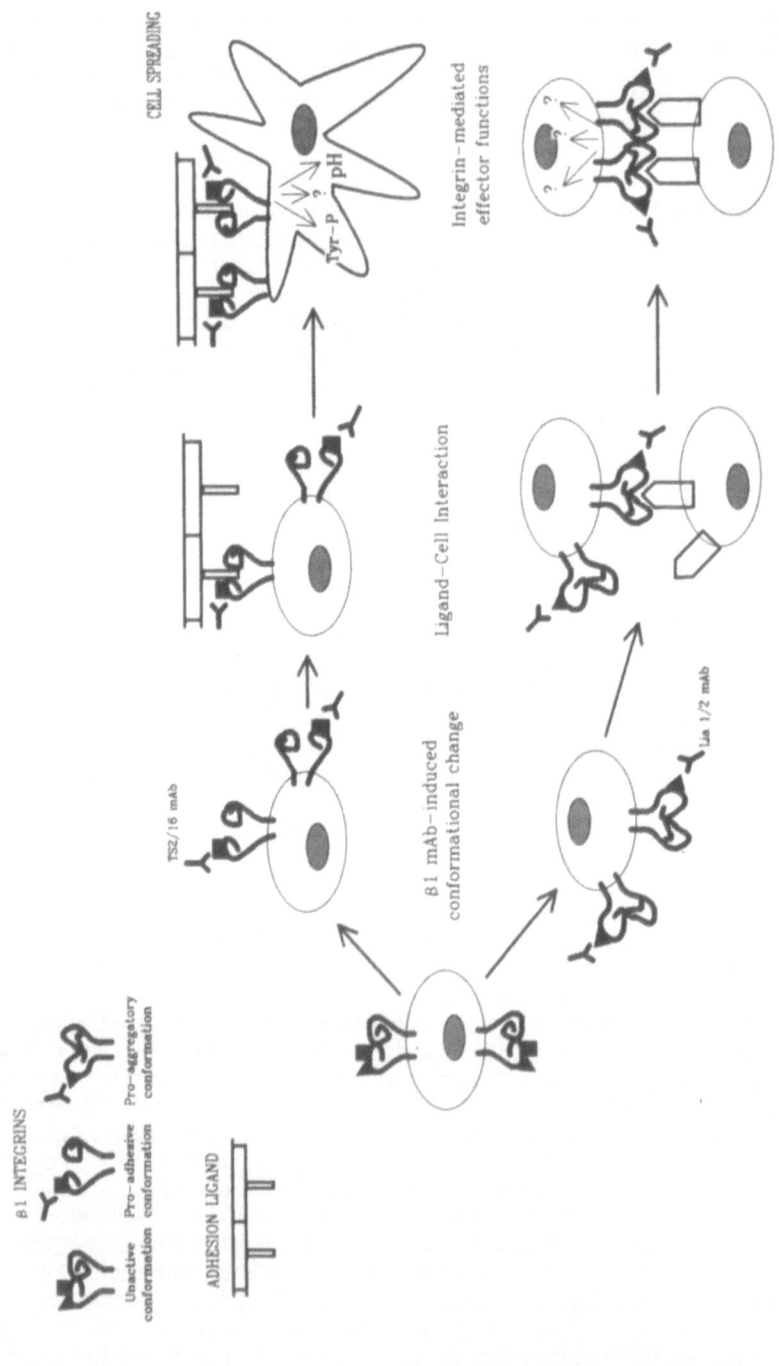

Figure 5. Sequential Steps Involved in β1-mediated Regulation of Cell Adhesive Functions. β1 integrins, in a Non-activated State, Express both Pro-adhesive (*squares*) and Pro-aggregatory (*triangles*) Epitopes on the β1 Chain. Binding of specific mAb to these epitopes would lead to a differential conformational change on the integrin, thus allowing cellular interaction with either adhesion or aggregation ligands. In a subsequent step, ligand-induced integrin cross-linking on the cell surface and further triggering of intracellular signals could take place.

54

ACKNOWLEDGEMENTS

We thank Dr. A. García-Pardo, R.R. Lobb and M.E. Hemler for their kind supply of fibronectin proteolytic fragments, recombinant soluble endothelial cell ligands, and α2- and α 4-K562 transfected cells, respectively.

This work was supported by grants from Insalud FISS 91/0259, and Fundación Ramón Areces.

REFERENCES

Akiyama, S.K., Yamada, S.S., Chen, W.T., and Yamada, K.M., 1989, Analysis of fibronectin receptor function with monoclonal antibodies: Roles in cell adhesion, migration, matrix assembly, and cytoskeletal organization. *J. Cell Biol.* 109:863-875.

Arroyo, A.G., Sánchez-Mateos, P., Campanero, M.R., Martin-Padura, I., Dejana, E., and Sánchez-Madrid, F., 1992, Regulation of the VLA integrin-ligand interactions through the β1 subunit. *J. Cell Biol.* 177:659-670.

Bauer, J.S., Schreiner, C.L., Giacotti, F.G., Ruoslahti, E., and Juliano, R.L., 1992, Motility of fibronectin receptor-deficient cells on fibronectin and vitronectin: Collaborative interactions among integrins. *J. Cell Biol.* 116:477-487.

Bednarczyk, J.L. and McIntyre, B.W., 1990, Induction of lymphocyte aggregation by mAb binding to a member of the integrin supergene family. *J. Immunol.* 144:777-784.

Burkly, L.C., Jakubowski, B.M., Newman, M.D., Roda, G., Chi-Rosso, G., and Lobb. R.R., 1991, Signaling by vascular cell adhesion molecule-1 (VCAM-1) through VLA-4 promotes CD3-dependent T cell proliferation. *Eur. J. Immunol.* 21:2871-2875.

Burn, P., Kupfer, A., and Singer, S.J., 1988, Dynamic membrane-cytoskeletal interactions: Specific association of integrin and talin arises *in vivo* after phorbol ester treatment of peripheral blood lymphocytes. *Proc. Natl. Acad. Sci. USA* 85:497-501.

Burridge, K., Fath, K., Nuckolls, G., Kelly, T., and Turner, C., 1988, Focal adhesions: Transmembrane junctions between the extracellular matrix and the cytoskeleton. *Annu. Rev. Cell. Biol.* 4:487-525.

Campanero, M.R., Pulido, R., Ursa, M.A., Rodriguez-Moya, M., de Landazuri, M.O., and Sánchez-Madrid, F.,1990, An alternative leukocyte homotypic adhesion mechanism, LFA1/ICAM-1-independent, triggered through the human VLA4 integrin. *J. Cell. Biol.* 110:2157-2163.

Campanero, M.R., Arroyo, A.G., Pulido, R., Ursa, A., de Matias, M.S., Sánchez-Mateos, P., Kassner, P.D., Chan, B.M.C., Hemler, M.E., de Landázuri, M.O., and Sánchez-Madrid, F., 1992, Functional role of α2/β1 and α4/β1 integrins in leukocyte intercellular adhesion induced through the common β1 subunit. *Eur. J. Immunol.* 22:3111-3119.

Carter, W.G., Wayner, E.A., Bouchard, T.S., and Kaur, P., 1990, The role of integrins α2β1 and α3β1 in cell-cell and cell-substrate adhesion of human epidermal cells. *J. Cell Biol.* 110:1387-1404.

Clayberger, C., Krensky, A.M., McIntyre, B.W., Koller, T.S., Parham, P., Brodsky, F., Linn, D.J., and Evans. E.L., 1987, Identification and characterization of two novel lymphocyte function-associated antigens, L24 and L25. *J. Immunol.* 138:1510-1514.

Damle, N.K. and Aruffo, A., 1991, Vascular cell adhesion molecule 1 induces T-cell antigen receptor-dependent activation of CD4+ T lymphocytes. *Proc. Natl. Acad. Sci. USA* 88:6403-6407.

Dang, N.H., Torimoto, Y., Schlossman, S.F., and Morimoto, C., 1990, Human CD4 helper T cell activation functional involvement of two distinct collagen receptors, 1F7 and VLA integrin family. *J. Exp. Med.* 172:649-652.

Darribière, T.,Guida, K., Larjava, H., Johnson, K.E., Yamada, K.M., Thiery, J.P., and Boucaut. J.C., 1990, *In vivo* analyses of integrin β1 subunit function in fibronectin matrix assembly. *J. Cell Biol.* 110:1813-1823.

Davis, L.S., Oppenheimer-Marks, N., Bednarcyzk, J.L., McIntyre, B.W., and Lipsky, P.E., 1990, Fibronectin promotes proliferation of naive and memory T cells by signaling through both the VLA-4 and VLA-5 integrin molecules. *J. Immunol.* 145:785-793.

Dobrina, A., Menegazzi, R., Carlos, T.M., Nardon, E., Cramer, R., Zacchi, T., Harlan, J.M., and Patriarca, P., 1991, Mechanisms of eosinophil adherence to cultured vascular endothelial cells. Eosinophils bind to the cytokine-induced endothelial ligand vascular cell adhesion molecule-1 via the very late activation antigen-4 integrin receptor. *J. Clin. Invest.* 88:20-26.

Du, X., Plow, E.F., Frelinger, A.L., O'Toole, T.E., Loftus, J.C., and Ginsberg. M.H., 1991, Ligands "activate" integrin $\alpha_{IIb}\beta_3$ (platelet GPIIb-IIIa). *Cell* 65:409-416.

Dustin, M.L. and Springer, T.A., 1991, Role of lymphocyte adhesion receptors in transient interactions and cell locomotion. *Annu. Rev. Immunol.* 9:27-66.

Dustin, M.L. and Springer, T.A., 1989, T-cell receptor cross-linking transiently stimulates adhesiveness through LFA-1. *Nature* 341:619-624.

Elices, M.J. and Hemler, M.E., 1989, The human integrin VLA2 is a collagen receptor on some cells and a collagen/laminin receptor on others. *Proc. Natl. Acad. Sci. USA.* 86:9906-9910.

Elices, M.J., Osborn, L., Takada, Y., Crouse, C., Luhowskyj, S., Hemler, M.E., and Lobb, R.R., 1990, VCAM-1 on activated endothelium interacts with the leukocyte integrin VLA4 at a site distinct from the VLA4/fibronectin binding site. *Cell* 60:577-584.

Figdor, C.G., Van Kooyk, Y., and Keizer, G.D., 1990, On the mode of action of LFA-1. *Immunol. Today* 11:277-280.

Freedman, A.S., Munro, J.M., Rice, G.G., Bevilacqua, M.P., Morimoto, C., McIntyre, B.W., Rhynhart, K., Pober, J.S., and Nadler, L.M., 1990, Adhesion of human B cells to germinal centers *in vitro* involves VLA-4 and INCAM-110. *Science* 249:1030-1033.

García-Vicuña, R., Humbría, A. Postigo, A.A. López-Elzardua, C. de Landázuri, M.O., Sánchez-Madrid, F., and Laffón, A., 1992, VLA family in rheumatoid arthritis: Evidence of "*in vivo*" regulated adhesion of synovial fluid T cells to fibronectin through VLA-5 integrins. *Clin. Exp. Immunol. (in press)*.

Groux, H., Huet, S., Valentin, H., Pham, D., and Bernard, A., 1989, Supressor effects and cyclic AMP accumulation by the CD29 molecule on CD4+ lymphocytes. *Nature* 339:152-154.

Guan, J.L., Trevithick, J.E., and Hynes, R.O., 1991, Fibronectin/integrin interaction induces tyrosine phosphorylation of a 120-kDa protein. *Cell. Regul.* 2:951-964.

Gulino, D., Ryckwaert, J.J., Andrieux, A., Rabiet, M.J., and Marguerie, G., 1990, Identification of a monoclonal antibody against platelet gpIIb that interacts with a calcium-binding site and induces aggregation. *J. Biol. Chem.* 265:9575-9581.

Hall, D.E., Reichardt, L.F., Crowley, E., Holley, B., Moezzi, H., Sonnenberg, A., and Damsky, C.H., 1990, The $\alpha1\beta1$ and $\alpha6\beta1$ integrin heterodimers mediate cell attachment to distinct sites on laminin. *J. Cell Biol.* 110:2175-2184

Haskill, S., Beg, A.A., Trompkins, S.M., Morris, J.S., Yurochko, D., Sampson-Johannes, A., Mondal, K., Ralph, P., and Baldwin, Jr. A., 1991, Characterization of an immediate-early gene induced in adherent monocytes that encodes IkB-like activity. *Cell* 65:1281-1289.

Hemler, M.E, 1990, VLA proteins in the integrin family: Structures, functions and expression on leukocytes. *Annu. Rev. Immunol.* 8:365-400.

Hermanowski-Vosatka, A., van Strijp, J.A.G., Swiggard, W.J., and Wright, S.D., 1992, Integrin modulating factor-1: A lipid that alters the function of leukocyte integrins. *Cell* 68:341-352.

Hibbs, M.L., Xu, H., Stacker, S.A., and Springer, T.A., 1991, Regulation of adhesion of ICAM-1 by the cytoplasmic domain of LFA-1 integrin beta subunit. *Science* 251:1611-1613.

Holzmann, B. and Weissman, I.L., 1989, Peyer's patch-specific lymphocyte homing receptors consist of a VLA-4 like α chain associated with either of two integrin β chain, one of which is novel. *EMBO. J.* 8:1735-1741.

Horwitz, A., Duggan, K., Buck, C., Beckerte, M.C., and Burridge, K., 1986, Interaction of plasma membrane fibronectin receptor with talin-a transmembrane linkage. *Nature* 320:531-533.

Hynes, R.O, 1992, Integrins: Versatility, modulation and signaling in cell adhesion. *Cell.* 69:11-25.

Keizer, G.D., Visser, W., Vleim, M., and Figdor, C., 1988, A monoclonal antibody (NK1-L16) directed against a unique epitope on the alpha chain of LFA-1 induces homotypic cell-cell interaction. *J. Immunol.* 140:1393-1400.

Kirchhofer, D., Languino, L.R., Ruoslahti, E., and Pierschbacher, M.D., 1990, α2β1 integrins from different cell types show different binding specificities. *J. Biol. Chem.* 265:615-618.

Koopman, G., Parmentier, K.H., Schuurman, H-J., Newman, W., Meijer, C.J.L.M., and Pals, S.T., 1991, Adhesion of human B cell to follicular dendritic cells involve both the lymphocyte function-associated antigen 1/intercellular adhesion molecule 1 and very late antigen 4/vascular cell adhesion molecule 1 pathway. *J. Exp. Med.* 173:1297-1304.

Kornberg, L.J., Earp, H.S., Turner, C.E., Prockop, C., and Juliano, R.L., 1991, Signal transduction by integrins: Increased protein tyrosine phosphorylation caused by clustering of β1 integrins. *Proc. Natl. Acad. Sci. USA* 88:8392-8396.

Kovach, N.L., Carlos, T.M., Yee, E., and Harlan, J.M., 1992, A monoclonal antibody to β1 integrin (CD29) stimulates VLA-dependent adherence of leukocytes to human umbilical vein endothelial cells and matrix components. *J. Cell Biol.* 116:499-509.

Kramer, R.H., McDonald, K.A., and Vu, M.P., 1989, Human melanoma cells express a novel integrin receptor for laminin. *J. Biol. Chem.* 264:15642-15649.

Laffón, A., García-Vicuña, R., Humbría, A., Postigo, A.A. Corbí, A.L. de Landázuri, M.O., and Sánchez-Madrid, F., 1991, Upregulated expression and function of VLA-4 fibronectin receptor on human activated T cells in rheumatoid arthritis. *J. Clin. Invest.* 88:546-552.

Maher, P.A., Pasquale, E.B., Wang, J.Y.J., and Singer, S.J.,1985, Phosphotyrosine-containing proteins are concentrated in focal adhesions and intercellular junctions in normal cells. *Proc. Natl. Acad. Sci. USA* 82:6576-6580.

Matsuyama, T., Yamada, A., Kay, J., Yamada, K.M., Akiyama, S.K., Schlossman, S.F., and Morimoto, C., 1989, Activation of CD4 cells by fibronectin and anti-CD3 antibody. A synergistic effect mediated by the VLA-5 fibronectin receptor complex. *J. Exp. Med.* 170:1133-1148.

Miyake, K., Weissman, I.L., Greenberger, J.S., and Kincade, P.W., 1991, Evidence for a role of the integrin VLA-4 in lymphohemopoiesis. *J. Exp. Med.* 173:599-607.

Mould, A.P., Komoriya, A., Yamada, K.M., and Humphries, M.J., 1991, The CS5 is a second site in the IIICS region of fibronectin recognized by the integrin α4β1. Inhibition of α4β1 function by RGD peptide homologues. *J. Biol. Chem.* 266:3579-3585.

Mueller, S.C., Kelly, T., Dai, M., Dai, H., and Chen, W.T., 1989, Dynamic cytoskeleton-integrin associations induced by cell binding to immobilized fibronectin. *J. Cell Biol.* 109:3455-3464.

Neugebauer, K.M. and Reichardt. L.F.,1991, Cell-surface regulation of β1-integrin activity on developing retinal neurons. *Nature* 350:68-71.

Nojima, Y., Humphries, M.J., Mould, A.P., Komoriya, A., Yamada, K.M., Schlossman, S.F., and Morimoto, C., 1990, VLA-4 mediated CD3-dependent CD4+ T cell activation via the CS1 alternatively spliced domain of fibronectin. *J. Exp. Med.* 172:1185-1192.

Nojima, Y., Rothstein, D.M., Sugita, K., Schlossman, S.F., and Morimoto, C., 1992, Ligation of VLA-4 on T cells stimulates tyrosine phosphorylation of a 105-kD protein. *J. Exp. Med.* 175:1045-1053.

O'Toole, T.E., Loftus, J.C., Du, X., Glass, A.A., Ruggeri, Z.M., Shattil, S.J., Plow, E.F., and Ginsberg, M.H., 1990, Affinity modulation of the αIIbβ3 integrin (platelet GPIIb-IIIa) is an intrinsic property of the receptor. *Cell Regul.* 1:883-893.

Osborn, L., Hession, C., Tizard, R., Vassallo, C., Luhowskyj, S., Chi-Rosso, G., and Lobb, R.R.,1989, Direct expression of vascular cell adhesion molecule 1, a cytokine-induced endothelial protein that binds to lymphocytes. *Cell* 59:1203-1211.

Otey, C., Pavalko, F.M., and Burridge, K., 1990, An interaction between α-actinin and the β1 integrin subunit *in vitro. J. Cell Biol.* 111:721-729.

Postigo, A., Pulido, R., Campanero, M.R., Acevedo, A., García-Pardo, A., Corbí, A., Sánchez-Madrid, L.F., and de Landázuri, M.O., 1991, Differential expression of VLA-4 integrin by resident and peripheral blood B lymphocytes. Acquisition of functionally active α4β1-fibronectin receptors upon B cell activation. *Eur. J. Immunol.* 21:2437-2445.

Postigo, A.A., García-Vicuña, R., Díaz-González, F., Arroyo, A.G., de Landázuri, M.O., Chi-Rosso, G., Lobb, R.R., Laffón, A., and Sánchez-Madrid., F., 1992, Increased binding of synovial T lymphocytes from rheumatoid arthritis to ELAM-1 and VCAM-1 endothelial adhesion molecules. *J. Clin. Invest.* 89:1445-1457.

Pulido, R., Elices, M.J., Campanero, M.R., Osborn, L., Schiffer, S., García-Pardo, A., Lobb, R.R., Hemler, M.E., and Sánchez-Madrid, F., 1991, Functional evidence for three distinct and independently inhibitable adhesion activities mediated by the human integrin VLA4. *J. Biol. Chem.* 266:10241-10245.

Robinson, M.K., Andrew, D., Rosen, H., Brown, D., Ortlepp, S., Stephens, P., and Butcher, E.C., 1992, Antibody against the Leu-CAM-β chain (CD18) promotes both LFA-1 and CR3-dependent adhesion events. *J. Immunol.* 148:1080-1085.

Ruoslahti, E, 1991, Integrins. *J. Clin. Invest.* 87:1-6.

Ruoslahti, E. and Pierschbacher, M.D., 1987, New perspectives in cell adhesion: RGD and integrins. *Science* 238:491-497.

Schaller, M.D., Borgman, C.A., Cobb, B.S., Vines, R.R., Reynolds, A.B., and Parsons, J.T., 1992, pp125[FAK], a structurally distinctive protein-tyrosine kinase associated with focal adhesions. *Proc. Natl. Acad. Sci. USA* 89:5192-5196.

Schleimer, R.P., Sterbinsky, S.A., Kaiser, J., Bickel, C.A., Klunk, D.A., Tomioka, K., Newman, W., Luscinskas, F.W., Gimbrone, M.A., McIntyre, B.W., and Bochner, B.S., 1992, IL-4 induces adherence of human esosinophils and basophils but not neutrophils to endothelium. 148:1086-1092.

Schwartz, M.A., Lechene, C., and Ingber, D.E., 1991, Insoluble fibronectin activates the Na/H antiporter by clustering and immobilizing integrin α5β1, independent of cell shape. *Proc. Natl. Acad. Sci. USA* 88:7849-7853.

Shimizu, Y., van Seventer, G.A., Horgan, K.J., and Shaw. S., 1990, Costimulation of proliferative responses of resting CD4+ T cells by the interaction of VLA-4 and VLA-5 with fibronectin or VLA-6 with laminin. *J. Immunol.* 145:59-67.

Shimizu, Y., Van Seventer, G.A., Horgan, K.J., and Shaw. S., 1990, Regulated expression and binding of three VLA (β1) integrin receptors on T cells. *Nature* 345:250-253.

Sonnenberg, A., Linders, C.J.T., Modderman, P.W., Damsky, C.H., Aumailley, M., and Timpl, R., 1990, Integrin recognition of different cell-binding fragments of laminin (P1,E3, E8) and evidence that α6β1 but not α6β4 functions as a major receptor for fragment E8. *J. Cell Biol.* 110:2145-2155

Springer, T.A, 1990, Adhesion receptors of the immune system. *Nature* 346:425-434.

Takada, Y., Elices, M.J., Crouse, C., and Hemler, M.E., 1989, The primary structure of the α4 subunit of VLA-4: Homology to other integrins and a possible cell-cell adhesion function. *EMBO J.* 8:1361-1368.

van de Wiel-van Kemenade, E., van Kooyk, Y., de Boer, A.J., Huijbens, R.J.F., Weder, P., de van de Kasteele, W., Melief, C.J.M., and Figdor, C.G., 1992, Adhesion of T and B lymphocytes to extracellular matrix and endothelial cells can be regulated through the β subunit of VLA. *J. Cell Biol.* 117:461-470.

Van Kooyk, Y., van de Wiel-van Kemenade, P., Weber, P., Kuijpers, T.W., and Figdor, C.G., 1989, Enhancement of LFA-1 mediated cell adhesion by triggering through CD2 or CD3 on T lymphocytes. *Nature* 342:811-813.

van Seventer, G.A., Newman, W., Shimizu, Y., Nutman, T.B., Tanaka, Y., Horgan, K.J., Gopal, T.V., Ennis, E., O'Sullivan, D., Grey, H., and Shaw, S., 1991, Analysis of T cell stimulation by superantigen plus major histocompatibility complex class II molecules or by CD3 monoclonal antibody: Costimulation by purified adhesion ligands VCAM-1, ICAM-1, but not ELAM-1. *J. Exp. Med.* 174:901-913.

van-Kooyk, Y., Weder, P., Hogervorts, F., Verhoeven, A.J., van Seventer, G., te Velde, A.A., Borst, J., Keizer, G.D., and Figdor, C.G., 1991, Activation of LFA-1 through a Ca²⁺ dependent epitope stimulates lymphocyte adhesion. *J. Cell Biol.* 112:345-354.

Vonderheide, R.H. and Springer, T.A., 1992, Lymphocyte adhesion through very late antigen 4: Evidence for a novel binding site in the alternatively spliced domain of vascular cell adhesion molecule 1 and an additional integrin counter-receptor on stimulated endothelium. *J. Exp. Med.* 175:1433-1440.

Walsh, G.M., Mermod, J-J., Hartnell, A., Kay, A.B., and Wardlaw, A.J., 1991, Human eosinophil, but not neutrophil, adherence to IL-1-stimulated human umbilical vascular endothelial cells is α4β1 (very late antigen-4) dependent. *J. Immunol.* 146:3419-3423.

Wayner, E.A., García-Pardo, A., Humphries, M.J., McDonald, J.A., and Carter, W.G., 1990, Identification and characterization of the T lymphocyte adhesion receptor for an alternative cell attachment domain (CS-1) in plasma fibronectin. *J. Cell Biol.* 109:1321-1330

Weller, P.F., Rand, T.H., Goelz, S.E., Chi-Rosso, G., and Lobb, R.R., 1991, Human eosinophil adherence to vascular endothelium mediated by binding to vascular cell adhesion molecule 1 and endothelial leukocyte adhesion molecule 1. *Proc. Natl. Acad. Sci. USA* 88:7430-7433.

Werb, Z., Tremble, P.M., Behrendtsen, O., Crowley, E., and Damsky, C.H., 1989, Signal transduction through the fibronectin receptor induces collagenase and stromelysin gene expression. *J. Cell Biol.* 109:877-889.

Yamada, A., Nojima, Y., Sugita, K., Dang, N.H., Schlossman, S.F., and Morimoto, C., 1991, Cross-linking of VLA/CD29 molecule has a comitogenic effect with anti-CD3 on CD4 cell activation in serum-free system. *Eur. J. Immunol.* 21:319-325.

Yamada, A., Nikaido, T., Nojima, Y., Schlossman, S.F., and Morimoto, C., 1991, Activation of human CD4 T lymphocytes. Interaction of fibronectin with VLA-5 receptor on CD4 cells induces the AP-1 transcription factor. *J. Immunol.* 146:53-56.

DISCUSSION

C. FIGDOR

There are additional phosphorylated polypeptides in your Western Blot experiment.

F. SANCHEZ-MADRID

Yes, in addition to the ones I have described, an enhancement in the phosphorylation of several polypeptides of 69, 65, and 55 Kd, was also detected after β_1-stimulated cell attachment to both FN80 and VCAM-1.

N. HOGG

OK. Question number 2. Co-localization of cytoskelatal proteins, the colocalization of talin. So, on long spikes like that you would expect to see reorganization of the cytoskeleton, or at least presence of microtubules. There has been a report in the *European Journal of Immunology* about LFA-1 antibodies showing the organization. So, have you done co-precipitation studies showing that?

F. SANCHEZ-MADRID

This is something very interesting to do. As you know, there is a recent article on *Journal of Cell Biology* describing the interaction of LFA-1 integrin with cytoskeletal activation after cell activation.

N. HOGG

Yes, because just colocalization does not prove anything. Finally, out of interest, Have you been able to discriminate between binding to VCAM-1 and fibronectin?

F. SANCHEZ-MADRID

We have been unable to discriminate by using antibodies. However, it is possible to do it by selective inhibition with soluble FN40.

N. HOGG

So you mean soluble VCAM will block binding to VCAM but not to fibronectin?

F. SANCHEZ-MADRID

No, I mean that cell binding to rsVCAM-1 can be inhibited by anti-VCAM-1 but not by anti-FN antibodies or soluble FN40 fragment. In reciprocal experiments, cell binding to FN40 is inhibited by anti-FN but not by anti-VCAM-1 antibodies.

59

G. DOUGHERTY

I think it important to distinguish between affinity and avidity. I assume from what you presented today that you believe that your antibody acts to increase the affinity of β_1 binding. I would be interested to know if it also altered in any way the distribution of β_1 integrins on the cell surface. In addition, have you looked at the question of affinity quantitatively, by doing for example Scatchard analysis on the binding of radiolabeled VCAM-1 to cells with or without stimulation with antibody?

F. SANCHEZ-MADRID

Yes, we have tried Scatchard analyses with radiolabeled purified VCAM-1 and FN fragments but so far we have been unable to quantitatively estimate the increased VLA integrin-ligand interaction triggered by the anti-β_1 TS2/16 mAb.

G. DOUGHERTY

When you stimulate with F(ab)' fragments do you see any capping or other changes in the distribution of the β_1 integrin molecules on the cell surface?

F. SANCHEZ-MADRID

We have not tested that point.

S. SHAW

I think that this is a very nice systematic analysis that you are putting together. My question is what are the molecular mechanisms used by the many different activation inducing pathways such as anti-integrin mAb, CD3, CD7, calcium and so on? Are each going through the same final common pathway or are there many changes in the integrins?

F. SANCHEZ-MADRID

Many receptors are using tyrosine kinase coupling, but probably there are different kinases dependent on each receptor.

S. SHAW

The question could perhaps be approached by exploring additivity. Does the antibody-induced conformational change act additively with the augmentation induced by PMA, CD3 or CD7 or the like?

F. SANCHEZ-MADRID

We have not done combination of different agents. I think that it is possible that physiologic ligands may exist working in the same fashion that this β_1 antibody, and thus, modulate the function of VLA integrins.

T. CAREY

A couple of clarifications -- FAB is monovalent?

F. SANCHEZ-MADRID
Yes.

T. CAREY

And, secondly, you said the resting T cells don't express GLA2, is that right?

F. SANCHEZ-MADRID

Yes.

T. CAREY

The third part of my question is the following -- there has been a lot of discussion about anti-β_1 blocking attachment to laminin. I was wondering whether you have looked at laminin attachment and what the effect of any of these fragments or antibody might be on laminin attachment in this model.

F. SANCHEZ-MADRID

Peripheral blood T cells as well as melanoma cells are susceptible to be regulated by the TS2/16 anti-β_1 in their binding to LN.

G. MARGUERIE

Comment on new concept about integrin ligand interaction. Most antibodies react in same ways as you recognize epitope already present in receptor. Binding on ligand or peptide will increase binding of antibody itself. Hypothesis of allosteric system. New idea, we tried in platelets to see if antibody by itself can induce phosphorylation and we were unsuccessful antibody plus ligand, then phosphorylation, activation of receptor?

G. DOUGHERTY

Final question -- What effects do inhibitors of tyrosine phosphorylation have on the adhesive events induced by your anti-β_1 antibody?

F. SANCHEZ-MADRID

In experiments of inhibition of cell spreading with genistein, there is not a clear answer, and thus I cannot make a clear statement due to problems of solubilization of genistein in DMSO.

R. JULIANO

I guess I just want to get something straight about the stimulation of tyrosine phosphorylation. Do you ever see an increase in tyrosine phosphorylation just using the antibodies alone without allowing cells to attach to a substratum or to interact with other cells? In other words, will the antibody on cells do the trick?

F. SANCHEZ-MADRID

We have a very weak signal with the antibody alone. Our working hypothesis is that the antibody reinforces the interaction of the integrin with the ligand. After receptor occupancy there is a signal transmission.

ELEMENTS FOR A STRUCTURAL/FUNCTIONAL MODEL OF HUMAN PLATELET PLASMA MEMBRANE FIBRINOGEN RECEPTOR, THE GLYCOPROTEIN IIb/IIIa COMPLEX (INTEGRIN αIIb/ß3)

Juan José Calvete

Instituto de Química-Física "Rocasolano" CSIC
c/Serrano 119
28006-Madrid, SPAIN

EVIDENCE FOR THE ROLE OF GLYCOPROTEIN IIb/IIIa AS THE PLATELET FIBRINOGEN RECEPTOR

The observation that Glanzmann thrombasthenia, a rare autosomal recessive congenital bleeding disorder characterized by the absence of macroscopic platelet aggregation in response to all physiologic stimuli (Caen et al., 1966; reviewed by Nurden, 1989), was associated with the absence, or marked reduction of the content, of two platelet surface glycoproteins, GPIIb (135 kDa) and GPIIIa (95 kDa), (Nurden and Caen, 1974) provided the first evidence to assign a given platelet function to specific membrane glycoproteins (Nurden and Caen, 1975; Phillips et al., 1975). Moreover, the well-documented role of fibrinogen as an essential cofactor for platelet aggregation (Caen and Inceman, 1963; McLean et al., 1964; reviewed by Peerschke, 1985; Plow et al., 1986), together with the demonstration of a direct association of fibrinogen with a single, or predominant, class of inducible, (~mM) Ca^{2+}-dependent, specific and saturable platelet receptor (Mustard et al., 1978; Marguerie et al., 1979; Bennett and Vilaire, 1979), and the failure of thrombasthenic platelets to bind fibrinogen and aggregate upon activation (Mustard et al., 1979), suggested a role for GPIIb and/or GPIIIa as part of the platelet fibrinogen receptor system (reviewed by Marguerie and Plow, 1983; Plow et al., 1984; Peerschke, 1985).

Subsequently, it was shown that GPIIb and GPIIIa form a Ca^{2+}-dependent complex with a 1:1 stoichiometry both in solution, after solubilization of platelet membranes with buffers containing Triton X-100, (Hagen et al., 1980; Kunicki et al., 1981; Pidard et al., 1982; Gogstad et al., 1982; Howard et al., 1982; Jennings and Phillips, 1982), and in the intact, resting platelet membrane (McEver et al., 1983; Fujimura and Phillips, 1983; Pidard et al., 1983; Coller et al., 1983; Bennett et al., 1983). Further studies from different laboratories provided important evidence supporting the concept that GPIIb/IIIa is the platelet fibrinogen receptor:

1) A direct interaction between GPIIb/IIIa and fibrinogen was demonstrated using an enzyme-linked immunosorbent assay (ELISA) (Nachman and Leung, 1982; Nachman et al., 1984), by crossed-immunoelectrophoresis (Gogstad et al., 1982a), as well as by photoaffinity labeling of fibrinogen to ADP-activated platelets (Bennett et al., 1982; Marguerie et al., 1984)

2) Several monoclonal antibodies against GPIIb and/or GPIIIa blocked fibrinogen binding and platelet aggregation (Bennett et al., 1983; Coller et al., 1983; Kornecki et al., 1983; Melero and González-Rodríguez, 1984, and references therein)

3) Purified glycoprotein IIb/IIIa complex, once incorporated into phospholipid vesicles (Parise and Phillips, 1985), bound fibrinogen with properties quantitatively similar to those of whole platelets or isolated platelet membranes (Baldassare et al., 1985; Parise and Phillips, 1985a)

4) Expression of recombinant GPIIb/IIIa in human embryonic kidney cells resulted in a functional fibrinogen-binding complex (Bodary et al., 1989)

Since under physiological conditions the stability of the GPIIb/IIIa complex depends upon micromolar calcium concentration (Fujimura and Phillips, 1983; Brass et al., 1985; Fitzgerald and Phillips, 1985; Steiner et al., 1991; Rivas et al., 1991), and the complexed form of these glycoproteins is necessary for the expression of fibrinogen receptors (Nachman and Leung, 1982; Gogstad et al., 1982; Phillips and Baughan, 1983), it was concluded that extracellular Ca^{2+}-binding sites on unstimulated platelet GPIIb/IIIa heterodimers are responsible for the stability of the complexes, and that only the GPIIb/IIIa complex and not the separate subunits constitutes the functional receptor unit.

Fibrinogen binding to intact platelets requires activation of its receptor for induction of the fibrinogen binding site. However, fibrinogen binding to proteoliposomes occurred in the absence of, and was not enhanced by, ADP. Moreover, the stoichiometry of the interaction was low, 0.1 mol fibrinogen bound per mol of GPIIb/IIIa (Baldassare et al., 1985; Parise and Phillips, 1985), indicating that, once isolated from its natural membrane, only a small subpopulation of GPIIb/IIIa complexes (around 18% of the right-side oriented) became unmasked and retained functional activity. Similarly, recombinant GPIIb/IIIa promoted adhesion of cells to substrates coated with adhesive proteins, but did not bind soluble fibrinogen (Bodary et al., 1989).

Altogether, this emphasizes the requirement of auxiliary molecules (i.e. other proteins, or a certain lipid environment) for induction and full expression of the receptor capability of GPIIb/IIIa. In this respect, using a cross-linking approach, Slupsky et al. (1989) have reported that platelet-aggregating concentrations of anti-CD9 monoclonal antibodies promote specific interaction of the CD9 antigen with the GPIIb/IIIa complex, and Torti et al. (1991) found physical association between GPIIb/IIIa and a 21 kDa GTP-binding protein. In addition, Conforti et al. (1990) showed that the membrane lipid composition modulates the receptor capability of the highly GPIIb/IIIa analogous vitronectin receptor (see below), and Cheresh et al. (1987a) found that vitronectin receptor on the surface of human melanoma cells exits in a divalent cation-dependent functional complex with the disialoganglioside GD2. The characterization of the relationships between these, and possibly other, molecules and GPIIb/IIIa is likely to provide a new insight into the expression and functioning of the platelet fibrinogen receptor.

The number and affinity of fibrinogen binding sites exposed at the surface of a single activated platelet have been derived by Scatchard plot analysis. Several laboratories have provided figures consistent with a single class of fibrinogen binding site ($K_a = 10^6$-10^7 M^{-1}; $n = 20$-50×10^3), while others have obtained curvilinear plots, which were interpreted in terms of a high-affinity, low capacity site ($K_a = 3 \times 10^7$ M^{-1}; $n = 1300$) and a low-affinity, high-capacity site ($K_a = 1.8 \times 10^6$ M^{-1}; $n = 80 \times 103$) (reviewed by Plow et al., 1984; Peerschke, 1985). Although these differences remain to be solved, the latter set of data may be explained in the context of our current understanding considering that: a) 80,000 to 100,000 copies of GPIIb/IIIa are exposed at the surface of intact, resting platelets (Calvete et al., 1986); and b) platelets also possess a minor vitronectin receptor (several thousand copies per platelet), a promiscuous heterodimeric receptor which contains a β-subunit almost identical to GPIIIa (Ginsberg et al., 1987; Troesch et al., 1990), and capable of interacting with an arginine-glycine-aspartic acid (RGD)-containing site near the C-terminus of the fibrinogen α chain (Smith et al., 1990; Cheresh et al., 1989).

GPIIB/IIIA BELONGS TO THE INTEGRIN SUPERFAMILY OF CELL ADHESION RECEPTORS

The presence of immunologically- and structurally-related GPIIb/IIIa-like molecules in endothelial cells was first reported by Thiagarajan et al. (1985) and Fitzgerald et al. (1985). Further immunological work demonstrated the presence of GPIIb/IIIa-related heterodimers on a wide variety of cells (reviewed by Hynes, 1987; Phillips et al., 1988; Plow and Ginsberg, 1989). Elucidation of the primary structures of human GPIIb (Poncz et al., 1987;

Fitzgerald et al., 1987; Uzan et al., 1988; Frachet et al., 1990) and GPIIIa (Fitzgerald et al., 1987a; Zimrin et al., 1988; Rosa et al., 1988; Frachet et al., 1990) led to the realization that GPIIb/IIIa is actually a member of a superfamily of structurally, immunologically, and functionally related adhesion receptors, for which the term "integrin" has been proposed (Hynes, 1987) to signify the role of these receptors in integrating extracellular ligand binding with the intracellular cytoskeleton.

Integrins are α/β noncovalent-associated heterodimeric protein complexes present on the surface of all adherent cells. Analysis of the human cDNA-derived primary structures of 12 different α-subunits ($\alpha^{1-6,8}$, α^{IIb}, α^v, α^M, α^L, and α^X) (Takada et al., 1991; Bossy et al., 1991; and references cited) and 8 unique β-subunits (β^{1-8}) (Moyle et al., 1991; and references therein) revealed that, although there is no structural relationship between both subunits, any pair of α- or β-subunits share 35-60% amino acid sequence identity. Depending of the cell type, the α- and β-subunits associate in various combinations. This, together with possible differential posttranscriptional processing and/or alternative mRNA splicing (van Kuppevelt et al., 1989; Altruda et al., 1990; Troesch et al., 1990; Bray et al., 1990; Tamura et al., 1990; Languino and Ruoslahti, 1992), creates functionally distinct receptors, which provide cells with varied capabilities to mediate a wide variety of cell-cell and cell-matrix interactions. The obvious importance of integrins in a number of biological processes have brought many investigators with diverse backgrounds (cell biology, molecular biology, protein chemistry, etc.) to the field, and led to a virtual explosion of work done on integrins during the last years. The discovery of integrins also integrated a large body of separate observations made during many years on platelet aggregation, immune functions, cell-extracellular matrix interactions during development, tissue repair, tumor invasion, etc., and brought, thus, new insights into the molecular mechanisms of cell adhesion (reviewed by Hynes, 1987; Ginsberg et al., 1988; Ruoslahti and Giancotti, 1989; Hemler, 1990; Humphries, 1990; Springer, 1990; Albelda and Buck, 1990; Dustin and Springer, 1991; Ruoslahti, 1991; Hynes, 1992).

Integrins may be subgrouped by various structural and/or functional criteria. GPIIb/IIIa, also denominated integrin $\alpha^{IIb}\beta_3$, the most extensively characterized integrin, shares with the endothelial vitronectin receptor (integrin $\alpha^v\beta_3$) the β-subunit, and together constitute the cytoadhesin, or β_3 family (Plow et al., 1986; Ginsberg et al., 1988).

INTEGRIN $\alpha^{IIb}\beta_3$: BIOSYNTHESIS, EXPRESSION, AND ANALYSIS OF CDNA-DERIVED AMINO ACID SEQUENCE

The genes coding for GPIIb and GPIIIa have been located to the q21-22 region on the long arm of human chromosome 17 (Sosnoski et al., 1988; Bray et al., 1988). The gene coding for GPIIb spans a region of 17.2 kb, contains 30 exons (Heidenreich et al., 1990), exists as a single copy per human genome (Prandini et al., 1988), and its expression is tissue-specific and restricted to cells of the megakaryocytic lineage (Uzan et al., 1991). On the other hand, the gene for GPIIIa spans 46 kb and is divided into 14 exons (Lanza et al., 1990; Zimrin et al., 1990).

Examination of the biosynthetic processing and assembly of the platelet GPIIb/IIIa complex in human megakaryocytes (Duperray et al., 1987) and HEL cells (Bray et al., 1986; Rosa and McEver, 1989) showed that both GPIIb and GPIIIa are synthesized as single-chain precursors to which N-linked mannose-rich oligosaccharides are co-translationally added in the endoplasmic reticulum. A 5-fold excess of pre-GPIIb is synthesized relative to GPIIIa. Assembly of the GPIIb and GPIIIa precursors require 4-6 hours for completion. All GPIIIa molecules are assembled into heterodimers, while the excess of pre-GPIIb is intracellularly degraded. Following assembly, the heterodimers are rapidly transported to the Golgi apparatus where the oligosaccharide chains of pre-GPIIb are processed into the complex-type, O-linked oligosaccharides are added, and the polypeptide chain undergoes proteolytic cleavage to yield mature GPIIb (135 kDa) made up of disulphide-linked heavy (GPIIbH) and light (GPIIbL) chains (114 and 22 kDa, respectively, Phillips and Agin, 1977; Calvete and González-Rodríguez, 1986; Usobiaga et al., 1987). Cleavage of pre-GPIIb/IIIa seems to be important for maturation into a functioning receptor, since expression of uncleaved GPIIb message produces an abnormal GPIIb/IIIa and a platelet thrombasthenic phenotype (Jung et al., 1988; Moroi et al., 1991).

Investigation on the biogenesis of the vitronectin receptor in human endothelial cells

(Polack et al., 1989) evidence that cytoadhesins $\alpha^v\beta_3$ and $\alpha^{IIb}\beta_3$ are synthesized by a common mechanism.

Analysis of the cDNA-derived sequences of GPIIb and GPIIIa (Poncz et al., 1987; Fitzgerald et al., 1987), and comparison with those of other integrins, provided evidence for the existence of a number of (evolutionarily conserved) structural features:

a) Both subunits, GPIIb and GPIIIa, each contain a single putative transmembrane domain near their carboxyl termini followed by a short (20-40 residues) cytoplasmic tail.

b) The cytoplasmic tail of GPIIIa contains a tyrosine and several threonine and serine residues within potential phosphorylation sites.

c) The putative intracellular portion of GPIIbL, coded entirely by exon 30 (Heidenreich et al., 1990), contains the sequence GFF(K/D)R, which has been absolutely conserved in all other integrin α-subunits.

d) GPIIbH, which may be completely extracellular, contains four stretches of 12 amino acids near the midportion of the molecule (residues 243-254; 297-308; 365-376; and 426-437) (Figure 1A) that are similar to calcium-binding sites found in other proteins.

e) The extracellular side of GPIIIa is characterized by the presence of 56 cysteine residues, all of them involved in intrachain disulphide bonding (Eirín et al., 1986) (Figure 1B). These residues are mainly distributed into two cysteine-rich domains (N-terminal: 1-50, 7 cysteines; and C-terminal: 400-655: 41 cysteines). The C-terminal cysteine-rich domain can be subdivided, depending on the amino acid sequence alignment, into three (Calvete et al., 1991) or four (Fitzgerald et al., 1987) cysteine-rich tandem repeats of about 40 amino acids long. The number and position of cysteine residues are highly, but not absolutely, conserved among the different integrin β-subunits, indicating that disulphide bonds may play an important role in stabilizing the functional active conformation of the protein.

f) Beside cysteine residues, certain polypeptide stretches of GPIIIa within the region between the cystine cores (positions 108-128; 142-164; 191-197; 217-228; 236-240; 245-252; 259-274; 307-311; 337-348) have been also highly (70-95%) conserved in all known integrin β-subunits during evolution, suggesting that they are of structural and/or functional relevance. The fact that some of these highly homologous regions are contained within individual exons (Lanza et al., 1990), provides genetic validation to this hypothesis.

g) Several glycine residues within the C-terminal cysteine-rich core are absolutely conserved among all integrin β-subunits. This may reflect that at those positions, probably due to the compactness of the folded core, only residues without side-chain are sterically allowed.

The structure and binding properties of GPIIb/IIIa indicate the presence of structural domains involved in at least three major functions: subunit association, extracellular ligand recognition and binding, and cytoskeletal interactions. In the following I will review our present knowledge on structure-function relationships in this platelet integrin.

PROTEIN-STRUCTURAL AND BIOCHEMICAL CHARACTERIZATION OF GPIIb/IIIa

In agreement with, and complementing, the molecular biology and cell biology approaches, protein chemical analyses showed that:

a) Only the heavy chain of GPIIb contains O-linked oligosaccharide chains (Eirín et al., 1986). Glycosylation of Ser^{847} seems to be essential for expression of the Bak (=Lek) alloantigen system (Take et al., 1990; Calvete, in preparation).

b) All the putative extracellular N-glycosylation sites of GPIIb (asparagines 15, 249, 570, 680 in GPIIbH, and 72 in GPIIbL1) (Figure 1A) and GPIIIa (positions 99, 320, 371, 452, 559, and 654) (Figure 1B) are actually occupied (Calvete et al., 1989a; Calvete, unpublished data).

c) The two chains of GPIIb are covalently-linked by a single disulphide bond (Calvete and González-Rodríguez, 1986) between the last cysteine of GPIIbH (Cys^{826}) and the first cysteine of GPIIbL1 (Cys^{21}) (Calvete et al.,1989) The intrachain disulphide bridges of GPIIb (7 S-S bonds in GPIIbH and 1 S-S bond in GPIIbL) form loops between nearest-neighbor cysteine residues (Calvete et al., 1989a) (Figure 1A). The importance of disulphide bridges in stabilizing the GPIIb folding pattern is suggested by the observation that a mutant GPIIb,

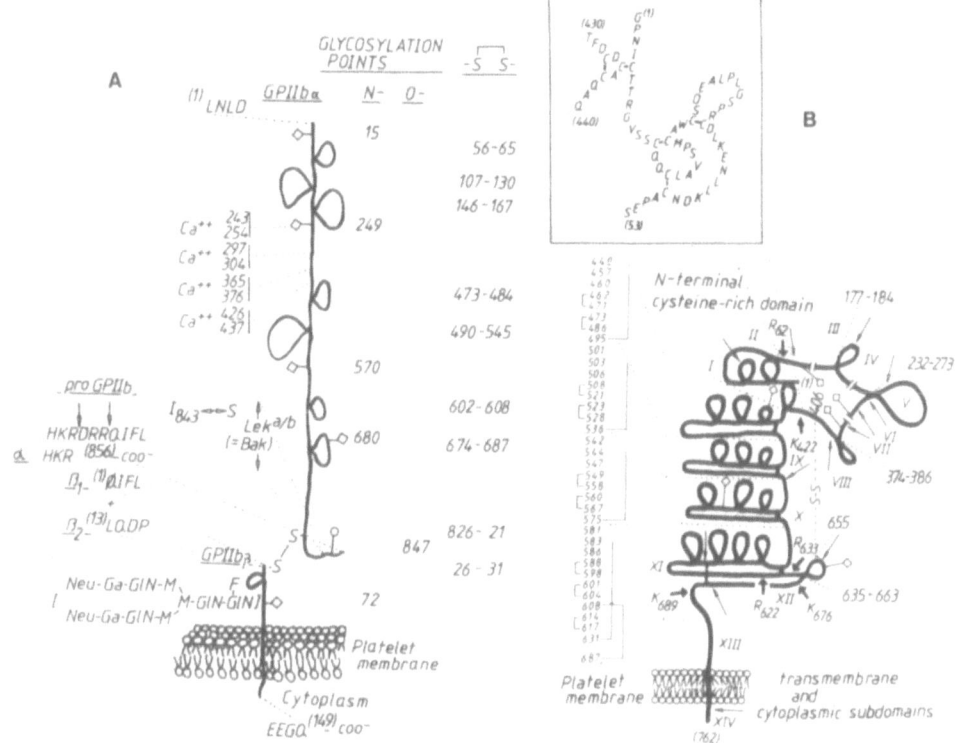

Figure 1. A) Schematic drawing of glycoprotein IIb showing some of its main structural features, like the position of: N- and O-glycosylation points; intra- and inter-molecular disulphide bridges; the four putative calcium-binding sites; the Bak$^{a/b}$ (Lek$^{a/b}$) alloantigen system; and the proteolytic cleavage sites which give rise to mature two-chain GPIIb made up of an homogeneous heavy chain (1-856) and an alternatively N-terminal cleaved light chain (β_1, 1-149, with pyroglutamic acid as its N-terminal residue; β_2, 13-149). The oligosaccharide structure attached to GPIIbβ is also shown. Neu, neuraminic acid; Ga, galactose; GIN, N-acetyl glucosamine; M, mannose; F, fucose.

B) A model of GPIIIa showing the tentative assigned disulphide bond pattern within the N-terminal cysteine-rich domain (*inset*) and the C-terminal cystine core. Thin, long arrows point to exon boundaries; the Roman nomenclature marks the position of the fourteen exons. The short, thick arrows show the position of the trypsin cleavage sites which produce the 35 kDa proteolytically-resistant GPIIIa core, K, lysine; R, arginine. The position of glycosylation points (*represented by diamonds attached to the polypeptide chain*) at positions 99, 320, 371, 452, 559, and 654 are also indicated.

in which a 13-base deletion in exon 4 produces a 6-amino acid deletion including Cys[107], is unable to form heterodimers with GPIIIa and result in a thrombasthenic phenotype in the Arab population in Israel (Newman et al., 1991).

    d) Proteolytic cleavage of pre-GPIIb occurs at two or three sites to give rise to an homogeneous heavy chain (GPIIbH, 1-856) disulphide-bonded to one of two light chains: GPIIbL$_1$, 860-1008, with pyroglutamic acid as its blocked N-terminus, and GPIIbL$_2$, 872-1008 (Calvete et al., 1990) (Figure 1A). This explains the observed heterogeneity of GPIIbL (Calvete and González-Rodríguez, 1986; Loftus et al., 1988). Its biological significance, if any, remains to be established.

    e) The GPIIIa sequence at positions 623-627, initially published as EPYMT (Fitzgerald et al., 1987; Zimrin et al., 1988), is actually GALHD (Calvete et al., 1990a) in accordance with Rosa et al. (1988).

f) The positions of the C-termini of GPIIIa (762) and GPIIbL (GPIIbL1 149, GPIIbL$_2$ 137) correspond with those deduced from the cDNAs, although glutamine instead of glutamic acid was found in GPIIbL (Calvete et al., 1990a). The potential significance of this finding requires further investigation, however.

g) The single oligosaccharide chain of GPIIbL is a disialylated biantennary structure substituted with fucose on the chitobiose core (Reason et al., 1991) (Figure 1A). The structures of the oligosaccharide chains of GPIIbH and GPIIIa remain to be established.

h) Glycoprotein IIIa contains both short- and long-range cystine links (Calvete et al., 1988; Beer and Coller, 1989; Niewiarowski et al., 1989; Calvete et al., 1991; Kouns et al., 1991b). The disulphide bridge between Cys^5-Cys^{435} defines a large loop containing the integrin β-subunit highly evolutionary conserved polypeptide stretches; the bond between Cys^{406}-Cys^{655} joins the N- and the C-terminal parts of the C-terminal cysteine-rich core (Figure 1B). These bridges are probably major determinants for the overall folding of the molecule. This hypothesis is supported by the identification of the molecular genetic basis of Glanzmann thrombasthenia in the Iraqi-Jewish population in Israel as a 11-base deletion within exon 12 of the GPIIIa gene (Newman et al., 1991). This mutation produces a frameshift leading to protein termination shortly before Cys^{655}. No GPIIIa message could be detected in these thrombasthenic platelets. The altered form of GPIIIa, lacking the Cys^{406}-Cys^{655} bond, may be highly unstable and rapidly degraded after biosynthesis. Protein unstability may be due to the absence of the long-range cystine residue, since the transmembrane and cytoplasmic domains of GPIIIa seem not to be essential for the association with pro-GPIIb, a correct cellular transit, and exposure of the complex at the cell surface (Frachet et al., 1992). The complete pattern of disulphide bridges of GPIIIa awaits elucidation.

i) Both subunits of GPIIb/IIIa contain single class of low-affinity Ca^{++}-binding sites (Rivas and González-Rodríguez, 1991). The isoterms at 21° C showed: GPIIbH, n = 5.4 +/- 0.9, K$_d$ = 0.17 +/- 0.03 mM; GPIIIa, n = 2.0 +/- 0.4, K$_d$ = 0.30 +/- 0.07 mM. The existence of calcium-binding sites in GPIIIa could not be deduced from cDNA-derived sequence analysis, although it was anticipated by Gogstad et al. (1983). The GPIIb/IIIa complex contains, at 21° C, both low- and high-affinity binding sites: n$_1$ = 0.9 +/- 0.2, K$_{d1}$ = 80 +/- 30 nM; n$_2$ = 3.4 +/- 0.6, K$_{d2}$ = 40 +/- 15 μM. The affinity of the single high-affinity site increased one order of magnitude when the GPIIb/IIIa complex was incorporated into egg-yolk lecithin liposomes, and occupancy of this Ca^{2+}-binding site correlated with stability of the heterodimer (Rivas et al., 1991). Gulino et al. (1992) have express a recombinant fragment of GPIIbH (171-464) which contains the four putative Ca2+-binding sites predicted from the cDNA-derived sequence (Figure 1A). In overall agreement with the data of Rivas et al. (1991), the authors found two classes of binding sites: two sites with K$_d$ of 30 μM and two sites with K$_d$ of 120 μM. Localization of the other Ca^{2+}-binding sites within GPIIb/IIIa requires further investigation.

What also remains to be clarified is the molecular basis, and the physiological significance, of the observed ability of GPIIb/IIIa to function as a voltage-independent, passive Ca^{2+} channel in the platelet plasma membrane (Brass, 1985), when reconstituted in liposomes (Rybak et al., 1988), as well as incorporated into a planar phospholipid bilayer (Fujimoto et al., 1991).

MOLECULAR CHARACTERIZATION OF ISOLATED GPIIb/IIIa

The glycoprotein IIb/IIIa has been isolated in buffers containing Triton X-100, and characterized using size-exclusion chromatography or laser-light scattering in combination with analytical ultracentrifugation (Jennings and Phillips, 1982; Carrell et al., 1985; Rivas et al., 1991a; 1991b). The Stokes radius of the GPIIb/IIIa-detergent complex calculated from hydrodynamic parameters (0.38 g TX-100 bound/g GPIIb/IIIa; s°$_{20}$ = 8.9S; f/f$_{min}$ = 1.5) was 62-74 Å. Carrell et al. (1985) reported that the morphology of individual complexes, determined by electron microscopy of rotary-shadowed specimens, consisted of two domains: an oblong head of ~8 x 10 nm with two rodlike tails (~14-17 x 2-3 nm) extending from one side of the head. Based on the morphology of the dissociated subunits (Carrell et al., 1985) and on hydrodynamic measurements (Jennings and Phillips, 1982), the authors concluded that GPIIb accounted for the formation of the head and GPIIIa for both tails of the complex. In the light of our current knowledge, it becomes clear that both the high frictional ratio (f/f_0 = 2.1) obtained for GPIIIa (Jennings and Phillips, 1982) and the filamentous structures reported for dissociated GPIIIa (Carrell et al., 1985) reflected the high tendency of GPIIIa to self-associate (Calvete et al., 1987; Rivas et al., 1991b).

The organization of GPIIb/IIIa has been subsequently reinterpreted based on the cDNA-deduced protein sequences. It has been proposed that GPIIb/IIIa has two sites of membrane insertion located at the ends of the filamentous domains (which in this model may correspond to the carboxy terminal portions of GPIIIa and GPIIbL), with the N-terminal domains of both subunits being extracellular and forming the bulk of the oblong head (see Figure 3 in Phillips et al., 1988). This figure is similar to the structural model of the human fibronectin receptor (integrin $\alpha^5\beta_1$) reported by Nermut et al. (1988). However, proteolytic dissection of isolated GPIIb/IIIa (Figure 2) has shown that sequences of the GPIIIa loop between the cysteine-rich regions (GPIIIa (217-235)-S-S-(262-298); 324-366; 403-421) and regions of the C-terminal half of GPIIbH (486-553; 696-734; 780-817) and the midregion of GPIIbL (GPIIbL$_2$ 30-75) are involved in the heterodimer intersubunit surface (Calvete et al., 1992). Thus, not only GPIIbH forms a complex with GPIIIa (Lam et al., 1989) but also GPIIbL is involved. This model of GPIIb/IIIa (Figure 2) does not fit the current accommodation of the amino acid sequences of GPIIb and GPIIIa in the head/two-tails image (Phillips et al., 1988). In addition, it must be noticed that electron micrographs of individual GPIIb/IIIa-detergent complexes show a great variety of shapes: filled globular, empty oval, head-two-tails, bilobular shapes, etc. (Rivas et al., 1991b). This fact may reflect inherent molecular flexibility of the polypeptide chains of the glycoprotein IIb/IIIa complex. Therefore, further image analysis, together with the use of monoclonal antibodies whose epitopes are well assigned, are needed to identify molecular domains and to be able to carry out a low-resolution three-dimensional reconstruction of the heterodimer images.

INDUCTION OF THE RECEPTOR CAPABILITY OF GPIIb/IIIa

Though the nature of the conformational change(s) in GPIIb/IIIa leading to induction of its receptor capability remains elusive, it has been shown that it does not involve phosphorylation (Hilery et al., 1991), that activation is mainly a property of the protein itself rather than of the surrounding cell membrane environment (O'Toole et al., 1990; Du et al., 1991), and that spatial separation or reorientation of extracellular domains within GPIIb and/or GPIIIa may serve to convert this integrin into a functional adhesion receptor (Sims et al., 1991). In agreement with the energy transfer experiments from Sims et al. (1991), Heath et al. (1991) and Kouns et al. (1991a) have showed that although inactive GPIIb/IIIa is highly resistant to proteolysis with Arg-C or thrombin, GPIIb within active GPIIb/IIIa becomes a substrate for these enzymes.

Investigation of the chymotryptic degradation pattern of GPIIb/IIIa in washed, resting platelets showed that a 95 kDa product, containing GPIIbH17-(551), was released to the supernatant at the time the GPIIIa region 100-348 was degraded, and that a 66 kDa GPIIIa proteolysis-resistant product, corresponding to (1-100)-S-S-(349-C-terminal), remained membrane-bound and noncovalently associated with a 25 kDa product of GPIIbH (encompassing residues 558-747) (Calvete et al., 1991b; Niewiarowski et al., 1989). However, analysis of a chymotryptic degradation product of GPIIb/IIIa bound to an affinity matrix of GRGDSPK-Sepharose (Lam, 1992) showed that GPIIb (1-485) and GPIIIa (1-727/739) remained noncovalently-associated as a functional divalent cation-dependent heterodimer. A possible explanation to reconcile these apparently contradictory data could be that upon receptor induction the N-terminal half of GPIIbH changes its relative orientation protecting itself and the GPIIIa 100-348 region from proteolysis, while the C-terminal half of GPIIbH becomes more exposed.

In agreement with the above-stated hypothesis, immunochemical studies on the topography of GPIIb (Calvete et al., 1991b) (Figure 3A) and GPIIIa (Calvete et al., 1991c; Andrieux et al., 1991) (Figure 3B) have provided evidence for assigning regions within GPIIb/IIIa involved in the receptor expression-mediated conformational change.

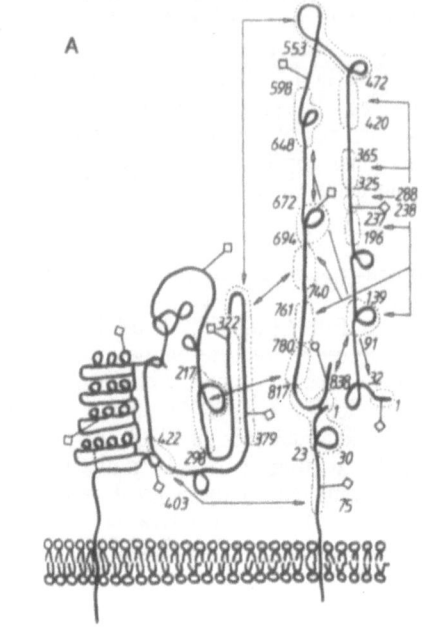

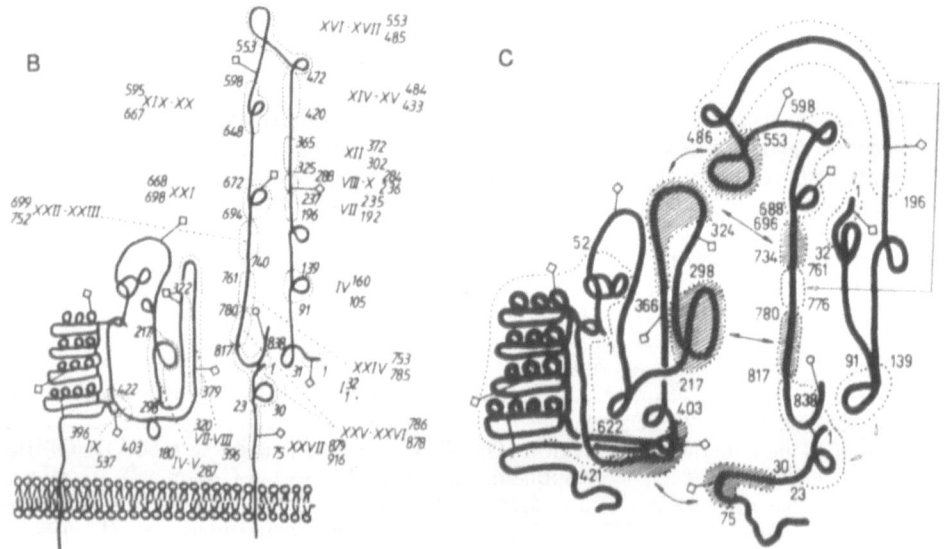

Figure 2. A) Schematic representation of intra- and inter-molecular domain connectivities within GPIIb/IIIa deduced by analysis of proteolytically-derived structures (Calvete et al., 1992).
B) Comparison of the position of GPIIb/IIIa interacting regions and exon boundaries (*in Roman nomenclature*) shows that there is a close relationship between protein domains and the genetic organization of the GPIIb and GPIIIa genes.
C) A rudimentary connectivity map of GPIIb/IIIa. Shadowed regions represent protein domains participating in the inter-subunit surface within the heterodimer.

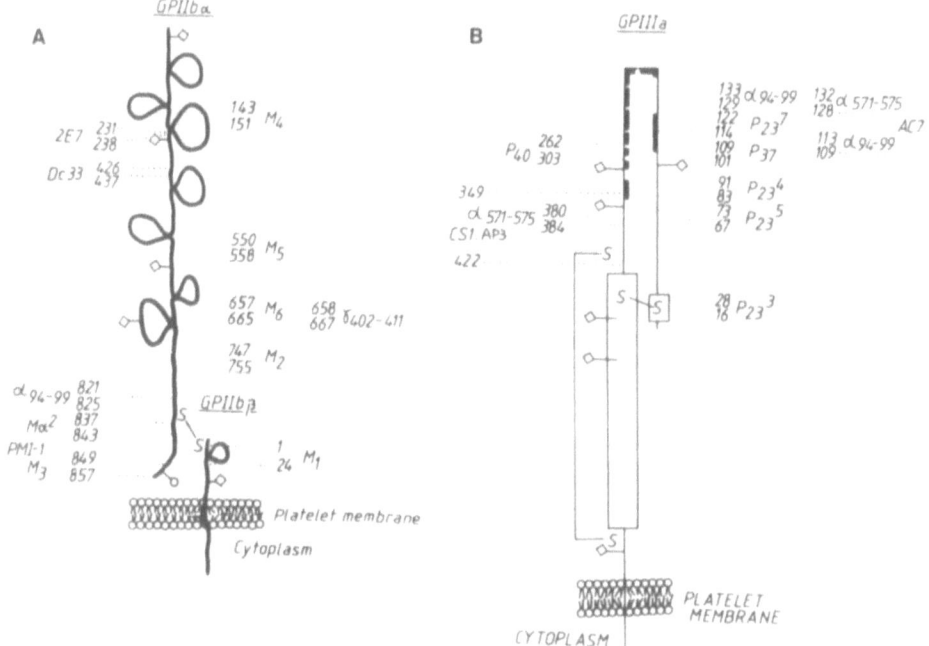

Figure 3. Schematic representation of the localization of monoclonal antibody epitopes (*described in the text*) within the linear primary structure of: **A)** GPIIb; and **B)** GPIIIa. α 94-99, α 571-575, and γ 402-411 indicate the theoretical hydropathic complementary sites for those human fibrinogen amino acid stretches (see Calvete et al., 1991 b, c). S-S, disulphide bond. The cysteine-rich regions of GPIIIa are represented as boxes. The thick lines within the large GPIIIa loop between the two cysteine-rich regions indicate the approximate position of the high evolutionary conserved polypeptide stretches within all known integrin β-subunits (positions 108-128; 142-164; 191-197; 217-228; 236-240; 245-252; 259-274; 307-311; 337-348).

Thus, binding of monoclonal antibodies $P_{23}{}^{7}$ (GPIIIa 114-122; Calvete et al., 1991c) and M_6 (GPIIbH 657-665; Calvete et al., 1991b) to their epitopes is EDTA- or thrombin-dependent. Similarly, monoclonal antibody AC7 (GPIIIa 109-128; Andrieux et al., 1991) interacts only with activated platelets. Once bound to their respective epitopes all three antibodies block fibrinogen binding to GPIIb/IIIa. Altogether this may imply that the exposure of those regions on GPIIb/IIIa (Figure 2) correlates with expression of receptor capability.

Gulino et al. (1990) have found that the monoclonal antibody D33C binds to a surface-exposed calcium-binding site on resting GPIIb/IIIa (GPIIbH 426-437; see Figure 3A), thereby inducing platelet aggregation and the exposure of the fibrinogen binding site. Similarly, Kouns et al. (1990) have reported that monoclonal antibody D3GP3 (or its Fab fragment), recognizing a conformational epitope (5000 sites/platelet; Kd = 80 nM) within the cyteine-rich core of GPIIIa (tentatively assigned within the region 422-692, Kouns et al., 1991b), induces fibrinogen binding in a limited population of GPIIb/IIIa within the platelet membrane and causes platelet aggregation. Receptor occupancy with RGD-peptides or soluble fibrinogen, but not platelet stimulation with ADP in the absence of ligand, led to an increase in the number of D3GP3-binding sites. Thus, the data suggest that GPIIb/IIIa can exist in several conformations, and that the antibody may recognize a region of GPIIIa involved in conformational changes following receptor occupation.

Analysis of molecular biology-derived GPIIb/IIIa constructs showed that while the cytoplasmic and transmembrane domains of GPIIIa do not seem to play a prominent role either in subunit association specificity or in receptor function (Solowska et al., 1991), the presence of the corresponding domains in the α subunit are necessary for expression of

GPIIb/IIIa at the cell surface (Frachet et al., 1992). Moreover, O'Toole et al. (1991) have shown that the cytoplasmic domain of GPIIbL controls ligand binding affinity, suggesting thus the existence of a mechanism for inside-out transmembrane signaling. The suggestive hypothesis put forward by these authors, namely that the cytoplasmic domain of GPIIbL may bind to an intracellular moiety to maintain the receptor in a low affinity conformation, and that modification of this interaction during platelet activation could induce the high affinity state, awaits further detailed investigation.

Limited proteolysis of GPIIb/IIIa by α-chymotrypsin has been shown to expose activation-independent fibrinogen binding sites on the surface of resting platelets (reviewed in Peerschke, 1985). Pidard et al. (1991) have studied the relationship between discrete proteolytic events within GPIIb/IIIa and the development of fibrinogen-binding function. These authors found that a single cleavage at the GPIIbH (tentatively assigned at Phe[827] or Trp[839]) released its C-terminal peptide and correlated with acquisition of permanently available fibrinogen binding sites. Interestingly, the released peptide contains the epitope for monoclonal antibody PMI-1 (GPIIbH 842-857, Loftus et al., 1987) (Figure 3A) whose surface expression under physiological concentration of divalent cations is restricted to less than 2000 sites per platelet (Ginsberg et al., 1986). Since interaction of RGD-containing ligands with GPIIb/IIIa alters the conformation of this region, resulting in increased exposure of the PMI-1 epitope (Loftus et al., 1987), it seems reasonable to propose that its proteolytic release from GPIIbH leads to a similar, and irreversible, conformational change within GPIIb/IIIa. The nature of this structural modification induced by proteolysis requires further elucidation, however.

GPIIb/IIIa ACTS AS A COMMON RECEPTOR FOR PLATELET ADHESIVE PLASMA PROTEINS. IDENTIFICATION OF ADHESIVE RECOGNITION SEQUENCES WITHIN THE LIGAND

Besides its role as fibrinogen receptor, GPIIb/IIIa has been implicated as a receptor for other adhesive proteins found in plasma. Specifically, GPIIb/IIIa has been shown to possess capability to interact with fibronectin (Ginsberg et al., 1983; Gardner and Hynes, 1985; Parise and Phillips, 1986), von Willebrand factor (Ruggeri et al., 1982; Pietu et al., 1984), and vitronectin (Pytela et al., 1986), although it is not the exclusive platelet receptor for these proteins.

The fact that the binding of the four adhesive proteins was mutually exclusive suggested that the ligands either share a common binding site within the receptor or bind to a site that affects the binding site(s) of the other adhesive proteins (Pietu et al., 1984; reviewed by Peerschke, 1985; Plow et al., 1986; Phillips et al., 1988). Because of the large molecular size of the adhesive proteins, it was unexpected that small synthetic peptides containing the amino acid sequence Arg-Gly-Asp (RGD), a polypeptide originally determined to represent a major cell-binding site on fibronectin (Pierschbacher and Ruoslahti, 1984; Yamada and Kennedy, 1984), inhibited the binding of fibrinogen, fibronectin, and von Willebrand factor to activated platelets (Haverstick et al., 1985; Ginsberg et al., 1985; Gartner and Bennett, 1985; Plow et al., 1985). A direct interaction of RGD-peptides with GPIIb/IIIa was demonstrated by:

a) The binding of a population of GPIIb-IIIa to immobilized Arg-Gly-Asp peptides (Pytela et al., 1986; Lam et al., 1987);

b) Chemical cross-linking of RGD peptides to GPIIb/IIIa in intact or activated platelets (Santoro and Lawling, 1987; D'Souza et al., 1988);

c) Equilibrium dialysis: the binding of RGD-containing peptides to GPIIb/IIIa showed many features in common with fibrinogen binding: it was specific to a single site, saturable, reversible (Kd = 1.0 +/- 0.2 µM), and Ca^{2+}-dependent (Steiner et al., 1989);

d) Biochemical analyses of natural GPIIb/IIIa antagonists isolated from viper venoms (called "disintegrins" (for a review see: Gould et al., 1990; Scarborough et al., 1991) or from the leech (called "ornatins or decorsin", reviewed in Mazur et al., 1991), suggest that the conserved RGD (or KGD) sequence, in the spatial configuration determined by the appropriate pairing of the cysteine residues (see Calvete et al., 1991a), functions as a cell recognition site.

Since the Arg-Gly-Asp sequence is present within the primary structure of many adhesive proteins, including fibrinogen, von Willebrand factor, fibronectin, and vitronectin (reviewed by Ruoslahti and Pierschbacher, 1986; 1987), the promiscuous receptor capability of GPIIb/IIIa may be due to its ability to bind to properly-exposed RGD sequences contained within the adhesive proteins. Thus, although fibrinogen contains two RGD sequences in its α chain (residues 95-97 and 572-574) (Doolittle et al., 1979), only the C-terminal RGD sequence does support cell adhesion to the macromolecule (Cheresh et al., 1989). This probably reflects the fact that $A\alpha^{95-97}$ is located in a coiled-coil region and may not be accessible for receptor recognition.

This rather simple view of the interaction of RGD-containing adhesive proteins with the glycoprotein IIb/IIIa complex is complicated by the fact that besides the RGD motif many adhesive proteins contain other receptor-binding domains (Plow et al., 1987; Yamada, 1991; Humphries, 1990). In the case of fibrinogen, a dodecapeptide derived from the C-terminus of the fibrinogen γ chain (HHLGGAKQAGDV), which does not contained the RGD sequence and which is not found in adhesive proteins other than fibrinogen, competes with RGD-containing peptides for binding to GPIIb/IIIa, and is capable of inhibiting the binding of each of the other three adhesive proteins to GPIIb/IIIa (Kloczewiak et al., 1984; Timmons et al., 1984; Plow et al., 1985; Gartner and Bennett, 1985; Lam et al., 1987; Santoro and Lawling, 1987; D'Souza et al., 1988; Steiner et al., 1989).

The relationship between the gamma chain and RGD binding sites on GPIIb/IIIa is still a matter of controversy. Thus, based on the effect of the γ chain peptide on the binding of an RGD peptide to platelets and by cross-elution of GPIIb/IIIa from affinity matrices containing either the RGD or the γ chain peptide, Lam et al. (1987) concluded that both peptides share a common binding site on platelets. On the other hand, Santoro and Lawing (1987) showed that upon platelet activation an RGD peptide cross-linked primarily to GPIIIa, whereas a γ chain peptide cross-linked predominantly to GPIIb, though each peptide inhibited cross-linking of the other to the respective subunit. The authors concluded that the peptides may compete for related, mutually exclusive, but nonidentical binding sites on glycoprotein IIb/IIIa complex. This latter interpretation does not take into account the possibility that both peptides bind actually to the same site though they cross-link to different receptor regions.

Relevant to this point is the finding that anti-idiotypic antibodies against PAC1, a monoclonal antibody anti-GPIIb/IIIa which contain in its hypervariable region μ-CDR3 the sequence RYD that mimics the RGD sequence(s) within fibrinogen (Taub et al., 1989), bind to fibrinogen fragments containing either or both the $A\alpha^{572-574}$ and the $\gamma^{400-411}$ stretches (Abrams et al., 1992). Since electron microscopy of fibrinogen antibody complexes confirmed that each antibody bound independently to sites on both the $A\alpha$ and the γ chains, the authors suggest that these regions may share a similar conformation within the native fibrinogen molecule. These data speak in favor of the identical-receptor-binding-site hypothesis, but are difficult to reconcile with the experimental evidences showing that integrin $\alpha^{v}\beta_3$, structurally similar to GPIIb/IIIa, and which also mediates cell adhesion to fibrinogen (Charo et al., 1987; Cheresh et al., 1987), does not interact with the fibrinogen γ 400-411 peptide (Cheresh et al., 1990; Smith et al., 1990). Ultimately, comparison of high resolution structural data of GPIIb/IIIa- and vitronectin receptor-peptide complexes would be needed to definitively settle down this point. Since the ligand-binding sites of the anti-idiotypic antibodies may bear an internal image of the ligand-binding site within GPIIb/IIIa, analysis of the three-dimensional structure of anti-PAC1 Fab fragment-peptide complexes may also provide relevant structural information.

Finally, it must be noticed that the synthetic peptide approach, although useful for designing platelet aggregation competitive antagonists (i.e. to define structural constrains and functional groups required for inhibitory potency and specificity), may overlook other functional conformations of the macromolecules (Yamada, 1991). Central to understanding integrin function is the basis of their natural ligand recognition, however. Therefore, a true understanding of the molecular basis of fibrinogen, or other adhesive protein, binding to GPIIb/IIIa requires fine structural characterization of the interaction surfaces of macromolecular ternary complexes.

THE LIGAND BINDING SITE WITHIN THE EXTRACELLULAR GPIIb/IIIa DOMAIN: MORE SITES THAN INSIGHTS!

Biochemical, immunological, and molecular-genetic analyses of the extracellular domain of GPIIb/IIIa have led to the identification of a variety of regions putatively involved in ligand-binding (Figure 4).

Several lines of evidence have implicated the N-terminal and central regions of GPIIIa in integrin receptor function:

a) Antibodies which bind to GPIIIa 101-109,114-122, 262-302 (P37, $P_{23}7$, P40, respectively; Calvete et al., 1991c), 109-128 (AC7, Andrieux et al., 1991), 133-136 (Pasqualini et al., 1989), and 349-422 (CS-1, Ramsamooj et al., 1991) inhibit both fibrinogen-binding to GPIIb/IIIa and platelet aggregation.

b) The human platelet alloantigen systems PLA[1/2] and Pen[a/b] are associated with a Leu (33)-> Pro and Arg (143)-> Gln polymorphisms, respectively (Newman et al., 1989; Newman, 1991; Wang et al., 1991; Bowditch et al., 1992). Anti-Pen and anti-PLA antibodies completely inhibit ADP-induced platelet aggregation (Furihata et al., 1987).

However, antibodies are large molecules which may bind to epitopes spatially distant from the ligand-binding site and inhibit ligand binding by steric hindrance. Alternatively, interaction of the antibody with its epitope can change directly or indirectly the conformation of the ligand-binding domain within the receptor (see above and Gulino et al., 1990; Kouns et al.,1990).

c) A point mutation within GPIIIa that results in an Asp[119]-> Tyr substitution produces a mutant $\alpha^{IIb}\beta_3$ integrin that lacks ligand recognition and shows perturbed interaction with divalent cations (Loftus et al., 1990). The authors noticed that the linear spacing of oxygenated residues in the surrounding of Asp[119] (Ser[121], Ser[123], Asp[126], Asp[127], and Ser[130]) approximates that of the residues in the calcium-binding loop of EF hand proteins, and postulated that this clustering could provide coordination sites for divalent cation essential for fibrinogen binding.

Other point mutations at GPIIIa Arg[214], resulting either in Gln (Bajt et al., 1992) or Trp (Lanza et al., 1992), result in thrombasthenic platelets with reduced ability of mutant GPIIb/IIIa to bind ligands, failure to aggregate, and increased sensitivity of these complexes to dissociation by EDTA.

Although these data point to essential residues in a region of GPIIIa involved in ligand binding and in the Ca^{2+}-dependent stability of the GPIIb/IIIa complex at the platelet surface, the response to a single-site modification can only be interpreted in a speculative fashion in the absence of high-resolution structural data. Thus, it is hazardous to interpret the alteration solely in terms of local interactions of the mutated residue, since the functional defect may be mediated by changes of the overall conformation or by specific long-range effects.

d) Bajt et al. (1991) have characterized a mutant GPIIb/IIIa in which the polypeptide stretch GPIIIa 129-133 was replaced by the corresponding amino acid sequence from integrin β_1 subunit (FNVKS). Cell lines expressing this mutant GPIIb/IIIa were hyperadherent and manifested spontaneous aggregation when fibrinogen was added to the medium. The basis for the capacity of this substitution to augment the ligand binding function of integrin $\alpha^{IIb}\beta_3$ requires further structural characterization.

e) Expression of a truncated heterodimeric form of GPIIb/IIIa (GPIIbH/GPIIIa[1-469]) in insect cells indicate that it exists as a functionally active (*properly folded?*) complex, and thus that the cysteine-rich core of GPIIIa may not be necessary for the formation of the ligand-binding site (Wippler et al., 1991).

f) D'Souza et al. (1988a) have identified a region within GPIIIa (residues 109-171) to which the peptide KYGRGDS becomes chemically cross-linked upon addition of bis(sulfosuccinimidyl) suberate (BS) to thrombin-stimulated platelets. Using a different azido-peptide (GRGDSPK) coupled to the cross-linker aryl azide (sulfosuccinimidyl) 2-(p-salicylamido)-1,3'-dithiopropionate (SASD), Smith and Cheresh (1988) have identified a similar region (61-203) within the vitronectin receptor β-subunit.

It must be noticed, however, that in both cases the (searched) binding- and the (found) cross-linking sites may be as much as 30 Å apart! (Calvete et al., 1992a). Since the folding pattern of GPIIb/IIIa is not known, and the average radius of the heterodimer is around 50 Å, any conclusion based solely on this approach must be *per se* quite risky.

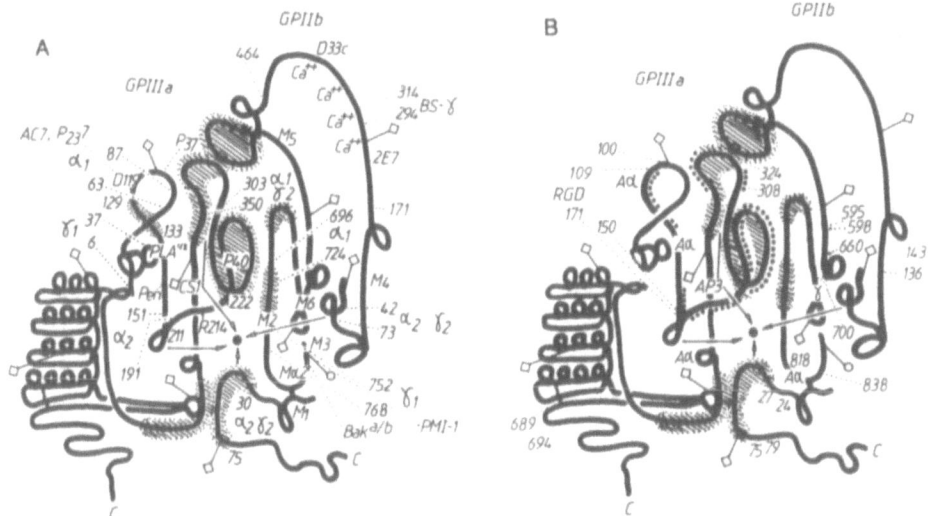

Figure 4. A Model of GPIIb/IIIa Based on the Data Presented in This Review. It incorporates most of the features of GPIIb and GPIIIa shown in more detail in Figures 1-3, and intends to correlate, in a speculative manner, known structural data with defined functional epitopes described within the text.

A) Ca++, putative calcium binding sites; M1, M2, Mα2, M3, M4, M5, M6, 2E7, D33c, PMI-1, and Bak[a/b], are epitopes on GPIIb recognized by antibodies defined in the text. PLA[1/2], P_{23}^4, P_{23}^5, P37, P_{23}^7, AC7, Pen, CS1, and P40 are antibody epitopes on GPIIIa whose exact locations are described in the text. D[119] and R[214], aspartic acid and arginine at positions 119 and 214, respectively, whose mutation have been found to be correlated with Glanzmann thrombasthenia phenotypes. BS-γ, the polypeptide stretch of GPIIbH where the fibrinogen γ-chain peptide was found covalently-linked. The polypeptide stretches GPIIIa 211-222 and GPIIbH 294-314, from where synthetic peptides have been derived which interact with fibrinogen and inhibit platelet aggregation, also are indicated. Shadowed regions represent GPIIb and GPIIIa domains involved in the formation of the heterodimer interface (*see Figure 2C*). α1, α2, γ1, and γ2, indicate the regions of GPIIb and GPIIIa where the synthetic peptides CGRGDF, CKRKRKRKRRGDV, CYHHLGGAKQAGDV, and CGAKQAGDV, respectively, were cross-linked (*see the text for details*). The black-filled circle represents a hypothetical ligand-binding site on GPIIb/IIIa, and the arrows pointing to it show those GPIIb and GPIIIa domains which may be closer than 10 Å to each other and to the ligand-binding site.

B) Regions of high susceptibility to proteolysis within the primary structures of GPIIIa (100-150, 308-324, 689-694), GPIIbH (136-143, 595-598, 660-700, 818-838) and GPIIbL2 (24-27, 75-89) are indicated. Aα and γ, hydropathic complementary amino acid sequences for the fibrinogen RGD-containing stretches Aα 571-575 or Aα 94-98, and γ 400-411, respectively (*see Figure 3*). RGD, the GPIIIa region where the synthetic peptide KYGRGDS was found, cross-linked. AP3, epitope recognized by the monoclonal antibody AP3. Filled circles indicate the position of high evolutionarily conserved regions among all integrin β-subunits (*Figure 3*).

g) Calvete et al. (1992a) have localized the cross-linking sites of RGD and KQAGDV, of different length and derivatized with the cross-linker N-hydroxysuccinimidyl-4-azidosalicylic acid (NHS-ASA) at the N-terminal α-amino group, to isolated GPIIb/IIIa in solution (Figure 4). Within GPIIIa, all the cross-linking sites where found within the first half of the molecule. It was concluded that some segments of the GPIIIa 151-191 and 303-350 stretches are very close (~10 Å) to each other and to the ligand-binding site.

h) Adhesive protein-binding to GPIIb/IIIa was blocked with two overlapping synthetic peptides encompassing residues 204-229 of GPIIIa ([211]SVSRNRDAPEGG[222], IC50 = 0.5 mM; Charo et al. (1991) or [217]DAPEGGFDAIMQATV[231]; Cook et al., 1992). Although fibrinogen binding to its receptor is Ca[2+]-dependent, the GPIIIa-derived peptides bound to fibrinogen in a cation-independent manner. Moreover, the interaction of GPIIIa 211-222 or

217-231 with fibrinogen was not inhibited by RGDS. Thus, though this region of GPIIIa could represent an important portion of the GPIIb/IIIa ligand-binding domain, there is no experimental evidence that it is actually involved in adhesive protein-binding within the structure of the native GPIIb/IIIa conformation.

i) The sequences of GPIIIa (128-132; 380-384) and (109-113; 128-133) were found hidropathically complementary to fibrinogen Aα 571-575 and 94-99, respectively (Calvete et al., 1991) (Figure 3B).

On the other hand, regions of GPIIb forming part of the ligand-binding pocket are not yet so well defined:

a) Lam et al. (1989) reported that GPIIbH contains sufficient information for maintenance of a functional complex with GPIIIa. Furthermore, after partial chymotryptic digestion of GPIIb/IIIa bound to an affinity matrix of GRGDSPK-coupled Sepharose 4B and washing of unbound peptides, two Ca++-dependent associated polypeptides of 55kDa (GPIIbH 1-[~485]) and 85 kDa (GPIIIa 1-[~730]) were specifically eluted by soluble HHLGGAKQAGDV or by GRGDSP (Lam, 1992).

b) The fibrinogen γ-chain peptide K(I^{125})YGGHHLGGAKQAGDV was covalently bound with BS to ADP- or thrombin-activated GPIIb/IIIa and its cross-linking site was localized to the 21 amino acid stretch GPIIb 294-314 (D`Souza et al., 1990) (Figure 4A).

Using a different approach, Calvete et al. (1992a) (Figure 4A) showed that the peptides CYHHLGGAKQAGDV (γ1), CKRKRKRKRRGDV (α1), CGAKQAGDV (γ2), and CGRGDF (α2) (carrying a photoactivatable cross-linker and an eosin molecule attached to the N-terminal cysteine residue) were covalently bound to GPIIbH 752-768; 696-724; 42-73 and GPIIbL$_2$ 30-75; and GPIIbH 42-73 and GPIIbL$_2$ 30-75, respectively, within isolated GPIIb/IIIa. It was concluded that the cross-linking regions for the short synthetic peptides (GPIIbH 42-73 and GPIIbL$_2$ 30-75) may be close to each other and at less than 10 Å from the peptide binding site.

c) A recombinant fragment corresponding to GPIIb 171-464 (Figure 4) bound to a fibrinogen-coated surface, with an estimated recombinant GPIIb:fibrinogen ratio of 2.2:1, and this interaction was optimal at 0.5 mM Ca^{2+}, when the four Ca^{++}-binding sites of recombinant GPIIb were occupied (Gulino et al., 1992). The recombinant GPIIb fragment-fibrinogen association was inhibited by soluble fibrinogen as well as by the synthetic peptides HLGGAKQAGDV and RGDF (IC50 = ~0.1 mM).

d) Synthetic peptides corresponding to the GPIIb regions 296-306 (B12: TDVNGDGRHDL; D`Souza et al., 1991), 309-312 (GAPL; Gartner et al., 1991), at milimolar concentration, inhibit platelet aggregation. Fibrinogen binding to B12 was dependent on divalent ions (Ca^{2+} or Mg^{2+}), and this interaction was inhibited by the fibrinogen γ-chain peptide HHLGGAKQAGDV and by RGDS (IC50 = 200 μM).

e) The epitope of platelet aggregation inhibitory monoclonal antibodies M5 and M6 were mapped to GPIIbH 550-558 and 657-665, respectively (Calvete et al., 1991b) (Figures 3A and 4A). The sequence where the epitope for M6 was located is, in addition, hydropathically complementary to the γ402-411 peptide of fibrinogen (Calvete et al., 1991b) (Figures 3A and 4B). The overlap between the thrombin-dependent M6 epitope and the putative fibrinogen binding site would imply that the unmasking of the region around the GPIIbH 657-665 peptide stretch could be one of the structural changes in GPIIb/IIIa required for the induction of the fibrinogen receptor in activated platelets.

In addition, analytical ultracentrifugation studies have recently showed that peptide GPIIbH 657-665 binds specifically to soluble fibrinogen in a Ca^{2+}-independent manner, and that this interaction was competitively inhibited by the fibrinogen γ402-411 peptide but not by RGDS (Rivas and Calvete, in preparation).

f) A high hydropathic complementarity between the region GPIIbH 821-825 and fibrinogen Aα 94-98 was noticed (Calvete et al., 1991b) (Figures 3A and 4B).

THE DARK SIDE OF GPIIb/IIIa FUNCTION: INTRACELLULAR INTERACT-IONS WITH THE PLATELET CYTOSKELETON

Resting platelet GPIIb/IIIa complexes are uniformly distributed along the plasma membrane (Isenberg et al., 1987), and are not associated with the platelet cytoskeleton (Phillips et al., 1980). Platelet activation and receptor occupancy lead to conformational changes in both GPIIb and GPIIIa (Shattil et al., 1985; Parise et al., 1987; Frelinger et al., 1988; 1990; 1991; O'Toole et al., 1990; Du et al., 1991; Sims et al., 1991), cause cytoskeletal reorganization-independent clustering of GPIIb/IIIa molecules within the plane of the membrane (Isenberg et al., 1987; Kouns et al., 1991) followed by platelet aggregation and clot retraction. These latter events are accompanied by the attachment of GPIIb/IIIa complexes to the cytoskeleton (Phillips et al., 1980; reviewed by Fox and Phillips, 1986; Kouns et al., 1991). The linkage to the cytoskeleton could provide a mechanism by which extracellular fibrin clots attach to intracellular actin filaments resulting in retraction of fibrin clots.

Painter et al. (1985) have provided evidence suggesting that the GPIIb/IIIa complex interacts directly with actin filaments. Neither the protein domains involved nor the mechanisms regulating the association are known, however.

The antigenic determinant of monoclonal antibody AP3, whose binding to GPIIb/IIIa inhibits platelet aggregation but not fibrinogen binding (Newman et al., 1987), have been tentatively assigned to GPIIIa 348-421 (Kouns et al., 1991b). It has been shown that AP3 (+ Tab)-induced inhibition of platelet aggregation is actually related to blockade of transmembrane signaling through the GPIIb/IIIa complex following receptor occupancy and leading to the cytoskeletal rearrangements required for the formation of platelet pseudopods important for platelet aggregation (Isenberg et al., 1990) and clot retraction (Gartner and Ogilvie, 1988).

It is tentative to speculate that the epitope recognized by AP3 is within, or spatially close to, a GPIIIa region involved in heterodimer patching. The finding that the epitope of monoclonal antibody C5GP3, which resides within the same region as the AP3 determinant, is cryptic in intact GPIIb/IIIa and become exposed following GPIIb/IIIa receptor occupancy (Kouns et al., 1991b), is in keeping with the hypothesis that this region of GPIIIa may be involved in conformational changes that occur after ligand binding, exposing neoepitopes which could be involved in interactions with other platelet surface and/or plasma components.

OTHER GPIIb/IIIa EPITOPES OF STRUCTURAL AND/OR FUNCTIONAL RELEVANCE (SEE FIGURE 4)

The epitope of monoclonal antibody PMI-1 (GPIIbH 842-857, Loftus et al., 1987), is essentially hidden within inactive GPIIb/IIIa (Ginsberg et al., 1986), but it is flanked by the surface-exposed antibody epitopes BaK$^{a/b}$ (GPIIbH Ile/Ser^{843}, Lyman et al., 1990), Mα2 (GPIIbH 837-843, Calvete et al., 1991b) and M3 (GPIIbH 849-857, Calvete ct al., 1991b).

Similarly, the epitope for the inhibitory monoclonal antibody P37 (GPIIIa 101-109) may be located within a surface-exposed loop (containing glycosylated Asp^{99}) between the surface inaccessible polypeptide stretches containing the hidden epitopes for $P_{23}5$ (67-73) and $P_{23}4$ (83-91), and the activation-dependent epitope of $P_{23}7$ (114-122) (Calvete et al., 1991c).

On the other hand, the epitopes for monoclonal antibodies M1 (GPIIbL$_2$ 2-24), M2 (GPIIbH 747-755) and M4 (GPIIbH 143-151) are well exposed to the platelet external medium (Calvete et al., 1991b).

The epitope recognized by the human monoclonal autoantibody 2E7 is stable to protein denaturation or reduction of disulphide bonds. Upon treatment of intact platelets with EDTA, the number of binding sites of 2E7 per GPIIb molecule increases four-fold while the dissociation constant decreases ten-fold. It specifically recognizes the peptide sequence GPIIb 231-238 (Kunicki et al., 1991). Tryptophan residue at position 235 is the immunodominant residue of the 2E7 epitope. Although the structural basis for expression of autoantigens on GPIIb/IIIa remains to be elucidated, it is reasonable to postulate that Trp^{235} may be permanently hidden within the native GPIIb/IIIa structure and only exposed upon partial unfolding of the molecule.

The patterns of proteolytic cleavage of isolated GPIIb/IIIa with different proteolytic enzymes indicate the presence of particular regions which show high susceptibility to proteolysis by trypsin, chymotrypsin and V8 endoproteinase (Calvete et al., 1992). These regions (GPIIIa 100-150, 308-324, 689-694; GPIIbL$_2$ 24-27, 75-89; GPIIbH 136-143, 595-598, 660-700, 818-838) may correspond to polypeptide extensions on the protein surface or surface-exposed segments connecting structural domains.

Greco et al. (1991) have identified an $(^{32}S)ATP$-α-S binding site on a GPIIb 18 kDa domain beginning at Tyr^{198}. This label could be competed by the agonist ADP. Since interaction of ADP with purified GPIIb/IIIa did not expose fibrinogen receptors (Parise and Phillips, 1985a), the biological role of the nucleotide-binding site on GPIIb awaits further investigation.

Cierniewski et al. (1989) reported that both subunits of GPIIb/IIIa in freshly isolated blood platelets contain covalently-bound palmitate, and that acylation was 4-5 fold intensified upon platelet activation with thrombin or ADP. The estimated stoichiometry was 0.14-0.33 mol palmitic acid/mol GPIIIa, and much lower for GPIIb. At present, the possible function(s), if any, of palmitylation are not clear.

OTHER GPIIb/IIIa PUTATIVE FUNCTIONS AND OCCUPANCY-DEPENDENT SIGNALING EVENTS

Banga et al. (1986) reported that apparent occupancy of GPIIb/IIIa by fibrinogen was correlated with activation of phospholipases A and C and was necessary to maintain Na^+/H^+ exchange in epinephrine-stimulated platelets. Recently, it has been shown that thrombin-induced tyrosine phosphorylation of several platelet proteins appears to require ligand binding to GPIIb/IIIa (Ferrell and Martin, 1989) as well as platelet aggregation (Golden et al., 1990), although GPIIb/IIIa itself is not tyrosine-phosphorylated (Wyler et al., 1986). These aspects of GPIIb/IIIa function are, at the molecular level, virtually unexplored.

REFERENCES

Abrams, C.S., Ruggeri, Z.M., Taub, R., Hoxie, J.A., Nagaswani, C., Weisel, J.W., and Shattil, S.J., 1992, Anti-idiotypic antibodies against an antibody to the platelet glycoprotein (GP) IIb-IIIa complex mimic GPIIb-IIIa by recognizing fibrinogen. *J. Biol. Chem.* 267:2775-2785.

Albelda, S.M. and Buck, C.A., 1990, Integrins and other cell adhesion molecules. *FASEB J.* 4:2868-2888.

Altruda, F., Cervella, P., Tarone, G., Botta, C., Balzac, F., Stefanuto, G., and Silengo, L., 1990, A human integrin β1 subunit with a unique cytoplasmic domain generated by alernative mRNA processing. *Gene* 95:261-266.

Andrieux, A., Rabiet, M-J., Chapel, A., Concord, E., and Marguerie, G., 1991, A highly conserved sequence of the arg-gly-asp-binding domain of the integrin β3 subunit is sentitive to stimulation. *J. Biol. Chem.* 266:14202-14207.

Bajt, M.L., Loftus, J.C., Gawaz, M.P., Plow, E.F., and Ginsberg, M.H., 1991, Characterization of a mutant GPIIb-IIIa superintegrin. *Thromb. Haemost.* 65:717 (Abstr. 210).

Bajt, M.L., Ginsberg, M.H., Frelinger, III, A.L., Berndt, M.C., and Loftus, J.C., 1992, A spontaneous mutation of integrin αIIbβ3 (platelet glycoprotein IIb-IIIa) helps define a ligand binding site. *J. Biol. Chem.* 267:3789-3794.

Baldassare, J.J., Kahn, R.A., Knipp, M.A., and Newman, P.J., 1985, Reconstitution of platelet proteins into phospholipid vesicles. Functional proteoliposomes. *J. Clin. Invest.* 75:35-39.

Banga, H.S., Simons, E.R., Brass, L.F., and Rittenhouse, S.E., 1986, Activation of phospholipases A and C in human platelets exposed to epinephrine: Role of glycoproteins IIb/IIIa and dual role of epinephrine. *Proc. Natl. Acad. Sci. USA* 83:9197-9201.

Beer, J. and Coller, B.S., 1989, Evidence that platelet glycoprotein IIIa has a large disulphide-bonded loop that is susceptible to proteolytic cleavage. *J. Biol. Chem.* 264:17564-17573.

Bennett, J.S. and Vilaire, G., 1979, Exposure of platelet fibrinogen receptors by ADP and epinephrine. *J. Clin. Invest.* 64:1393-1400.

Bennett, J.S., Vilaire, G., and Cines, D.B., 1982, Identification of the fibrinogen receptor on human platelets by photoaffinity labeling. *J. Biol. Chem.* 257:8049-8054.

Bennett, J.S., Hoxie, J.A., Leitman, S.F., Vilaire, G., and Cines, D.B., 1983, Inhibition of fibrinogen binding to stimulated human platelets by a monoclonal antibody. *Proc. Natl. Acad. Sci. USA* 80:2417-2421.

Bodary, S.C., Napier, M.A., and McLean, J.W., 1989, Expression of recombinant platelet glycoprotein IIbIIIa results in a functional fibrinogen-binding complex. *J. Biol. Chem.* 264:18859-18862.

Bossy, B., Bossy-Wetzel, E., and Reichardt, L.F., 1991, Characterization of the integrin α8 subunit: A new integrin β1-associated subunit, which is prominently expressed on axons and on cells in contact with basal laminae in chick embryos. *EMBO J.* 10:2375-2385.

Bowditch, R.D., Tani, P.H., Halloran, C.E., Frelinger III, A.L., McMillan, R., and Ginsberg, M.H., 1992, Localization of a PLA1 epitope to the amino terminal 66 residues of platelet glycoprotein IIIa. *Blood* 79:559-562.

Brass, L.F., 1985, Ca^{2+} transport across the platelet plasma membrane. A role for membrane glycoproteins IIb and IIIa. *J. Biol. Chem.* 260:2231-2236.

Brass, L.F., Shattil, S.J., Kunicki, T.J., and Bennett, J.S., 1985, Effect of calcium on the stability of the platelet membrane glycoprotein IIb-IIIa complex. *J. Biol. Chem.* 260:7875-7881.

Bray, P.F., Rosa, J-P., Lingappa, V.R., Kan, Y.W., McEver, R.P., and Shuman, M.A., 1986, Biogenesis of the platelet receptor for fibrinogen: Evidence for separate precursors for glycoproteins IIb and IIIa. *Proc. Natl. Acad. Sci. USA* 83:1480-1484.

Bray, P.F., Barsh, G., Rosa, J-P., Juo, X.Y., Magenis, E., and Shuman, M.A., 1988, Physical linkage of the genes for platelet membrane glycoproteins IIb and IIIa. *Proc Natl. Acad. Sci. USA* 85:8683-8687.

Bray, P.F., Leung, C.S-I., and Shuman, M.A., 1990, Human platelets and megakaryocytes contain alternatively spliced glycoprotein IIb mRNAs. *J. Biol. Chem.* 265:9587-9590.

Caen, J.P. and Inceman, S., 1963, Considerations sur l`allongement du temps de saignement dans l afibrinogenemie congenitale. *Nouv. Rev. Fr. Hematol.* 3:614-615.

Caen, J.P., Castaldi, P.A., Leclerc, J.C., Inceman, S., Larrieu, M.J., Probst, M., and Bernard, J., 1966, Congenital bleeding disorders with long bleeding time and normal platelet count. I. Glanzmann's thrombasthenia (report of fifteen patients). *Am. J. Med.* 41:4-26.

Calvete, J.J. and González-Rodríguez, J., 1986, Isolation and biochemical characterization of the α- and β-subunits of glycoprotein IIb of human platelet plasma membrane. *Biochem. J.* 240:155-161.

Calvete, J.J., Alvarez, M.V., and González-Rodríguez, J., 1986, Quantitation and subcellular distribution of GPIIb and GPIIIa in human platelets and in the external platelet surface. Characterization of these glycoproteins isolated from the different subcellular fractions, in: "Monoclonal Antibodies and Human Blood Platelets," INSERM Symposium N° 27, J.L. McGregor, ed., Elsevier, Amsterdam, pp 179-190.

Calvete, J.J., McGregor, J.L., Rivas, G., and González-Rodríguez, J., 1987, Identification of a glycoprotein IIIa dimer in polyacrylamide gel separations of human platelet membranes. *Thromb. Haemost.* 58:694-697.

Calvete, J.J., Rivas, G., Maruri, M., Alvarez, M.V., McGregor, J.L., Hew, C-L., and González-Rodríguez, J., 1988, Tryptic digestion of human GPIIIa. Isolation and biochemical characterization of the 23 kDa N-terminal glycopeptide carrying the antigenic determinant for a monoclonal antibody (P37) which inhibits platelet aggregation. *Biochem. J.* 250:697-704.

Calvete, J.J., Alvarez, M.V., Rivas, G., Hew, C-L., Henschen, A., and González-Rodríguez, J., 1989, Interchain and intrachain disulphide bonds in human platelet glycoprotein IIb. Localization of the epitopes for several monoclonal antibodies. *Biochem. J.* 261:551-560.

Calvete, J.J., Henschen, A., and González-Rodríguez, J., 1989a, Complete localization of the intrachain disulphide bonds and the N-glycosylation points in the α-subunit of human platelet glycoprotein IIb. *Biochem. J.* 261:561-568.

Calvete, J.J., Schäfer, W., Henschen, A., and González-Rodríguez, J., 1990, Characterization of the β-chain N-terminus heterogeneity and the α- chain C-terminus of human platelet GPIIb. Posttranslational cleavage sites. *FEBS Lett.* 272:37-40.

Calvete, J.J., Schäfer, W., Henschen, A., and González-Rodríguez, J., 1990a, C-terminal amino acid determination of the transmembrane subunits of the human platelet fibrinogen receptor, the GPIIb/IIIa complex. *FEBS Lett.* 263:43-46.

Calvete, J.J., Henschen, A., and González-Rodríguez, J., 1991, Assignment of disulphide bonds in human platelet GPIIIa. A disulphide pattern for the β-subunits of the integrin family. *Biochem. J.* 274:63-71.

Calvete, J.J., Schäfer, W., Soszka, T., Lu, W., Cook, J.J., Jameson, B.A., and Niewiarowski, S., 1991a, Identification of the disulphide bond pattern in albolabrin, an RGD-containing peptide from the venom of *trimeresurus albolabris*: Significance for the expression of platelet aggregation inhibitory activity. *Biochemistry* 30:5225-5229.

Calvete, J.J., Arias, J., Alvarez, M.V., Lopez, M.M., Henschen, A., and González-Rodríguez, J., 1991b, Further studies on the topography of human platelet glycoprotein IIb. Localization of monoclonal antibody epitopes and the putative glycoprotein IIIa- and fibrinogen-binding regions. *Biochem. J.* 273:767-775.

Calvete, J.J., Arias, J., Alvarez, M.V., Lopez, M.M., Henschen, A., and González-Rodríguez, J., 1991c, Further studies on the topography of the N-terminal region of human platelet glycoprotein IIIa. Localization of monoclonal antibody epitopes and the putative fibrinogen-binding sites. *Biochem. J.* 274:457-463.

Calvete, J.J., Mann, K., Alvarez, M.V., López, M.M., and González-Rodríguez, J., 1992, Proteolytic dissection of the isolated platelet fibrinogen receptor, integrin GPIIb/IIIa. Localization of GPIIb and GPIIIa sequences putatively involved in the subunit interface and in intra-subunit and intrachain contacts. *Biochem. J.* 282:523-532.

Calvete, J.J., Schäfer, W., Mann, K., Henschen, A., and González-Rodríguez, J., 1992a, Localization of the cross-linking sites of RGD and KQAGDV peptides to the isolated fibrinogen receptor, the human platelet integrin glycoprotein IIb/IIIa. Influence of peptide length. *Eur. J. Biochem.* 206:759-765.

Carrell, N.A., Fitzgerald, L.A., Steiner, B., Erickson, H.P., and Phillips, D.R., 1985, Structure of human platelet membrane glycoproteins IIb and IIIa as determined by electron microscopy. *J. Biol. Chem.* 260:1743-1749.

Charo, I.F., Bekeart, L.S., and Phillips, D.R., 1987, Platelet glycoprotein IIb-IIIa-like proteins mediate endothelial cell attachment to adhesive proteins and the extracellular matrix. *J. Biol. Chem.* 262:9935-9938.

Charo, I.F., Nannizzi, L., Phillips, D.R., Hsu, M.A., and Scarborough, R., 1991, Inhibition of fibrinogen binding to GPIIb-IIIa by a GPIIIa peptide. *J. Biol. Chem.* 266:1415-1421.

Cheresh, D.A., 1987, Human endothelial cells synthesize and express an arg-gly-asp-directed adhesion receptor involved in attachment to fibrinogen and von Willebrand factor. *Proc. Natl. Acad. Sci. USA* 84:6471-6475.

Cheresh, D.A., Pytela, R., Pierschbacher, M.D., Klier, F.D., and Ruoslahti, E, 1987a, An Arg-Gly-Asp-directed receptor on the surface of human melanoma cells exists in a divalent cation-dependent functional complex with the disialoganglioside GD2. *J. Cell Biol.* 105:1163-1173.

Cheresh, D.A., Berliner, S.A., Vicente, V., and Ruggeri, Z.M., 1989, Recognition of distinct adhesive sites on fibrinogen by related integrins on platelets and endothelial cells. *Cell* 58:945-953.

Cierniewski, C.S., Krzeslowska, J., Pawlowska, Z., Witas, H., and Meyer, M., 1989, Palmitylation of the glycoprotein IIb/IIIa complex in human blood platelets. *J. Biol. Chem.* 264:12158-12164.

Coller, B.S., Peerschke, E.I., Scudder, L.E., and Sullivan, C.A., 1983, A murine monoclonal antibody that completely blocks the binding of fibrinogen to platelets produces a thrombasthenic-like state in normal platelets and binds to glycoproteins IIb and/or IIIa. *J. Clin. Invest.* 72:278-286.

Conforti, G., Zanetti, A., Pasquali-Ronchetti, I., Quaglino, D. Jr., Neyroz, P., and Dejana, E., 1990, Modulation of vitronectin receptor binding by membrane lipid composition. *J. Biol. Chem.* 265:4011-4019.

Cook, J.J., Trybulek, M., Lasz, E.C., Khan, S., and Niewiarowski, S., 1992, Binding of glycoprotein IIIa-derived peptide 217-231 to fibrinogen and von Willebrand factor and its inhibition by platelet glycoprotein IIb/IIIa complex. *Biochim. Biophys. Acta* 1119:312-321.

Doolittle, R.F., Watt, K.W.K., Colltrell, B.A., Strong, D.D., and Riley, M., 1979, The amino acid sequence of the α-chain of human fibrinogen. *Nature* 280:464-468.

D'Souza, S.E., Ginsberg, M.H., Lam, C-T., and Plow, E.F., 1988, Chemical cross-linking of arginyl-glycyl-aspartic acid peptides to an adhesion receptor on platelets. *J. Biol. Chem.* 263:3943-3951.

D'Souza, S.E., Ginsberg, M.H., Burke, T.A., Lam, C-T., and Plow, E.F., 1988a,

Localization of an arg-gly-asp recognition site within an integrin adhesion receptor. *Science* 242:91-93.

D'Souza, S.E., Ginsberg, M.H., Burke, T.A., and Plow, E.F., 1990, The ligand binding site of the platelet integrin receptor GPIIb-IIIa is proximal to the second calcium binding domain of its α subunit. *J. Biol. Chem.* 265:3440-3446.

D'Souza, S.E., Ginsberg, M.H., Matsueda, G.R., and Plow, E.F., 1991, A discrete sequence in a platelet integrin is involved in ligand recognition. *Nature* 350:66-68.

Du, X., Plow, E.F., Frelinger, III,A.L., O'Toole, T.E., Loftus, J.C., and Ginsberg, M.H., 1991, Ligands "activate" integrin αIIbβ3 (platelet GPIIb-IIIa). *Cell* 65:409-416.

Duperray, A.., Berthier, R., Chagnon, E., Ryckwaert, J-J., Ginsberg, M., Plow, E.F., and Marguerie, G., 1987, Biosynthesis and processing of platelet GPIIb-IIIa in human megakaryocytes. *J. Cell Biol.* 104:1665-1673.

Dustin, M.L. and Springer, T.A., 1991, Role of lymphocyte adhesion receptors in transient interactions and cell locomotion. *Annu. Rev. Immunol.* 9:27-66.

Eirín, M.T., Calvete, J.J., and González-Rodríguez, J., 1986, New isolation procedure and further biochemical characterization of glycoproteins IIb and IIIa from human platelet plasma membrane. *Biochem. J.* 240:147-153.

Ferrell, J.E. and Martin, G.S., 1989, Tyrosine-specific protein phosphorylation is regulated by glycoprotein IIb-IIIa in platelets. *Proc. Natl. Acad. Sci. USA* 86:2234-2238.

Fitzgerald, L.A. and Phillips, D.R., 1985, Calcium regulation of the platelet membrane glycoprotein IIb-IIIa Complex. *J. Biol. Chem.* 260:11366-11374.

Fitzgerald, L.A., Charo, I.F., and Phillips, D.R., 1985, Human and bovine endothelial cells synthesize membrane proteins similar to human platelet glycoproteins IIb and IIIa. *J. Biol. Chem.* 260:10893-10896.

Fitzgerald, L.A., Poncz, M., Steiner, B., Rall, S.C., Jr., Bennett, J.S., and Phillips, D.R., 1987, Comparison of the cDNA-derived protein sequences of the human fibronectin and vitronectin receptor α-subunits and platelet glycoprotein IIb. *Biochemistry* 26:8158-8165.

Fitzgerald, L.A., Steiner, B., Rall, S.C., Jr., Lo, S-S., and Phillips, D.R., 1987a, Protein sequence of endothelial glycoprotein IIIa derived from a cDNA clone. Identity with platelet glycoprotein IIIa and similarity to "integrin". *J. Biol. Chem.* 262:3936-3939.

Fox, J.E.B. and Phillips, D.R., 1986, Actin-membrane interactions in platelets, in: "Membrane Skeletons and Cytoskeletal-Membrane Associations," Alan R. Liss, Inc., pp. 281-292.

Frachet, P., Uzan, G., Thevenon, D., Denarier, E., Prandini, M-H., and Marguerie, G., 1990, GPIIb and GPIIIa amino acid sequences deduced from human megakaryocyte cDNAs. *Mol. Biol. Repr.* 14:27-33.

Frachet, P., Duperray, A., Delachanal, E., and Marguerie, G., 1992, Role of the transmembrane and cytoplasmic domains in the assembly and surface exposure of the platelet integrin GPIIb/IIIa. *Biochemistry* 31:2408-2415.

Frelinger, III, A.L., Lam, C-T., Plow, E.F., Smith, M.A., Loftus, J.C., and Ginsberg, M.H., 1988, Occupancy of an adhesive glycoprotein receptor modulates expression of an antigenic site involved in cell adhesion. *J. Biol. Chem.* 263:12397-12402.

Frelinger, III, A.L., Cohen, I., Plow, E.F., Smith, M.A., Roberts, J., Lam, C-T., and Ginsberg, M.H., 1990, Selective inhibition of integrin function by antibodies specific for ligand-occupied receptor conformers. *J. Biol. Chem.* 265:6346-6352.

Frelinger, III, A.L., Du, X., Plow, E.F., and Ginsberg, M.H., 1991, Monoclonal antibodies to ligand-occupied conformers of integrin αIIbβ3 (glycoprotein IIb-IIIa) alter receptor affinity, specificity, and function. *J. Biol. Chem.* 266:17106-17111.

Fujimoto, T., Fujimura, K., and Kuramoto, A., 1991, Functional Ca^{2+} channel produced by purified platelet membrane glycoprotein IIb-IIIa complex incorporated into planar phospholipid bilayer. *Thromb. Haemost.* 66:598-603.

Fujimura, K. and Phillips, D.R., 1983, Calcium cation regulation of glycoprotein IIb-IIIa complex formation in platelet plasma membranes. *J. Biol. Chem.* 258:10247-10252.

Furihata, K., Nugent, D.J., Bissonette, A., Aster, R.H., and Kunicki, T.J., 1987, On the association of the platelet-specific alloantigen, Pen[a], with glycoprotein IIIa. Evidence for heterogeneity of glycoprotein IIIa. *J. Clin. Invest.* 80:1624-1630.

Gardner, J.M. and Hynes, R.O., 1985, Interaction of fibronectin with its receptor on platelets. *Cell* 42:439-448.

Gartner, T.K. and Bennett, J.S., 1985, The tetrapeptide analogue of the cell attachment site of fibronectin inhibits platelet aggregation and fibrinogen binding to activated platelets. *J. Biol. Chem.* 260:11891-11894.

Gartner, T.K. and Ogilvie, M.L., 1988, Peptides and monoclonal antibodies which bind to platelet glycoprotein IIb and/or IIIa inhibit clot retraction. *Thromb. Res.* 49:49-56.

Gartner, T.K., Loudon, R., and Taylor, D.B., 1991, The peptides APLHK, EHIPA and GAPL are hydropathically equivalent peptide mimics of a fibrinogen binding domain of glycoprotein IIb/IIIa. *Biochem. Biophys. Res. Commun.* 180:1446-1452.

Ginsberg, M.H., Forsyth, J., Lightsey, A., Chediak, J., and Plow, E.F., 1983, Reduced surface expression and binding of fibronectin by thrombin-stimulated thrombasthenic platelets. *J. Clin. Invest.* 71:619-624.

Ginsberg, M., Pierschbacher, M.D., Ruoslahti, E., Marguerie, G., and Plow, E.F, 1985, Inhibition of fibronectin binding to platelets by proteolytic fragments and synthetic peptides which support fibroblast adhesion. *J. Biol. Chem.* 260:3931-3936.

Ginsberg, M.H., Lightsey, A., Kunicki, T.J., Kaufmann, A., Marguerie, G., and Plow, E.F., 1986, Divalent cation regulation of the surface orientation of platelet membrane glycoprotein IIb. Correlation with fibrinogen binding function and definition of a novel variant of Glanzmann's thrombasthenia. *J. Clin. Invest.* 78:1103-1111.

Ginsberg, M.H., Loftus, J., Ryckwaert, J-J., Pierschbacher, M.D., Pytela, R., Ruoslahti, E., and Plow, E.F., 1987, Immunochemical and amino-terminal sequence comparison of two cytoadhesins indicates they contain similar or identical β subunits and distinct α subunits. *J. Biol. Chem.* 262:5437-5440.

Ginsberg, M.H., Loftus, J.C., and Plow, E.F., 1988, Cytoadhesins, integrins, and platelets. *Thromb. Haemostas.* 59:1-6.

Gogstad, G.O., Hagen, I., Krutnes, M.B., and Solum, N.O., 1982, Dissociation of the glycoprotein IIb-IIIa complex in isolated human platelet membranes. Dependence of pH and divalent cations. *Biochim. Biophys. Acta* 689:21-30.

Gogstad, G.O., Brosstad, F., Krutnes, M-B., Hagen, I., and Solum, N.O., 1982a, Fibrinogen-binding properties of the human platelet glycoprotein IIb-IIIa complex: A study using crossed radioimmunoelectrophoresis. *Blood* 60:663-671.

Gogstad, G.O., Krutnes, M-B., and Solum, N.O., 1983, Calcium-binding proteins from human platelets. A study using crossed immunoelectrophoresis and $^{45}Ca^{2+}$. *Eur. J. Biochem.* 133:193-199.

Golden, A., Brugge, J.S., and Shattil, S.J., 1990, Role of platelet membrane glycoprotein IIb-IIIa in agonist-induced tyrosine phosphorylation of platelet proteins. *J. Cell Biol.* 111:3117-3127.

Gould, R.J., Polokoff, M.A., Friedman, P.A., Huang, T-F., Holt, J.C., Cook, J.J., and Niewiarowski, S., 1990, Disintegrins: A family of integrin inhibitory proteins from viper venoms. *Proc. Soc. Exp. Biol. Med.* 195:168-171.

Greco, N.J., Yamamoto, N., Jackson, B.W., Tandon, N.N., Moos, M., and Jamieson, G.A., 1991, Identification of a nucleotide-binding site on glycoprotein IIb. Relationship to ADP-induced platelet activation. *J. Biol. Chem.* 266:13627-13633.

Gulino, D., Ryckewaert, J-J., Andrieux, A., Rabiet, M-J., and Marguerie, G., 1990, Identification of a monoclonal antibody against platelet GPIIb that interacts with a calcium-binding site and induces aggregation. *J. Biol. Chem.* 265:9575-9581.

Gulino, D., Boudignon, C., Zhang, L., Concord, E., Rabiet, M-J., and Marguerie, G., 1992, Ca2+-binding properties of the platelet glycoprotein IIb ligand-interacting domain. *J. Biol. Chem.* 267:1001-1007.

Hagen, I., Nurden, A., Bjerrum, O.J., Solum, N.O., and Caen, J.P., 1980, Immunochemical evidence for protein abnormalities in platelets from patients with Glanzmann`s thrombasthenia and Bernard-Soulier syndrome. *J. Clin. Invest.* 65:722-731.

Haverstick, D.M., Cowan, J.F., Yamada, K.M., and Santoro, S.A., 1985, Inhibition of platelet adhesion to fibronectin, fibrinogen, and von Willebrand factor substrates by a synthetic tetrapeptide derived from the cell-binding domain of fibronectin. *Blood* 66:946-952.

Heath, T.L., Slack, S.M., White, M.M., Kouns, W.C., Robertson, J.T., Steiner, B., Turitto, V.T., and Jennings, L.K., 1991, Thrombin hydrolysis of human GPIIb-IIIa upon complex activation. *Blood* 78 (Suppl.I) Abstr. 542.

Heidenreich, R., Eisman, R., Surrey, S., Delgrosso, K., Bennett, J.S., Schwartz, E., and Poncz, M., 1990, Organization of the gene for platelet glycoprotein IIb. *Biochemistry* 29:1232-1244.

Hemler, M.E., 1990, VLA proteins in the integrin family: Structures, functions, and their role on leukocytes. *Annu. Rev. Immunol.* 8:365-400.

Hillery, C.A., Smyth, S.S., and Parise, L.V., 1991, Phosphorylation of human platelet glycoprotein IIIa (GPIIIa). Dissociation from fibrinogen receptor activation and phosphorylation of GPIIIa *in vitro. J. Biol. Chem.* 266:14663-14669.

Howard, L., Shulman, S., Sadanandan, S., and Karpatkin, S., 1982, Crossed immunoelectrophoresis of human platelet membranes. The major antigen consists of a complex of glycoproteins, GPIIb and GPIIIa, held together by Ca^{2+} and missing in Glanzmann's thrombasthenia. *J. Biol. Chem.* 257:8331-8338.

Humphries, M.J., 1990, The molecular basis and specificity of integrin-ligand interactions. *J. Cell Science* 97:585-592.

Hynes, R.O., 1987, Integrins: A family of cell surface receptors. *Cell* 48:549-554.

Hynes, R.O., 1992, Integrins. Versatility, modulation, and signaling in cell adhesion. *Cell* 69:11-26.

Isenberg, W.M., Bainton, D.F., and Newman, P.F., 1990, Monoclonal antibodies bound to subunits of the integrin GPIIb-IIIa are internalized and interfere with filipodia formation and platelet aggregation. *Blood* 76:1564-1575.

Isenberg, W.M., McEver, R.P., Phillips, D.R., Shuman, M.A., and Bainton, D.F., 1987, The platelet fibrinogen receptor: An immunogold-surface replica study of agonist-induced ligand binding and receptor clustering. *J. Cell Biol.* 104:1655-1663.

Jennings, L.K. and Phillips, D.R., 1982, Purification of glycoproteins IIb and III from human platelet plasma membranes and characterization of a calcium-dependent glycoprotein IIb-III complex. *J. Biol. Chem.* 257:10458-10466.

Jung, S.M., Yoshida, N., Aoki, N., Tanoue, K., Yamazaki, H., and Moroi, M., 1988, Thrombasthenia with an abnormal platelet membrane glycoprotein IIb of different molecular weight. *Blood* 71:915-922.

Kloczewiak, M., Timmons, S., Lukas, T.J., and Hawiger, J., 1984, Platelet receptor recognition site on human fibrinogen. Synthesis and structure-function relationship of peptides corresponding to the carboxyl-terminal segment of the γ chain. *Biochemistry* 23:1767-1774.

Kornecki, E., Tuszynski, G.P., and Niewiarowski, S., 1983, Inhibition of fibrinogen receptor-mediated platelet aggregation by heterologous anti-human platelet membrane antibody. Significance of an Mr = 66,000 protein derived from glycoprotein IIIa. *J. Biol. Chem.* 258:9349-9356.

Kouns, W.C., Wall, C.D., White, M.M., Fox, C.F., and Jennings, L.K., 1990, A Conformation-dependent epitope of human platelet glycoprotein IIIa. *J. Biol. Chem.* 265:20594-20601.

Kouns, W.C., Fox, C.F., Lamoreaux, W.J., Coons, L.B., and Jennings, L.K., 1991, The effect of glycoprotein IIb-IIIa receptor occupancy on the cytoskeleton of resting and activated platelets. *J. Biol. Chem.* 266:13891-13900.

Kouns, W.C., Hadváry, P., and Steiner, B., 1991a, Conformational modulation of purified GPIIb-IIIa allows proteolytic generation of active fragments from either active or inactive GPIIb-IIIa. *Blood* 78 (Suppl.I):Abstr. 543.

Kouns, W.C., Newman, P.J., Puckett, K.J., Miller, A.A., Wall, C.D., Fox, C.F., Seyer, J.M., and Jennings, L.K., 1991b, Further characterization of the loop structure of platelet glycoprotein IIIa: Partial mapping of functionally significant glycoprotein IIIa epitopes. *Blood* 78:3215-3223.

Kunicki, T.J., Pidard, D., Rosa, J-P., and Nurden, A.T., 1981, The formation of Ca^{++}-dependent complexes of platelet membrane glycoproteins IIb and IIIa in solution as determined by crossed immunoelectrophoresis. *Blood* 58:268-278.

Kunicki, T.J., Plow, E.F., Kekomaki, R., and Nugent, D.J., 1991, Human monoclonal autoantibody 2E7 is specific for a peptide sequence of platelet glycoprotein IIb. Localization of the epitope to IIb$_{231-238}$ with an immunodominant Trp$_{235}$. *J. Autoimmunol.* 4:415-431.

Lam, C-T., Plow, E.F., Smith, M.A., Andrieux, A., Ryckwaert, J-J., Marguerie, G., and Ginsberg, M.H., 1987, Evidence that arginyl-glycyl-aspartate peptides and fibrinogen γ chain peptides share a common binding site on platelets. *J. Biol. Chem.* 262:947-950.

Lam, C-T., Plow, E.F., and Ginsberg, M.H., 1989, Platelet membrane glycoprotein IIb heavy chain forms a complex with glycoprotein IIIa that binds arg-gly-asp peptides. *Blood* 73:1513-1518.

Lam, C-T., 1992, Isolation and characterization of a chymotryptic fragment of platelet Glycoprotein IIb-IIIa retaining arg-gly-asp binding activity. *J. Biol. Chem.* 267:5649-5655.

Languino, L.R. and Ruoslahti, E., 1992, An alternative form of the integrin $\beta 1$ subunit with a variant cytoplasmic domain. *J. Biol. Chem.* 267:7116-7120.

Lanza, F., Kieffer, N., Phillips, D.R., and Fitzgerald, L.A., 1990, Characterization of the human platelet glycoprotein IIIa gene. Comparison with the fibronectin receptor β-subunit gene. *J. Biol. Chem.* 265:18098-18103.

Lanza, F., Stierlé, A., Fournier, D., Morales, M., André, G., Nurden, A., and Cazenave, J-P., 1992, A new variant of Glanzmann's thrombasthenia: Platelets with functionally defective GPIIb-IIIa complexes and a GPIIIa Arg^{214}-> Trp^{214} mutation. *J. Clin. Invest.* 89:1995-2004.

Loftus, J.C., Plow, E.F., Frelinger, III, A.L., D'Souza, S.E., Dixon, D., Lacy, J., Sorge, J., and Ginsberg, M.H., 1987, Molecular cloning and chemical synthesis of a region of platelet glycoprotein IIb involved in adhesive function. *Proc. Natl. Acad. Sci. USA* 84:7114-7118.

Loftus, J.C., Plow, E.F., Jennings, L.K., and Ginsberg, M.H., 1988, Alternative proteolytic processing of platelet membrane glycoprotein IIb. *J. Biol. Chem.* 263:11025-11028.

Loftus, J.C., O'Toole, T.E., Plow, E.F., Glass, A., Frelinger, III, A.L., and Ginsberg, M.H., 1990, A $\beta 3$ integrin mutation abolishes ligand binding and alters divalent cation-dependent conformation. *Science* 249:915-918.

Lyman, S., Aster, R.H., Visentin, G.P., and Newman, P.J., 1990, Polymorphism of human platelet membrane glycoprotein IIb associated with the Bak^a/Bak^b alloantigen system. *Blood* 75:2343-2348.

Marguerie, G.A., Plow, E.F., and Edgington, T.S., 1979, Human platelets possess an inducible and saturable receptor specific for fibrinogen. *J. Biol. Chem.* 254:5357-5363.

Marguerie, G. and Plow, E.F., 1983, The fibrinogen-dependent pathway of platelet aggregation. *Ann. NY Acad. Aci.* 408:556-566.

Marguerie, G.A., Thomas-Maison, N., Ginsberg, M.H., and Plow, E.F., 1984, The platelet-fibrinogen interaction. Evidence for proximity of the Aα chain of fibrinogen to platelet membrane glycoproteins IIb/III. *Eur. J. Biochem.* 139:5-11.

Mazur, P., Henzel, W.J., Seymour, J.L., and Lazarus, R.A., 1991, Ornatins: Potent glycoprotein IIb-IIIa antagonists and platelet aggregation inhibitors from the leech *Placobdella ornata. Eur. J. Biochem.* 202:1073-1082.

McEver, R.P., Bennett, E.B., and Martin, M.N., 1983, Identification of two structurally and functionally distinct sites on platelet GPIIb-IIIa using monoclonal antibodies. *J. Biol. Chem.* 258:5269-5275.

McLean, J.R., Maxwell, R.E., and Hertler, D., 1964, Fibrinogen and adenosine diphosphate-induced aggregation of platelets. *Nature* 202:605-606.

Melero, J.A. and González-Rodríguez, J., 1984, Preparation of monoclonal antibodies against glycoprotein IIIa of human platelets. Their effect on platelet aggregation. *Eur. J. Biochem.* 141:421-427.

Moroi, M., Yamamura, J., Koga, H., Miyazaki, S., and Jung, S.M., 1991, Analysis of a variant form of platelet glycoprotein (GP) IIb: A second patient with abnormal molecular weight. *Thromb. Res.* 62:215-225.

Moyle, M., Napier, M.A., and McLean, J.W., 1991, Cloning and expression of a divergent integrin subunit β_8. *J. Biol. Chem.* 266:19650-19658.

Mustard, J.F., Packman, M.A., Kinlough-Rathbone, R.L., Perry, D.W., and Regoeczi, E., 1978, Fibrinogen and ADP-induced platelet aggregation. *Blood* 52:453-465.

Mustard, J.F., Kinlough-Rathbone, R.L., Packham, M.A., Perry, D.W., Harfenist, E.J., and Pai, K.R.M., 1979, Comparison of fibrinogen association with normal and thrombasthenic platelets on exposure to ADP or chymotrypsin. *Blood* 54:987-993.

Nachman, R.L. and Leung, L.L.K., 1982, Complex formation of platelet membrane glycoproteins IIb and IIIa with fibrinogen. *J. Clin. Invest.* 69:263-269.

Nachman, R.L., Leung, L.L.K., Kloczewiak, M., and Hawiger, J., 1984, Complex formation of platelet membrane glycoproteins IIb and IIIa with the fibrinogen D domain. *J. Biol. Chem.* 259:8584-8588.

Nermut, M.V., Green, N.M., Eason, P., Yamada, S.S., and Yamada, K.M., 1988, Electron microscopy and structural model of human fibronectin receptor. *EMBO J.* 7:4093-4099.

Newman, P.J., Derbes, R.S., and Aster, R.H., 1989, The human platelet alloantigens, PLA1 and PLA2, are associated with a leucine 33/proline 33 amino acid polymorphism in membrane glycoprotein IIIa, and are distinguishable by DNA typing. *J. Clin. Invest.* 83:1778-1781.

Newman, P.J., McEver, R.P., Doers, M.P., and Kunicki, T.J., 1987, Synergistic action of two murine monoclonal antibodies that inhibit ADP-induced aggregation without blocking fibrinogen binding. *Blood* 69:668-676.

Newman, P.J., 1991, Platelet GPIIb-IIIa: Molecular variations and alloantigens. *Thromb. Haemost.* 66:111-118.

Newman, P.J., Seligsohn, U., Lyman, S., and Coller, B.S., 1991, The molecular genetic basis of glanzmann thrombasthenia in the Iraqi-Jewish and Arab populations in Israel. *Proc. Natl. Acad. Sci. USA.* 88:3160-3164.

Niewiarowski, S., Norton, K.J., Eckardt, A., Lukasiewicz, H., Holt, J.C., and Kornecki, E., 1989, Structural and functional characterization of major platelet membrane components derived by limited proteolysis of glycoprotein IIIa. *Biochim. Biophys. Acta* 983:91-99.

Nurden, A.T. and Caen, J.P., 1974, An abnormal glycoprotein pattern in three cases of Glanzmann's thrombasthenia. *Br. J. Haematol.* 28:253-260.

Nurden, A.T. and Caen, J.P., 1975, Specific roles for platelet surface glycoproteins in platelet function. *Nature* 255:720-722.

Nurden, A.T., 1989, Congenital abnormalities of platelet membrane glycoproteins, in: "Platelet Immunobiology. Molecular and Clinical Aspects," T.J. Kunicki and J.N. George, eds., J.B. Lippincott Co., Philadelphia, pp 63-96.

O'Toole, T.E., Loftus, J.C., Du, X., Glass, A.A., Ruggeri, Z.M., Shattil, S.J., Plow, E.F., and Ginsberg, M.H., 1990, Affinity modulation of the αIIbβ3 integrin (platelet GPIIb-IIIa) is an intrinsic property of the receptor. *Cell Regulation* 1:883-893.

O'Toole, T.E., Mandelman, D., Forsyth, J., Shattil, S.J., Plow, E.F., and Ginsberg, M.H., 1991, Modulation of the affinity of integrin αIIbβ3 (GPIIb-IIIa) by the cytoplasmic domain of αIIb. *Science* 254:845-847.

Painter, R.G., Prodouz, K.N., and Gaarde, W., 1985, Isolation of a subpopulation of glycoprotein IIb-III from platelet membranes that is boudn to membrane actin. *J. Cell Biol.* 100:652-657.

Parise, L.V. and Phillips, D.R., 1985, Platelet membrane glycoprotein IIb-IIIa complex incorporated into phospholipid vesicles. Preparation and morphology. *J. Biol. Chem.* 260:1750-1756.

Parise, L.V. and Phillips, D.R., 1985a, Reconstitution of the purified platelet fibrinogen receptor. Fibrinogen binding properties of the glycoprotein IIb-IIIa complex. *J. Biol. Chem.* 260:10698-10707.

Parise, L.V. and Phillips, D.R., 1986, Fibronectin-binding properties of the purified platelet Glycoprotein IIb-IIIa complex. *J. Biol. Chem.* 261:14011-14017.

Parise, L.V., Helgerson, S.L., Steiner, B., Nannizzi, L., and Phillips, D.R., 1987, Synthetic peptides derived from fibrinogen and fibronectin change the conformation of purified platelet glycoprotein IIb-IIIa. *J. Biol. Chem.* 262:12597-12602.

Pasqualini, R., Chamone, D.F., and Brentani, R.R., 1989, Determination of the putative binding site for fibronectin on platelet glycoprotein IIb-IIIa complex through a hydropathic complementary approach. *J. Biol. Chem.* 264:14566-14570.

Peerschke, E.I.B., 1985, The platelet fibrinogen receptor. *Sem. Hematol.* 22:241-259.

Phillips, D.R., Jenkins, C.S.P., Lüscher, E.F., and Larrieu, M.J., 1975, Molecular differences of exposed surface proteins on thrombasthenic platelet plasma membranes. *Nature* 257:599-600.

Phillips, D.R. and Agin, P.P., 1977, Platelet plasma membrane glycoproteins. Evidence for the presence of nonequivalent disulphide bonds using nonreduced-reduced two-dimensional gel electrophoresis. *J. Biol. Chem.* 252:2121-2126 .

Phillips, D.R., Jennings, L.K., and Edwards, H.H., 1980, Identification of membrane proteins mediating the interaction of human platelets. *J. Cell Biol.* 86:77-86.

Phillips, D.R. and Baughan, A.K., 1983, Fibrinogen binding to human platelet plasma membranes. Identification of two steps requiring divalent cations. *J. Biol. Chem.* 258:10240-10246.

Phillips, D.R., Charo, I.F., Parise, L.V., and Fitzgerald, L.A., 1988, The platelet membrane glycoprotein IIb-IIIa complex. *Blood* 71:831-843.

Pidard, D., Rosa, J-P., Kunicki, T.J., and Nurden, A.T., 1982, Further studies on the interaction between human platelet membrane glycoproteins IIb and IIIa in triton X-100. *Blood* 60:894-904.

Pidard, D., Montgomery, R.R., Bennett, J.S., and Kunicki, T.J., 1983, Interaction of AP-2, a monoclonal antibody specific for the human platelet glycoprotein IIb-IIIa complex, with intact platelets. *J. Biol. Chem.* 258:12582-12586.

Pidard, D., Frelinger, A.L., Bouillot, C., and Nurden, A.T., 1991, Activation of the fibrinogen receptor on human platelets exposed to alpha chymotrypsin. Relationship with a major proteolytic cleavage at the carboxyterminus of the membrane glycoprotein IIb heavy chain. *Eur. J. Biochem.* 200:437-447.

Pierschbacher, M.D. and Ruoslahti, E., 1984, Cell attachment activity of fibronectin can be duplicated by small synthetic fragments of the molecule. *Nature* 309:30-33.

Pietu, G., Cherel, G., Marguerie, G.A., and Meyer, D., 1984, Inhibition of von Willebrand factor platelet interaction by fibrinogen. *Nature* 308:648-650.

Plow, E.F., Ginsberg, M.H., and Marguerie, G., 1984, Regulation of platelet aggregation by inducible receptors for fibrinogen, in: "The Receptors," Vol.1, P.M. Conn, ed., Academic Press, New York, pp 465-510.

Plow, E.F., Pierschbacher, M.D., Ruoslahti, E., Marguerie, G.A., and Ginsberg, M.H., 1985, The effect of arg-gly-asp-containing peptides on fibrinogen and von Willebrand factor binding to platelets. *Proc. Natl. Acad. Sci. USA* 82:8057-8061.

Plow, E.F., Ginsberg, M.H., and Marguerie, G., 1986, Expression and function of adhesive proteins on the platelet surface, in: "Biochemistry of Platelets," D.R. Phillips and M.A. Shuman, eds., Academic Press, Orlando, pp 225-256.

Plow, E.F., Marguerie, G., and Ginsberg, M.H., 1987, Fibrinogen, fibrinogen receptors, and the peptides that inhibit these interactions. *Biochem. Pharmacol.* 36:4035-4040.

Plow, E.F. and Ginsberg, M.H., 1989, Cellular adhesion: GPIIb-IIIa as a prototypic adhesion receptor. *Prog. Thromb. Haemost.* 9:117-156.

Polack, B., Duperray, A., Troesch, A., Berthier, R., and Marguerie, G., 1989, Biogenesis of the vitronectin receptor in human endothelial cell: Evidence that the vitronectin receptor and GPIIb-IIIa are synthesized by a common mechanism. *Blood* 73:1519-1524.

Poncz, M., Eisman, R., Heidenreich, R., Silver, S.M., Vilaire, G., Surrey, S., Schwartz, E., and Bennett, J.S., 1987, Structure of the platelet membrane glycoprotein IIb. Homology to the α subunits of the vitronectin and the fibronectin membrane receptors. *J. Biol. Chem.* 262:8476-8482.

Prandini, M.H., Denarier, E., Frachet, P., Uzan, G., and Marguerie, G., 1988, Isolation of the human platelet glycoprotein IIb gene and characterization of the 5' flanking region. *Biochem. Biophys. Res. Commun.* 156:595-601.

Pytela, R., Pierchbacher, M.D., Ginsberg, M.H., Plow, E.F., and Ruoslahti, E., 1986, Platelet membrane glycoprotein IIb/IIIa: Member of a family of arg-gly-asp-specific adhesion receptors. *Science* 231:1559-1562.

Ramsamooj, P., Lively, M.O., and Hantgan, R.R., 1991, Evidence that the central region of glycoprotein IIIa participates in integrin receptor function. *Biochem. J.* 276:725-732.

Reason, A.J., Dell, A., Morris, H.R., Rogers, M.E., Calvete, J.J., and González-Rodríguez, J., 1991, Characterization of the N-linked oligosaccharides of the light chain of human glycoprotein IIb by f.a.b.-m.s. *Carbohydrate Res.* 221:169-177.

Rivas, G.A. and González-Rodríguez, J., 1991, Calcium binding to human platelet integrin GPIIb/IIIa and to its constituent glycoproteins. Effects of lipids and temperature. *Biochem. J.* 276:35-40.

Rivas, G.A., Calvete, J.J., and González-Rodríguez, J., 1991a, A large-scale procedure for the isolation of integrin GPIIb/IIIa, the human platelet fibrinogen receptor. *Prot. Express. Purif.* 2:248-255.

Rivas, G.A., Aznárez, J.A., Usobiaga, P., Saiz, J.L., and González-Rodríguez, J., 1991b, Molecular characterization of the human platelet integrin GPIIb/IIIa and its constituent glycoproteins. *Eur. Biophys. J.* 19:335-345.

Rivas, G.A., Usobiaga, P., and González-Rodríguez, J., 1991, Calcium and temperature regulation of the stability of the human platelet integrin GPIIb/IIIa in solution: An analytical centrifugation study. *Eur. Biophys. J.* 20:287-292.

Rosa, J-P., Bray, P.F., Gayet, O., Johnston, G.I., Cook, R.G., Jackson, K.W., Shuman, M.A., and McEver, R.P., 1988, Cloning of glycoprotein IIIa cDNA from human erythroleukemia cells and localization of the gene to chromosome 17. *Blood* 72:593-600.

Rosa, J-P. and McEver, R.P., 1989, Processing and assembly of the integrin, glycoprotein IIb-IIIa, in HEL cells. *J. Biol. Chem.* 264:12596-12603.

Ruggeri, Z.M., Bader, R., and DeMarco, L., 1982, Glanzmann thrombasthenia. Deficient binding of von Willebrand factor to thrombin-stimulated platelets. *Proc. Natl. Acad. Sci. USA* 79:6038-6041.

Ruoslahti, E. and Pierschbacher, M.D., 1986, Arg-gly-asp: A versatile cell recognition signal. *Cell* 44:517-518.

Ruoslahti, E. and Pierschbacher, M.D., 1987, New perspectives in cell adhesion: RGD and integrins. *Science* 238:491-497.

Ruoslahti, E. and Giancotti, F.G., 1989, Integrins and tumor cell dissemination. *Cancer Cells* 1:119-126.

Ruoslahti, E., 1991, Integrins. *J. Clin. Invest.* 87:1-5.

Rybak, M.E., Renzulli, L.A., Bruns, M.J., and Cahaly, D.P., 1988, Platelet glycoproteins IIb and IIIa as a calcium channel in liposomes. *Blood* 72:714-720.

Santoro, S.A. and Lawling, W.J., Jr., 1987, Competition for related but nonidentical binding sites on the glycoprotein IIb-IIIa complex by peptides derived from platelet adhesive proteins. *Cell* 48:867-873.

Scarborough, R.M., Rose, J.W., Hsu, M.A., Phillips, D.R., Fried, V.A., Campbell, A.M., Nannizzi, L., and Charo, I.F., 1991, Barbourin. A GPIIb-IIIa-specific integrin antagonist from the venom of *sistrurus m. barbouri*. *J. Biol. Chem.* 266:9359-9362.

Shattil, S.J., Hoxie, J.A., Cunningham, M., and Brass, L.F., 1985, Changes in the platelet membrane glycoprotein IIb.IIIa complex during platelet activation. *J. Biol. Chem.* 260:11107-11114.

Sims, P.J., Ginsberg, M.H., Plow, E.F., and Shattil, S.J., 1991, Effect of platelet activation on the conformation of the plasma membrane glycoprotein IIb-IIIa complex. *J. Biol. Chem.* 266:7345-7352.

Slupsky, J.R., Seehafer, J.G., Tang, S-C., Masellis-Smith, A., and Shaw, A.R.E., 1989, Evidence that monoclonal antibodies against CD9 antigen induce specific association between CD9 and the platelet glycoprotein IIb-IIIa complex. *J. Biol. Chem.* 264:12289-12293.

Smith, J.W. and Cheresh, D.A., 1988, The arg-gly-asp binding domain of the vitronectin receptor. Photoaffinity cross-linking implicates amino acid residues 61-203 of the β subunit. *J. Biol. Chem.* 263:18726-18731.

Smith, J.W., Ruggeri, Z.M., Kunicki, T.J., and Cheresh, D.A., 1990, Interaction of integrins αvβ3 and glycoprotein IIb-IIIa with fibrinogen. Differential peptide recognition accounts for distinct binding sites. *J. Biol. Chem.* 265:12267-12271.

Solowska, J., Edelman, J.M., Albelda, S.M., and Buck, C.A., 1991, Cytoplasmic and transmembrane domains of integrin β1 and β3 subunits are functionally interchangeable. *J. Cell Biol.* 114:1079-1088.

Sosnoski, D.M., Emanuel, B.S., Hawkins, A.L., van Tuilen, P., Ledbetter, D.H., Nussbaum, R.L., Kaos, F-T., Schwartz, E., Phillips, D.R., Bennett, J.S., Fizgerald, L.A., and Poncz, M., 1988, Chromosomal localization of the genes for the vitronectin and the fibronectin receptors α subunits and for platelet glycoproteins IIb and IIIa. *J. Clin. Invest.* 81:1993-1998.

Springer, T.A., 1990, Adhesion receptors of the immune system. *Nature* 346:425-434.

Steiner, B., Cousot, D., Trzeciak, A., Gillessen, D., and Hadváry, P., 1989, Ca^{2+}-dependent binding of a synthetic arg-gly-asp (RGD) peptide to a single site on the purified platelet glycoprotein IIb-IIIa complex. *J. Biol. Chem.* 264:13102-13108.

Steiner, B., Parise, L.V., Leung, B., and Phillips, D.R., 1991, Ca^{2+}-dependent structural transitions of the platelet glycoprotein IIb-IIIa complex. Preparation of stable glycoprotein IIb and IIIa monomers. *J. Biol. Chem.* 266:14986-14991.

Takada, Y., Murphy, E., Pil, P., Chen, C., Ginsberg, M.H., and Hemler, M.E., 1991, Molecular cloning and expression of the cDNA for α3 subunit of human α3β1 (VLA-3), an integrin receptor for fibronectin, laminin, and collagen. *J. Cell Biol.* 115:257-266.

Take, H., Tomiyama, Y., Shibata, Y., Furubayashi, T., Honda, S., Mizutani, H., Nishiura, T., Tsubakio, T., Kurata, Y., Yonezawa, T., and Tarui, S., 1990, Demonstration of the heterogeneity of the platelet-specific alloantigen, Baka. *Br. J. Haematol.* 76:395-400.

Tamura, R.N., Cooper, H.M., Collo, G., and Quaranta, V., 1991, Cell type-specific integrin variants with alternative α chain cytoplasmic domains. *Proc. Natl. Acad. Sci. USA* 88:10183-10187.

Taub, R., Gould, R.J., Garsky, V.M., Ciccarone, T.M., Hoxie, J., Friedman, P.A., and Shattil, S.J., 1989, A monoclonal antibody against the platelet receptor contains a sequence that mimics a receptor recognition domain in fibrinogen. *J. Biol. Chem.* 264:259-265.

Thiagarajan, P., Shapiro, S.S., Levine, E., DeMarco, L., and Yalcin, A., 1985, A monoclonal antibody to human platelet glycoprotein IIIa detects a related protein in cultured human endothelial cells. *J. Clin. Invest.* 75:896-901.

Timmons, S., Kloczewiak, M., and Hawiger, J., 1984, ADP-dependent common receptor mechanism for binding of von Willebrand and fibrinogen to human platelets. *Proc. Natl. Acad. Sci. USA* 81:4935-4939.

Torti, M., Sinigaglia, F., Ramaschi, G., and Balduini, C., 1991, Platelet glycoprotein IIb-IIIa is associated with 21-kDa GTP-binding protein. *Biochim. Biophys. Acta* 1070:20-26.

Troesch, A., Duperray, A., Polack, B., and Marguerie, G., 1990, Comparative study of the glycosylation of platelet glycoprotein GPIIb/IIIa and the vitronectin receptor. Differential processing of their β-subunit. *Biochem. J.* 268:129-133.

Usobiaga, P., Calvete, J.J., Saíz, J.L., Eirín, M.T., and González-Rodríguez, J., 1987, Molecular characterization of human platelet glycoproteins IIIa and IIb and the subunits of the latter. *Eur. J. Biophys.* 14:211-218.

Uzan, G., Frachet, P., Lajmanovich, A., Prandini, M-H., Denarier, E., Duperray, A., Loftus, J., Ginsberg, M., Plow, E.F., and Marguerie, G., 1988, cDNA clones for human platelet GPIIb corresponding to mRNA from megakaryocytes and HEL cells. Evidence for an extensive homology to other arg-gly-asp adhesion receptors. *Eur. J. Biochem.* 171:87-93.

Uzan, G., Prenant, M., Prandini, M-H., Martin, F., and Marguerie, G., 1991, Tissue-specific expression of the platelet GPIIb gene. *J. Biol. Chem.* 266:8932-8939.

van Kuppevelt, T.H.M.S.M., Languino, L.R., Gailit, J.O., Suzuki, S., and Ruoslahti, E., 1989, An alternative cytoplasmic domain of the integrin β3 subunit. *Proc. Natl. Acad. Sci. USA* 86:5415-5418.

Wang, R.G., McFarland, J., Furihata, K., Friedman, K., and Newman, P.J., 1991, An amino acid polymorphism within the RGD binding domain of GPIIIa is associated with the Pen[a]/Pen[b] alloantigen system. *Blood* 78 (Suppl.I):281a.

Wippler, J., Steiner, B., and Hadváry, P., 1991, Expression of a truncated heterodimeric form of the platelet receptor GPIIb-IIIa in insect cells. *Throm. Haemost.* 65:655 (Abstr. 31).

Wyler, B., Bienz, D., Clemetson, K.J., and Lüscher, E.F., 1986, Glycoprotein Ibβ is the only phosphorylated major membrane glycoprotein in human platelets. *Biochem. J.* 234:373-379.

Yamada, K.M. and Kennedy, D.W., 1984, Dualistic nature of adhesive protein function: Fibronectin and its biologically active peptide fragments can autoinhibit fibronectin function. *J. Cell Biol.* 99:29-36.

Yamada, K.M., 1991, Adhesive recognition sequences. *J. Biol. Chem.* 266:12809-12812.

Zimrin, A.B., Eisman, R., Vilaire, G., Schwartz, E., Bennett, J.S., and Poncz, M., 1988, Structure of platelet glycoprotein IIIa. A common subunit for two different membrane receptors. *J. Clin. Invest.* 81:1470-1475.

Zimrin, A.B., Gidwitz, S., Lord, S., Schwartz, E., Bennett, J.S., White, II, G.C., and Poncz, M., 1990, The genomic organization of platelet glycoprotein IIIa. *J. Biol. Chem.* 265:8590-8595.

DISCUSSION

N. HOGG

I think there is a little bit further you could go with this model taking advantage of the location of the epitopes. You have mapped the disulphides, you have mapped epitopes along the "stick" model, and you have done limited proteolysis which has shown association betwen various parts of the molecule. Have you also done an epitope map of all the antibodies (i.e., their relationship one to the other), because if you could do that, you would then know how all those antibodies were docking onto your model and that would either confirm the model or show you that perhaps some of the associations were different. Lastly, you didn't mention how you actually mapped all those epitopes. I know some of them have been done by peptide blocking.

J. CALVETE

Answering your last question, all of the epitopes were assigned using synthetic peptides, because all of the monoclonals recognize linear epitopes within GPIIb or GPIIIa. We immunized mice with denatured antigen and we only identified those monoclonals which recognized the intact complex on platelets. So, we were lucky to have this collection but we also worked in that direction. Regarding the other part of your question, on the spatial relationship between the different epitopes, I may say that I would like to do these experiments and also experiments to see how the binding of one monoclonal affects the binding of a different one. This has been initiated in our laboratory. I would also like to take pairs of monoclonal antibodies as reference points for electron microscopy and to relate defined epitopes to the topography of the GPIIb/IIIa images.

N. HOGG

You can do a criss-cross experiment just looking at the antibodies themselves?

J. CALVETE

Yes, but you do not know if you have altered the conformation of the molecule. Thus what you conclude from pair AB cannot be related to pair AC if there is a conformational change between the binding of B and C.

N. HOGG

Why do you think your isolated molecule looks so different from the EM pictures?

J. CALVETE

The EM pictures we are used to seeing are just typical but arbitrary pictures of the molecule. If you look at hundreds of pictures in the same photograph you may find 10 or 15 different conformations. This might indicate that the molecule contains domains which are quite flexible.

M. HEMLER

I gather you are proposing two heresies. Not only are the regions near the membrane more compact but you had the putative divalent cation sites way out on the periphery, suggesting that they might not be very relevant. How do you reconcile that with the data suggesting that the divalent cation sites directly bind to ligand? Are you suggesting that the divalent cation sites are merely stabilizing the molecule?

J. CALVETE

If you measure the calcium-binding sites in GPIIb/IIIa you may find that GPIIIa also contains one calcium-binding site of low affinity. GPIIb contains 4 or 5 low affinity binding sites. At this stage we do not know whether the high affinity calcium-binding site responsible for the association of both subunits lies entirely in GPIIb, in GPIIIa, or is actually formed by dicarbosilic acids from both molecules. Before we answer this question, the relevance of the four well-assigned calcium binding sites in GPIIb α cannot be answered.

S. GOODMAN

Can I just ask what sort of quantities of GPIIb/IIIa did you need to carry out these studies?

J. CALVETE

I used to isolate tens of milligrams per week, but you do not need so much to carry out these analyses. Since I have enough material, I used to digest around half a milligram for each condition. I still have much of those peptides in the freezer.

A. SONNENBERG

I assume that you have used out-dated platelets, which may already have been activated.

J. CALVETE

The extent of activation of GPIIb/IIIa in outdated platelets may not be more than 15%, because if you assume that all the RGD binding GPIIb/IIIa population corresponds to activated GPIIb/IIIa, then 85% of the receptors are in the non-activated conformation.

G. LENAZ

In your model you have some repeated secondary structure. Was this based on predictions or on actual experimental data?

J. CALVETE

Usually you do not believe that this is the actual 2D structure of the molecule. What I did was to predict the secondary structure of the molecule with three different methods, compare the results, and just take the prediction in agreement with all three methods. This is just to have a model to think in other experiments. I have not published it and probably will never do it.

G. MARGUERIE

You mentioned the interactions of this molecule: how many molecules of peptide were bound? Was this binding inhibited by normal peptides? Was this binding or cross-linking sensitive to stimulation?

J. CALVETE

First I will answer your last question. It was done with isolated complex. We did not do the experiments at the platelet level. The peptides could be released by the non-labelled peptides. You need around three times more non-labelled peptides to release the same amount of labelled peptides. The population of GPIIb/IIIa complexes that bound RGD or γ-chain peptide was the same in both cases, around 20% of total GPIIb/IIIa complexes.

V. QUARANTA

Do you have any evidence for or against physical association of the cytoplasmic domains with each other?

J. CALVETE

No.

E. DEJANA

If you consider the β_3 chain and antibodies that inhibit the binding of fibrinogen or activate the molecule, would you expect that they would work when you have α_V instead of GPIIb bound to β_3. Would these mechanisms be influenced by the type of α-chain linked to the β?

J. CALVETE

The epitope for M5 is located on the α-chain and it is an inhibitory antibody of fibrinogen binding. It also binds to vitronectin receptor.

E. DEJANA

What I mean is if you consider an antibody that inhibits the binding of fibrinogen to GPIIb/IIIa and binds to β_3, would this antibody also inhibit the binding of a $\alpha_V\beta_3$ to the specific substrata?

J. CALVETE

Probably some of them may recognize the complex without activation, but I have no experimental data.

EXTRACELLULAR AND INTRACELLULAR FUNCTIONS OF VLA PROTEINS

Martin E. Hemler*, Akihide Masumoto*, Bosco M.C. Chan**,
Paul Kassner*, and Joaquin Teixidó***

*Dana-Farber Cancer Institute
  Boston, Massachusetts, 02115 U.S.A.

**Department of Chemistry and Biochemistry, University of Guelph
  Guelph, Ontario NIG 2W1, CANADA

***Seccion De Inmunologia, Hospital de la Princesa, Diego de Leon, 62
  28006 Madrid, SPAIN

INTRODUCTION

An extensive assortment of cell adhesion processes are mediated by receptors called integrins, a family composed of at least 20 distinct α-β heterodimers. Ligands bound by integrins include a variety of extracellular matrix molecules, cell surface Ig-like proteins, and some serum and complement proteins (Hemler, 1990; Hynes, 1992; Larson and Springer, 1990). Adhesion mediated by integrins is an essential feature of diverse processes such as tumor cell growth and metastasis, development, wound healing, and immunolocalization of leukocytes.

Recent studies have found that integrins not only mediate cell attachment, but also can transmit signals into the cell, thus influencing post-ligand-binding events such as cell migration (Akiyama et al., 1989; Grzesiak et al., 1992; Jaffredo et al., 1988; Letourneau et al., 1988; Perris et al., 1989; Smith et al., 1989; Yamada et al., 1990), proliferation (Brown et al., 1990; Pircher et al., 1986; Shimizu et al., 1990), gene induction (Blum et al., 1987; Li et al., 1987; Werb et al., 1989), and ultimately differentiation. It is assumed that integrins spanning the cell membrane might act as a critical interface between the intracellular and extracellular environments, but the precise biochemical mechanisms whereby different integrins might transmit different signals into the cell is not known. It is interesting that triggering through integrins can lead to an increase in intracellular pH (Schwartz et al., 1991), and to activation of FAK (focal adhesion kinase) (Guan et al., 1991; Kornberg et al., 1991; Schaller et al., 1992). However, these events are not very specific, since the same increase in pH can be achieved through a variety of different integrins (Schwartz et al., 1991), and FAK also can be activated by other mechanisms (Guan and Shalloway, 1992).

THE ROLE OF INTEGRIN CYTOPLASMIC DOMAINS

To understand how different integrins might translate adhesive signals into a diversity of post-ligand-binding events, we and others have focused on the role of cytoplasmic domains. Studies of β subunits have shown that deletion of cytoplasmic domains of β_1 (Solowska et al., 1989) or β_2 (Hibbs et al., 1991) resulted in diminished cell adhesion to immobilized ligands, whereas $\alpha^{IIb}\beta_3$ activity was retained in CHO cells despite deletion of the β_3 cytoplasmic domain (O'Toole et al., 1991). In other studies, the β_1 cytoplasmic domain was found to be essential for integrin localization to focal contacts (Hayashi et al., 1990; Marcantonio et al., 1990; Reszka et al., 1992; Solowska et al., 1989). In fact, even when the β_1 cytoplasmic domain was expressed as part of a hybrid molecule, it was sufficient to drive that molecule into focal contacts (LaFlamme et al., 1992). Whereas the cytoplasmic domains of β_1 and β_3 appear to be functionally interchangeable with respect to focal contact formation (Solowska et al., 1991), the cytoplasmic domain of β_5 subunit was less able to localize there (Wayner et al., 1992). The ability of the cytoplasmic domain of β_1 (and probably other β-chains) to directly associate with cytoskeletal proteins such as talin and α-actinin (Horwitz et al., 1986 Otey et al., 1990; Tapley et al., 1989) is probably a critical feature of localization to focal contacts.

Studies of the functional roles for α-chain cytoplasmic domains have so far been less extensive. Deletion of the cytoplasmic domain of the integrin α^L subunit caused no change in LFA-1 cell attachment functions (Hibbs et al., 1991). In contrast, exchange (with the α^5 cytoplasmic domain) or deletion of the cytoplasmic domain of α^{IIb} caused an increase in the ligand-binding function of $\alpha^{IIb}\beta_3$ (O'Toole et al., 1991). The integrin α^5-chain by itself did not localize to focal contacts, but it has been suggested that it could act as a negative regulator of the localization of β_1 to focal adhesions (LaFlamme et al., 1992).

Our studies have initially focused on the roles of integrin α-chains because they have not yet been well studied, and because their high degree of structural diversity predicts functional diversity. Aside from a consensus "KXGFFKR" motif found in most of the sequences, there is little other amino acid similarity among α-chain cytoplasmic domains. Although these α-chain cytoplasmic domains are highly dissimilar from one another, most are very highly conserved when compared among different species. This latter point suggests that α-chain cytoplasmic tails may mediate functions that are not only diverse, but also essential.

To begin to test the hypothesis that α-chain cytoplasmic domains have distinct functional roles, we have constructed a series of chimeric molecules, using the α^2, α^4, and α^5 cytoplasmic tails, and the α^2 and α^4 extracellular and transmembrane sequences. Constructs containing the extracellular and transmembrane sequences of α^2 (the X2 series) and the cytoplasmic tails of α^2, α^4 or α^5 (C2, C4, C5) were expressed in the rhabdomyosarcoma cell line RD. These mutant forms of VLA-2 were all expressed at equivalent levels at the cell surface, and each supported a similar level of VLA-2-dependent cell adhesion to collagen and laminin (Chan et al., 1992). These results suggested that ligand-binding was not influenced by exchange of cytoplasmic tails. However, examination of post-ligand-binding events (Table 1) indicated that the X2C4 construct was substantially better able to support random cell migration on surfaces coated with the VLA-2 ligands collagen and laminin. In contrast, the X2C2 and X2C5 constructs better supported VLA-2-dependent collagen gel contraction. These results provide strong evidence that different α-chain cytoplasmic domains can play critical but distinct roles in post-ligand-binding events.

To extend this concept, we analyzed another series of constructs (X4 series) expressed in the carcinoma cell line MIP-101. The X4C2, X4C4 and X4C5 constructs were expressed at similar levels, and mediated similar levels of cell adhesion to the VLA-4 ligands fibronectin and VCAM-1 (not shown). Analysis of cell migration (Table 1, top) revealed that the X4C4 construct supported a higher migration rate on fibronectin than did X4C2. Thus, in two different cell types, with two different extracellular domains, and migrating on different ligands, C4 supported migration better than did C2. However, MIP cells expressing the X4C5 construct showed a surprisingly high migration rate, in marked contrast to the low migration rate seen using the X2C5 construct in RD cells. This result emphasizes that unknown factors within different cellular environments may collaborate with α-chain cytoplasmic domains to yield variable results.

Having established that integrin α-subunit cytoplasmic domains have functional activities, we now have an insight into the extensive redundancy of ligand-binding functions within the integrin family (e.g., 3 different receptors for collagen, 4-6 receptors for laminin, 6-8 receptors for fibronectin). We hypothesize that although multiple receptors might recognize the same ligand, because these receptors have different cytoplasmic domains, it would allow a single ligand to stimulate a variety of post-ligand-binding events.

Table 1. Functional Differences Between α-chain Cytoplasmic Domains

Cells	Collagen	Migration on: Laminin	Fibronectin	Collagen Gel Contraction
MIP-Pf	--	--	Low	--
MIP-X4C2	--	--	Low	--
MIP-X4C4	--	--	Intermediate	--
MIP-X4C5	--	--	High	--
RD-Pf	Low	Low	--	No
RD-X2C2	Low	Low	--	Yes
RD-X2C4	High	High	--	No
RD-X2C5	Low	Low	--	Yes

Random migration of transfected MIP-101 and RD cells was carried out using plastic coated with 1-10 μg of laminin or collagen or 5 μg/ml of the FN-40 chymotryptic fragment of fibronectin using methods previously described (Chan et al., 1992). The low level of background migration seen for the RD cells is due to other integrins besides VLA-2. Collagen gel contraction was measured as determined elsewhere (Chan et al., 1992). Chimeric molecules consist of extracellular and transmembrane regions from α^2 and α^4 (X2 and X4) and cytoplasmic domains from α^2, α^4, and α^5 (C2, C4, C5). Cells mock-transfected with vector only are designated Pf.

DIFFERENT STATES OF FUNCTIONAL ACTIVITY FOR INTEGRINS

A remarkable feature of integrins is that they can assume multiple states of functional activation. On some cells, integrins are expressed in an inactive form, but then can be converted rapidly to an active (adhesive) state, without a requirement for new protein synthesis (Dustin and Springer, 1989; Hynes, 1992; Phillips et al., 1991; Plow and Ginsberg, 1989; Van Kooyk et al., 1989; Wright and Meyer, 1986). In other cases, functionally active integrins appear to lose function during cell differentiation, despite continued cell surface expression (Neugebauer and Reichardt, 1991; Watt, 1990). This ability of integrins to assume different states of functional activity provides cells with the flexibility to regulate their level of adhesiveness, consistent with their location and differentiation state. The mechanisms by which cells control the variable functional states of integrins are currently unknown, and at present this is an active area of research.

In our laboratory, we have discovered an added dimension of integrin ligand-binding flexibility for certain members of the β_1 family. Not only do they display different states of functional activity, but at the same time they display variable ligand-binding specificities in coordination with these different states of activation.

As shown in Figure 1, we have isolated a preparation of α^2-transfected K562 cells (called KA2) that showed little adherence to either collagen or laminin. Another preparation of α^2-transfected K562 cells was selected (by enrichment for attachment to matrix) which displayed adhesion to collagen, but not to laminin (KA2-M2 cells). In contrast, when the same α^2 cDNA that was used to transfect K562 cells was put into RD cells, adhesion to both collagen and laminin was observed. These results show that the same α^2 cDNA can be utilized to express VLA-2 in three functional states. We have named these "Form CL" (binds both collagen and laminin), "Form C" (binds collagen) and "Form 0" (binds neither). Because the same cDNA was utilized to obtain each of the functional forms, alternative splicing of α^2 is clearly not part of the mechanism responsible for cell type-specific variation in integrin function.

When the stimulatory anti-β_1 antibody TS2/16 (Arroyo et al., 1992; van de Wiel-van Kemenade et al., 1992) was added to K562 cells displaying either Form 0 or Form C activity, the VLA-2 was rapidly converted to form CL (Chan and Hemler, 1992). Thus, there does not appear to be an inherent deficiency in the VLA-2 molecule itself as it is expressed in K562 cells, and probably there are conformational differences between the different functional forms of VLA-2, rather than differences in primary sequence or in covalent post-translational modifications. Supporting the idea that TS2/16 may act by directly inducing a conformational change, it has been found that Fab fragments of the antibody can stimulate cell adhesion (Arroyo et al., 1992; van de Wiel-van Kemenade et al., 1992), thus ruling out a clustering mechanism. Also, we have found that TS2/16 appears to directly stimulate the binding of solubilized VLA-2 to collagen affinity columns (Chan and Hemler, 1992).

Analysis of VLA-4 functions also revealed similar functional variability. On certain cells, VLA-4 mediated adhesion to both VCAM-1 and to the CS1 region of fibronectin, whereas on some cells it bound to only VCAM-1, and on other cells it bound to neither CS1 or to VCAM-1 (Masumoto and Hemler, 1992). An example of VLA-4 with little adhesive function was seen in α^4-transfected K562 cells (Figure 2). Emphasizing again that this functional state is not irreversible, the addition of the stimulatory anti-β_1 antibody (TS2/16) caused an increase in VCAM-1 but not CS1 adhesion. The addition of 1 mM Mn^{2+} had a greater stimulatory effect, causing increased adhesion to both CS1 and VCAM-1, and the simultaneous addition of both TS2/16 and Mn^{2+} was even more stimulatory.

The results in Figures 1 and 2 can be summarized as shown schematically in Figure 3. In these hypothetical adhesion curves, VLA-2 (Part A) and VLA-4 (Part B) have different thresholds of activation for different ligands. Thus at the highest state of activation, the integrins acquire new ligand-binding activities not seen at intermediate or low states of activation. These findings with VLA-4 and VLA-2 are perhaps analogous to results observed for the $\alpha^{IIb}\beta_3$ integrin. When that structure is partly active, it binds only to insoluble fibrinogen, but when fully active it binds to many ligands, including soluble fibrinogen, fibronectin, and von Willebrand factor (Phillips et al., 1991; Plow and Ginsberg, 1989). Thus, the expansion of ligand specificity in parallel with increased activation may be

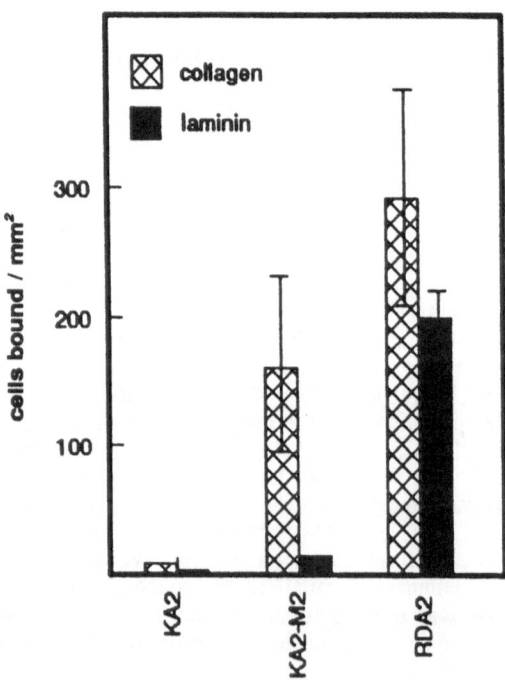

Figure 1. Multiple Functional Forms of VLA-2. Attachment to plastic surfaces coated with collagen or laminin was determined using a standard cell adhesion assay as described elsewhere (Chan and Hemler, 1992). Cells utilized were α^2-transfected K562 cells (KA2), a subclone of KA2 selected for attachment and growth on matrigel (KA2-M2), and α^2-transfected RD cells (RDA2) (Chan and Hemler, 1992).

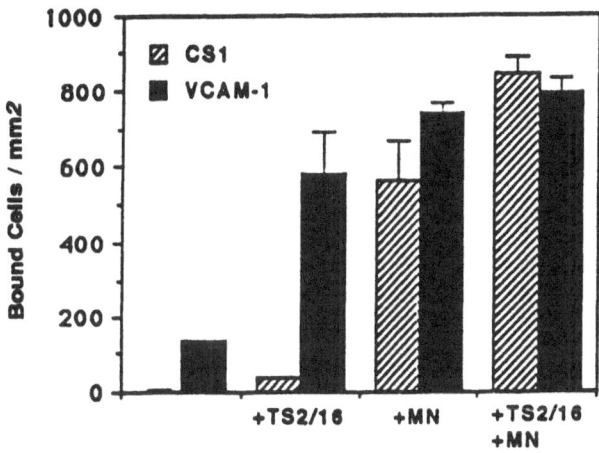

Figure 2. Stimulation of VLA-4-mediated Adhesion. Adhesion of α^4-transfected K562 cells (KA4) to the CS1 peptide derived from fibronectin or to VCAM-1 was determined as described elsewhere (Masumoto and Hemler, 1992), in the presence or absence of the stimulatory anti-β_1 antibody (TS2/16, ~3 µg/ml), or 1 mM Mn^{2+}.

a common theme among integrins, and provides a mechanism whereby additional flexibility and versatility can be achieved through each different receptor.

At present it is not clear why laminin would have a higher threshold of activation than collagen for binding to VLA-2. However for VLA-4, it perhaps makes sense that fibronectin binding would have a higher threshold of activation than VCAM-1. Because fibronectin is widely distributed as an extracellular matrix protein and also present in soluble form in circulating blood, it may be critical to limit the adhesive interaction to selected situations, only occurring when VLA-4 is highly activated. On the other hand, regulation of VCAM-1 at the level of expression on endothelial cells is perhaps sufficient to control its interaction with VLA-4.

We hypothesize that the stimulatory anti-β_1 antibody (TS2/16) may either bypass the normal regulatory pathway or mimic some naturally occurring stimuli. A current research challenge is to identify the unknown cellular factors, present in different cell types, which may regulate the appearance of these different functional states. Because Mn^{2+} can also convert VLA-4 to a higher state of activation, we speculate that integrin divalent cation binding sites may be critically involved in determining the activation state of this, and perhaps other, integrins.

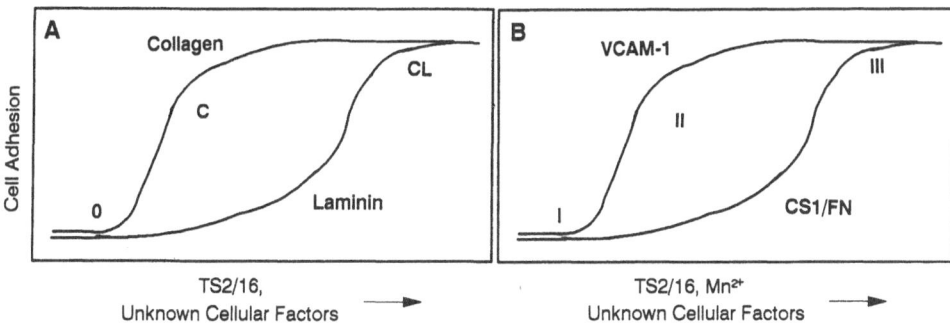

Figure 3. Schematic Diagram of VLA-2 (Part A) and VLA-4 (Part B) Functions.

MULTIPLE BIOCHEMICAL FORMS OF VLA-4: PROTEOLYTIC CLEAVAGE OF α^4

The integrin VLA-4 can be expressed in multiple biochemical forms, due to variable proteolytic cleavage of the α^4 subunit during biosynthesis (Teixidó et al., 1992). An attractive hypothesis was that the VLA-4 with cleaved α^4 ($\alpha^{4/80,70}$) might have different functional properties than VLA-4 with intact α^4 ($\alpha^{4/150}$). Recently we have identified the site of α^4 cleavage, and shown that mutations at that site can abolish cleavage (Teixidó et al., 1992). Thus we were able to prepare α^4-transfected cells expressing both cleaved and uncleaved forms of α^4 (Teixidó et al., 1992). Analysis of VLA-4 mediated adhesion to its two ligands (Table 2) revealed that both the cleaved and uncleaved forms of VLA-4 had essentially the same functional activities. Thus we conclude that proteolytic cleavage of α^4 is most likely not responsible for regulation of the state of functional activity displayed by VLA-4.

Table 2. Functional Differences Between Cleaved and Uncleaved VLA-4

Cell	α4 Status	Cells Bound/mm^2 to:	
		Fibronectin	**VCAM-1**
K-Pf	Absent	9	10
K-A4	Cleaved	222	611
K-A4 (KE/NA)	Cleaved	340	646
K-A4 (R/L)	Uncleaved	194	470
K-A4 (K/Q)	Uncleaved	291	578

Cell attachment was carried out as previously described (Chan et al., 1992), using plastic coated with 4 μg/ml of the FN-40 fragment of fibronectin or with 3.5 μg/ml of recombinant soluble VCAM-1.

CONCLUSIONS AND FUTURE DIRECTIONS

Here we have addressed two critical features of integrin-mediated cell adhesion. First, we have demonstrated and discussed how unique α-chain cytoplasmic domains can translate adhesive information into diverse post-ligand-binding events. As extension of these concepts, the wide diversity of α-chain cytoplasmic domain sequences leads to the prediction that a variety of specific biochemical interactions on the cytoplasmic side of the membrane remain to be discovered.

Second, we have described how certain integrins (VLA-2 and VLA-4) can have multiple activation states. These activation states arise due to unknown cell-type specific factors, are interconvertible, and may occur due to differing conformations in the vicinity of divalent cation sites. The challenge of future studies will be to discover and characterize the physiologically relevant biochemical signals that regulate these variable integrin activation states.

ACKNOWLEDGEMENTS

This work was supported by National Institutes of Health Grants GM38903 and GM46526 (to M.E.H.) and a Centennial fellowship from the MRC of Canada (to B.M.C.C.).

REFERENCES

Akiyama, S.K., Yamada, S.S., Chen, W.-T., and Yamada., K.M., 1989, Analysis of fibronectin receptor function with monoclonal antibodies: Roles in cell adhesion, migration, matrix assembly, and cytoskeletal organization. *J. Cell Biol.* 109:863-875.

Arroyo, A.G., Sánchez-Mateos, P., Campanero, M.R., Martín-Padura, I, Dejana, E., and Sánchez-Madrid, F., 1992, Regulation of the VLA integrin-ligand interactions through the β1 subunit. *J. Cell Biol.* 117:659-670.

Blum, J.L., Zeigler, M.E., and Wicha, M.S., 1987, Regulation of rat mammary gene expression by extracellular matrix components. *Exp. Cell Res.* 173:322-340.

Brown, E., Hooper, L., Ho, T., and Gresham, H., 1990, Integrin-associated protein: A 50 kD plasma membrane antigen physically and functionally associated with integrins. *J. Cell Biol.* 111:2785-2794.

Chan, B.M.C. and Hemler, M.E., 1993, Multiple functional forms of the integrin VLA-2 derived from a single α^2 cDNA clone: interconversion of forms induced by an anti-β_1 antibody. *J. Cell Biol. (in press).*

Chan, B.M.C., Kassner, P.D., Schiro, J.A., Byers, H.R., Kupper, T.S., and Hemler, M.E., 1992, Distinct cellular functions mediated by different VLA integrin α subunit cytoplasmic domains. *Cell* 68:1051-1060.

Dustin, M.L. and Springer, T.A., 1989, T-cell receptor cross-linking transiently stimulates adhesiveness through LFA-1. *Nature* 341:619-624.

Grzesiak, J.J., Davis, G.E., Kirchhofer, D., and Pierschbacher, M.D., 1992, Regulation of α_2 β_1-mediated fibroblast migration on type I collagen by shifts in the concentrations of extracellular Mg^{2+} and Ca^{2+}. *J. Cell Biol.* 117:1109-1117.

Guan, J-L. and Shalloway, D., 1992, Regulation of focal adhesion-associated protein tyrosine kinase by both cellular adhesion and oncogenic transformation. *Nature* 358:690-692.

Guan, J-L., Trevithick, J.E., and Hynes, R.O., 1991, Fibronectin/integrin interaction induces tyrosine phosphorylation of a 120 kDa protein. *Cell Regul.* 2:951-964.

Hayashi, Y., Haimovich, B., Reszka, A., Boettiger, D., and Horwitz, A., 1990, Expression and function of chicken integrin beta-1 subunit and its cytoplasmic domain mutants in mouse NIH 3T3 cells. *J. Cell Biol.* 110:175-184.

Hemler, M.E., 1990, VLA proteins in the integrin family: Structures, functions, and their role on leukocytes. *Ann. Rev. Immunol.* 8:365-400.

Hibbs, M.L., Xu, H., Stacker, S.A., and Springer, T.A., 1991, Regulation of adhesion to ICAM-1 by the cytoplasmic domain of LFA-1 integrin beta subunit. *Science* 251:1611-1613.

Horwitz, A., Duggan, K., Buck, C., Beckerle, M.C., and Burridge, K., 1986, Interaction of plasma membrane fibronectin receptor with talin-a transmembrane linkage. *Nature* 320:531-533.

Hynes, R.O., 1992, Integrins: Versatility, modulation and signalling in cell adhesion. *Cell* 69:11-25.

Jaffredo, T., Horwitz, A.F., Buck, C.A., Rong, P.M., and Dieterlen-Lievre, F., 1988, Myoblast migration specifically inhibited in the chick embryo by grafted CSAT hybridoma cells secreting an anti-integrin antibody. *Development* 103:431-446.

Kornberg, L.J., Earp, H.S., Turner, C.E., Prockop, C., and Juliano, R.L., 1991, Signal transduction by integrins: Increased protein tyrosine phosphorylation caused by clustering of β_1 integrins. *Proc. Natl. Acad. Sci. USA* 88:8392-8396.

LaFlamme, S.E., Akiyama, S.K., and Yamada, K.M., 1992, Regulation of fibronectin receptor distribution. *J. Cell Biol.* 117:437-447.

Larson, R.S. and Springer, T.A., 1990, Structure and function of leukocyte integrins. *Immunol. Rev.* 114:181-217.

Letourneau, P.C., Pech, I.V., Rogers, S.L., Palm, S.L., McCarthy, J.B., and Furcht. L.T., 1988, Growth cone migration across extracellular matrix components depends on integrin, but migration across glioma cells does not. *J. Neur. Res.* 21:286-297.

Li, M.L., Aggeler, J., Farson, D.A., Hatier, C., Hassell, J., and Bissell, M.J., 1987, Influence of a reconstituted basement membrane and its components on casein gene expression and secretion in mouse mammary epithelial cells. *Proc. Natl. Acad. Sci. USA* 84:136-140.

Marcantonio, E.E., Guan, J., Trevithick, J.E., and Hynes, R.O., 1990, Mapping of the functional determinants of the integrin beta-1 cytoplasmic domain by site-directed mutagenesis. *Cell Regul.* 1:597.

Masumoto, A. and Hemler, M.E., 1993, Multiple activation states of VLA-4: Mechanistic differences between adhesion to CS1/fibronectin and to VCAM-1. *J. Biol. Chem.* (*in press*).

Neugebauer, K.M. and Reichardt, L.F., 1991, Cell-surface regulation of β_1-integrin activity on developing retinal neurons. *Nature* 350:68-71.

O'Toole, T.E., Mandelman, D., Forsyth, J., Shattil, S.J., Plow, E.F., and Ginsberg. M.H., 1991, Modulation of the affinity of integrin $\alpha IIb\beta 3$ (GPIIb-IIIa) by the cytoplasmic domain of αIIb. *Science* 254:845-847.

Otey, C.A., Pavalko, F.M., and Burridge, K., 1990, An interaction between α-actinin and the $\beta 1$ integrin subunit *in vitro*. *J. Cell Biol.* 111:721-729.

Perris, R., Paulsson, M., and Bronner-Fraser, M., 1989, Molecular mechanisms of avian neural crest cell migration on fibronectin and laminin. *Dev. Biol.* 136:222-238.

Phillips, D.R., Charo, I.F., and Scarborough, R.M., 1991, GPIIb-IIa: The responsive integrin. *Cell* 65:359-362.

Pircher, H., Groscurth, P., Baumhütter, S., Aguet, M., Zinkernagel, R.M., and Hengartner, H., 1986, A monoclonal antibody against altered LFA-1 induces proliferation and lymphokine release of cloned T cells. *Eur. J. Immunol.* 16:172-181.

Plow, E.F. and Ginsberg, M.H., 1989, Cellular adhesion: GPIIb/IIIa as a prototypic adhesion receptor. *Prog. Hemost. Thromb.* 9:117-156.

Reszka, A.A., Hayashi, Y., and Horwitz, A.F., 1992, Identification of amino acid sequences in the integrin β_1 cytoplasmic domain implicated in cytoskeletal association. *J. Cell Biol.* 117:1321-1330.

Schaller, M.D., Borgman, C.A., Cobb, B.S., Vines, R.R., Reynolds, A.B., and Parsons, J.T., 1992, pp125^{FAK}, a structurally distinctive protein-tyrosine kinase associated with focal adhesions. *Proc. Natl. Acad. Sci. USA* 89:5192-5196.

Schwartz, M.A., Ingber, D.E., Lawrence, M., Springer, T.A., and Lechene, C., 1991, Multiple integrins share the ability to induce elevation of intracellular pH. *Exp. Cell Res.* 195:533-535.

Schwartz, M.A., Lechene, C., and Ingber, D.E., 1991, Insoluble fibronectin activates the Na/H antiporter by clustering and immobilizing integrin $\alpha_5\beta_1$, independent of cell shape. *Proc. Natl. Acad. Sci. USA* 88:7849-7853.

Shimizu, Y., Van Seventer, G.A., Horgan, K.J., and Shaw, S., 1990, Costimulation of proliferative responses of resting CD4+ T cells by the interaction of VLA-4 and VLA-5 with fibronectin or VLA-6 with laminin. *J. Immunol.* 145:59-67.

Smith, C.W., Marlin, S.D., Rothlein, R., Toman, C., and Anderson, D.C., 1989, Cooperative interactions of LFA-1 and Mac-1 with intercellular adhesion molecule-1 in facilitating adherence and transendothelial migration of human neutrophils *in vitro*. *J. Clin. Invest.* 83:2008-2017.

Solowska, J., Edelman, J.M., Albelda, S.M., and Buck, C.A., 1991, Cytoplasmic and transmembrane domains of integrin β_1 and β_3 subunits are functionally interchangeable. *J. Cell Biol.* 114:1079-1088.

Solowska, J., Guan, J-L., Marcantonio, E.E., Trevithick, J.E., Buck, C.A., and Hynes, R.O., 1989, Expression of normal and mutant avian integrin subunits in rodent cells. *J. Cell Biol.* 109:853-861.

Tapley, P., Horwitz, A., Buck, C., Duggan, K., and Rohrschneider, L., 1989, Integrins isolated from Rous sarcoma virus-transformed chicken embryo fibroblasts. *Oncogene* 4:325-333.

Teixidó, J., Parker, C.M., Kassner, P.D., and Hemler, M.E., 1992, Functional and structural analysis of VLA-4 integrin α^4 subunit cleavage. *J. Biol. Chem.* 267:1786-1791.

van de Wiel-van Kemenade, E., Van Kooyk, Y., de Boer, A.J., Huijbens, R.J.F., Weder, P., van de Kasteele, W., Melief, C.J.M., and Figdor, C.G., 1992, Adhesion of T and B lymphocytes to extracellular matrix and endothelial cells can be regulated through the β subunit of VLA. *J. Cell Biol.* 117:461-470.

Van Kooyk, Y., Van DeWiel-Van Kemenade, P., Weder, P., Kuijpers, T.W., and Figdor, C.G., 1989, Enhancement of LFA-1-mediated cell adhesion by triggering through CD2 or CD3 on T lymphocytes. *Nature* 342:811-813.

Watt, F.M., 1990, Changes in keratinocyte adhesion during terminal differentiation: reduction in fibronectin binding precedes $\alpha 5\beta 1$ integrin loss from the cell surface. *Cell* 63:425-435.

Wayner, E.A., Orlando, R.A., and Cheresh, D.A., 1992, Integrins αvβ3 and αvβ5 contribute to cell attachment to vitronectin but differentially distribute on the cell surface. *J. Cell Biol.* 113:919-929.

Werb, Z., Tremble, P.M., Behrendtsen, O., Crowley, E., and Damsky, C.H., 1989, Signal transduction through the fibronectin receptor induces collagenase and stromelysin gene expression. *J. Cell Biol.* 109:877-889.

Wright, S.D. and Meyer, B.C., 1986, Phorbol esters cause sequential activation and deactivation of complement receptors on polymorphonuclear leukocytes. *J. Immunol.* 136:1759-1764.

Yamada, K.M., Kennedy, D.W., Yamada, S.S., Gralnick, H., Chen, W-T., and Akiyama, S.K., 1990, Monoclonal antibody and synthetic peptide inhibitors of human tumor cell migration. *Cancer Res.* 50:4485-4496.

DISCUSSION

G. MARGUERIE

I agree with you that you have different forms, active versus not active, but, in your mind, are these four separate entities present *in vivo*?

M. HEMLER

They may be in equilibrium but the point is in different cells the equilibrium is totally different and while I tried to explain some of the differences between the different forms what I had not explained at all, and no one else has either, I do not think, is how can a cell stably and consistently make one form and not another? What is the difference in the cellular environment that produces an active, fully active or partly active form? That I do not know. Now presumably there is some sort of equilibrium which can be shifted by having manganese present or the antibody to this epitope on β_1, but the really key question which we have not addressed is why does one cellular environment give you one state and another cellular environment gives you another state?

D. LIVINGSTON

Is there a possibility that the activity, for example in a culture, is cell-cycle dependent and the reason the molecule appears to be inactive is that the growth properties of that culture are such that most of the cells there are in one particular stage of the cell cycle. When you test cells in that stage of the cell cycle are the molecules of interest active or inactive?

M. HEMLER

I know there is not much published on integrin activity and cell cycle dependence, because I presume that no one has found any obvious relationship. In Richard Hynes' laboratory a very nice story is coming out suggesting that at a certain point in the cycle of an attached cell, when it is about to detach or when it has detached, then you have phosphorylation of the β_1-chain, correlating with cell detachment. But that is in a case where you have a very specific situation. In our cells we are looking at cells that remain adherent throughout or they remain in suspension throughout and they certainly have not been synchronized in terms of cell-cycle and once the phenotype is established it persists whether the cells are going at high density or low density or whatever. So I do not think it is the answer.

D. LIVINGSTON

Perhaps a synchronization experiment would reveal a specific phase of the cycle when the VLA-μ molecules are active and other times when they are inactive.

M. HEMLER

That is an interesting idea, I will have to think about that.

A. ANICHINI

A recent paper in *Nature* described the gene called CAR (cell adhesion regulator), which is a gene that can modulate the function of integrins without changing their expression. Do you think that these or similar genes might be involved in the phenomena of inhibition of the function of your cells?

M. HEMLER

I have read that paper several times and I find it very curious, I mean I am not exactly sure what they are measuring because they describe an RGD-dependent adhesion to collagen, but they have not indicated if in fact it is an integrin or which particular integrin might be involved. So for that reason, I am a little puzzled. The fact that they have been able to change the apparent adhesive capability by transfecting this particular gene is interesting but I wonder how closely connected it is; maybe in some non-specific way they shifted the balance between phosphatase and kinase activity. They show at best a 2-3 fold difference, whereas I think some of the things we are seeing here are practically all or none.

D. LIVINGSTON

If you take ten clones of RD, let us say transfected with the same chimera with the same C terminal segment, is the phenotype fixed? Do all ten clones reveal the same phenotype?

M. HEMLER

We have not studied ten different clones of RD, but we have done assays with a number of uncloned cultures and usually the cells all show similar adhesion phenotypes. In addition we have studied two or three individual clones, but there is not really a notable difference in adhesion properties.

D. LIVINGSTON

It makes you wonder whether there might not be significant clonal variation in the cell line that might in the aggregate be wildly positive.

M. HEMLER

Right, but we stretched the system quite a bit, we really searched hard to find adhesive colonies. After a month we had four colonies out of perhaps billions of cells.

R. JULIANO

I have a question on the phorbol ester effects on cell adhesion. I was struck by your observation that the LA4 with a truncated cytoplasmic domain was still active but did not respond to phorbol ester. Now what is going on there? Are the cells adhering and then spreading, or is there cytoskeletal interaction of the truncated construct, or is there just basal adhesion in the cells?

M. HEMLER

In the case of the X4C0 construct they do not adhere very well, they adhere in fact a little bit less than the constitutively low level seen using X4C4. With T S2/16 stimulation, but not with phorbol ester stimulation, adhesion clearly goes up quite a bit for the X4CO construct. I do not know if we have waited long enough to see spreading in that assay but they look very similar if not identical to all the other cases where you have no cytoplasmic tail or where you have C2 or C5.

T. CAREY

I was wondering if you have looked at phosphorylation of the α tails in any of these cases. Do they get phosphorylated?

M. HEMLER

We have done the requisite phosphotyrosine blots and everything looks nearly the same. Regardless of which cell or which cytoplasmic tail, the same pattern of 10-20 weak to intermediate strength phosphotyrosine bands appear.

T. CAREY

But have you precipitated the α_2 integrins from that clone? At one point we noticed that α_4 cytoplasmic tail seemed to be constitutively phosphorylated but we have not really followed that up, we want to get some functional insights before we go looking for those things.

G. TARONE

You reported that when you transfected the α_2 cDNA in K562 the resulting $\alpha_2\beta_1$ receptor is inactive, but when you transfect RD cells with the same cDNA the $\alpha_2\beta_1$ receptor in these cells is active. Do you know whether these two cell types express different isoforms of β_1 subunit?

M. HEMLER

Well, the β_1 chains look the same and react with the same polyclonal serum to the cytoplasmic tail which only recognizes the common alternative splice form of β_1. Thus in all cases β_1 alternative splicing did not correlate with different functional capabilities. There might be some more subtle difference in β that we have not seen, but I do not think that our cell-type specific adhesion differences have anything to do with the known β alternative splicing.

R. BANKERT

I have two questions, first, have you looked at various blocking antibodies, and, second, do they still block your transfected cell line?

M. HEMLER

Yes, antibody 5E8, very kindly provided by yourself, was used and we carefully tested the inhibitory capability of the antibody to block each of our different chimeric constructs to show that the adhesive capability of each construct was similarly inhibited by the antibody. We have not done the most definitive experiment, which is to obtain some sort of collagen peptide fragment and do some sort of direct affinity measurement. That is very hard to do with an insoluble collagen triple helix which is our excuse for not having done it yet, but I think that comparison of sensitivity to antibody inhibition gives you some idea that their adhesion capability is not dramatically different and in fact it was very similar.

A. SONNENBERG

It is still puzzling for me that $\alpha_2\beta_1$ is a laminin receptor. You now say that you have conclusive evidence that if you transfected the α_2 cDNA clone into RD cells their binding to laminin is enchanced. It is not possible that by just transfecting them with the α_2 cDNA, you have activated $\alpha_6\beta_1$ or have enhanced its activity. In that case the effect that you have noticed would actually be mediated by $\alpha_6\beta_1$ and not by $\alpha_2\beta_1$.

M. HEMLER

I think I was at a Gordon conference a couple of years ago, we agreed that optimally you would like to have three independent criteria before you would say x integrin binds to y ligand. So in the case of VLA-2 we have blocking of adhesion to laminin by several antibodies to α_2, we have binding of solubulized receptors to collagen and laminin columns, and third, α_2-transfection confers on a cell the ability to adhere to both collagen and laminin in a way that is completely inhibited by antibodies to α_2. So, I suggest that there is a pretty definitive evidence for VLA-2 interaction with laminin.

A. SONNENBERG

I would still like to point out that it is possible that antibodies against $\alpha 2$ may produce a signal and thus increase the activity of $\alpha_6 \beta_1$.

M. HEMLER

If they signal α_6 that would be interesting. They certainly do not send a signal to alter adhesion to a control substance such as fibronectin because that does not change in the untransfected and transfected cells.

A. SONNENBERG

I have a question about the efficiency of isolating $\alpha_2 \beta_1$ on laminin columns. Some people claim that $\alpha_2 \beta_1$ was isolated, but they did not check the flow through and the percentage of $\alpha_2 \beta_1$ which really bound to the laminin column.

M. HEMLER

If you read very carefully a paper by M.J. Elvies and myself, from 1989, you would see 90% of the $\alpha_2 \beta_1$ in the specific eluate compared to the flow-through fraction.

A. SONNENBERG

I still find it hard to believe that $\alpha_2 \beta_1$ is a laminin receptor, because there is no structural basis for this function. Collagen and laminin have entirely different structures.

M. HEMLER

There is no structural similarity between VCAM and CS1 peptide, and yet they both are recognized by VLA-4.

SIGNAL TRANSDUCTION FROM LEUKOCYTE INTEGRINS

Eric J. Brown

Departments of Medicine, Cell Biology and Physiology, and Molecular   Microbiology

Washington University School of Medicine
St. Louis, Missouri, U.S.A.

ABBREVIATIONS USED

DMSO = Dimethylsulfoxide; **ECM** = Extracellular Matrix; **IAP** = Integrin-associated 50 kD Integral Membrane Protein; **LRI** = Leukocyte-response Integrin; M_r = Relative molecular mass as estimated from SDS-PAGE; **PMN** = Polymorphonuclear Neutrophil

INTRODUCTION

The activation of leukocytes at sites of inflammation is a topic of great medical importance. Leukocytes normally exist in the circulation unactivated. In response to specific inflammatory signals, they can migrate out of the bloodstream, into the site of inflammation. At the inflammatory site, they express a different phenotype than in the blood. Neutrophils and monocytes are easily stimulated to release reactive oxygen metabolites which act both as cytocidal agents and as metabolic intermediates (Durum and Oppenheim, 1989); monocytes and macrophages are more highly phagocytic or more competent for antigen presentation (Silverstein, et al., 1989; Unanue and Allen, 1987); lymphocytes are activated to proliferate, synthesize cytokines, or perform effector functions (Durum and Oppenheim, 1989). Inhibition of leukocyte migration into, or activation at, sites of inflammation can dramatically reduce tissue damage and alter survival (Ferguson et al., 1991; Rosen, 1989; Vedder et al., 1988; Vercellotti et al., 1983). Thus, understanding the mechanisms of leukocyte exit from the bloodstream and activation at inflammatory sites is presently a subject of intense effort in several laboratories.

We have hypothesized that a major component of leukocyte activation is likely to be recognition of extracellular matrix (ECM) components (Brown, 1986). In anthropomorphic terms, this is a mechanism whereby leukocytes can "learn" that they are out of the bloodstream, in an extravascular site of inflammation or infection. Using similar reasoning, leukocyte interaction with clotting factors might provide similar information, since clotting is an essentially invariable component of the inflammatory response. In recent years, many observations have been made which are consistent with this hypothesis, demonstrating both that leukocytes have receptors which recognize a variety of ECM and clotting proteins and that interaction with these proteins, alone or in combination with other inflammatory factors, can effect leukocyte activation (Bohnsack, 1992; Bohnsack et al., 1990; Bohnsack et al., 1985; Loike et al., 1991; Nathan and Sanchez, 1990; Nathan et al., 1989; Wright and Meyer, 1985).

A major focus in my laboratory for the past several years has been to understand the phenotypic changes which occur when leukocytes interact with extracellular matrix. Initially, we performed experiments which demonstrated that both fibronectin and laminin could increase the rate and extent of phagocytosis by both neutrophils and macrophages (Bohnsack et al., 1990b; Bohnsack et al., 1985; Pommier et al., 1983; Pommier et al., 1984; Pommier, O'Shea, et al., 1984). The enhancement of phagocytosis occurred because of leukocyte interaction with ECM independent of the phagocytic event and not because of fibronectin or laminin opsonization of the phagocytic target. Indeed, work first done by Wright and Silverstein showed that an interaction with fibronectin on the basal surface of the phagocyte could stimulate ingestion by receptors on the cell's apical surface (Wright et al., 1983). Thus, binding of these ECM ligands to the cells caused other receptors on the cells to ingest more efficiently. These observations suggested that the effect on phagocytosis was the result of signal transduction from one plasma membrane receptor to another. This "receptor-to-receptor" signal transduction was distinct from the sort of signal transduction involved in regulation of cell proliferation by growth factors -- "receptor-to nucleus" signal transduction -- because it was immediate and reversible (Wright, Licht, et al., 1984). Thus, one might predict that although "receptor-to-receptor" and "receptor-to-nucleus" signal transductions might involve some similar biochemical pathways, they were likely to be different in fundamental respects as well. Thus, we have set out to understand "receptor-to-receptor" signal transduction because of its importance in the process of leukocyte activation at sites of inflammation. Primarily, we have used ECM protein stimulation of phagocytosis as a model for this process, although we have recently begun to use adhesion and chemotaxis assays as well. In this paper I will summarize our recent work in dissection of this challenging and important problem.

RESULTS AND DISCUSSION

Leukocyte Response Integrin (LRI)

An important first step in understanding ECM activation of leukocytes is to determine the receptors necessary for this phenomenon. In theory, the interactions between leukocytes and ECM proteins may be complex events. Multiple receptors for laminin and fibronectin have been described (Hynes, 1992; Mecham, 1991); several of these are clearly present on leukocytes. How is it possible to distinguish which, if any, of these are required for ligand binding by intact cells? When more than one receptor for an ECM protein is expressed on a cell, it is possible that the receptor with highest ligand affinity, which should be most important for initial cell-ligand binding, is not the receptor which signals leukocyte activation. How might binding and activation be distinguished? To describe our approaches to these problems, it is necessary to summarize the literal explosion in understanding of cell-extracellular matrix interactions ushered in by the description of the integrin protein family.

Several years ago, in attempting to understand fibroblast interaction with fibronectin, Pierschbacher and Ruoslahti described that a tripeptide within fibronectin, arg-gly-asp (RGD), was necessary for adhesion (Pierschbacher and Ruoslahti, 1984). Remarkably, the RGD sequence alone could also mediate adhesion of many cells. Using the specificity of the adhesion receptor for this peptide, Pytela et al., purified the fibronectin receptor (Pytela et al., 1985). At the same time, several groups, most notably Buck and Horwitz, prepared monoclonal antibodies which inhibited cell adhesion to ECM substrates (Brown and Juliano, 1985; Neff et al., 1982). These monoclonal antibodies immunoprecipitated several polypeptide chains from cells, including ones with characteristics similar to those of the fibronectin receptor isolated by Ruoslahti's group. Interestingly, the monoclonal prepared by Buck and Horwitz inhibited interaction with fibronectin, collagen and laminin, while the receptor isolated by affinity chromatography apparently bound only to fibronectin. Concurrently, Hemler prepared a monoclonal antibody which immunoprecipitated several polypeptide chains from lymphocytes and other cells and which he correctly concluded represented several different heterodimeric complexes with one chain in common (Hemler et al., 1987). The common chain of the avian receptors was cloned and termed integrin because it was part of a receptor which associated the extracellular matrix with intracellular cytoskeletal molecules (Tamkun et al., 1986). Ultimately, integrin was renamed β1, and several polypeptide chains of extracellular matrix and cell adhesion receptors were cloned

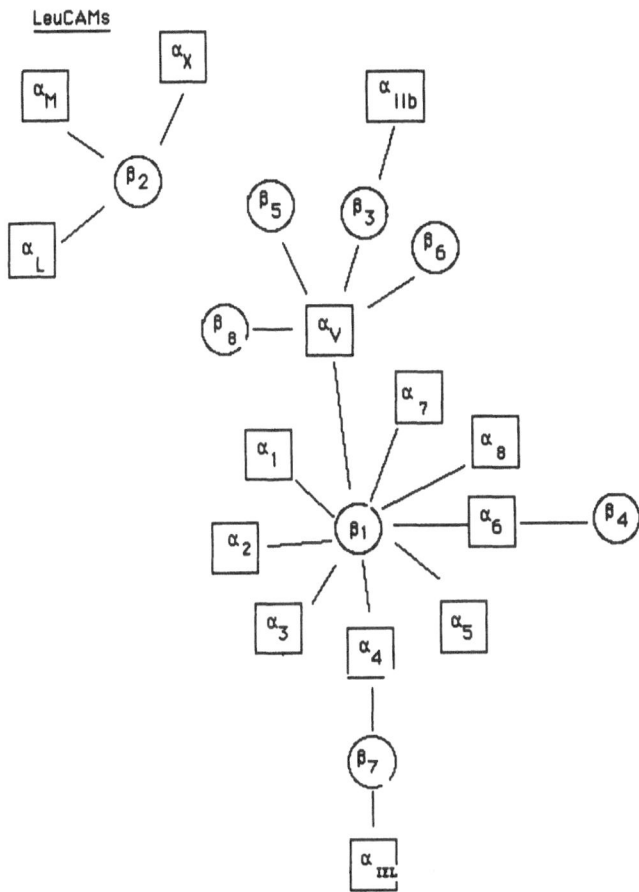

Figure 1. The Integrin Receptor Family. Integrin heterodimers which have been found to exist are indicated by solid lines connecting the α- and β-chains of each heterodimer.

and shown to have sequence homology to $\beta 1$. Now the entire superfamily of heterodimeric receptors is known as integrins. In recent years there has been an explosion of identification and cloning of polypeptide chains in this superfamily, and the extent of the diversity within the family, either in terms of the number of genes, or the ligands recognized, is still not known. Figure 1 shows an up-to-date version of the known receptors of the family. As recently discussed by Hynes (Hynes, 1992), many of the receptors of the family show the ability to recognize short linear peptide sequences with μM affinity, although only a few recognize the RGD sequence. This property has been extremely valuable to us in our characterization of the activating integrin of leukocytes.

RGD-binding Receptors of Neutrophils (PMN) and Monocytes. A first step in understanding the activation of leukocytes by extracellular matrix was the recognition that RGD-containing peptides could mimic these effects (Wright and Meyer, 1985). When RGD-binding receptors were isolated from surface-radiolabelled neutrophils and monocytes, they resembled receptors of the β_3 family in structure and immunologic reactivity (Brown and Goodwin, 1988). This led to the hypothesis that the activating receptor on these cells was $\alpha_v\beta_3$, and evidence has been presented that $\alpha_v\beta_3$ is present in macrophages and perhaps in monocytes and neutrophils as well (Krissansen et al., 1990; Singer et al., 1989). Indeed, even certain lymphocytes apparently express α_v-containing integrins (Maxfield et al., 1989; Moulder et al., 1991). Therefore, we prepared monoclonal and polyclonal antibodies against $\alpha_v\beta_3$ purified from placenta and examined their effect on phagocyte activation (Brown et al., 1990; Gresham, Goodwin, et al., 1989). The conclusion of these studies was that although polyclonal antibody to $\alpha_v\beta_3$ could inhibit stimulation of ingestion by a variety of extracellular matrix proteins and by synthetic RGD peptides, we could find no evidence for expression of α_v on neutrophils or freshly isolated monocytes. Furthermore, none of the anti-α_v monoclonal antibodies we prepared inhibited ingestion (although they did inhibit some functions of α_v on macrophages [Savill et al., 1990]), nor did LM609, a monoclonal antibody made by Cheresh, which inhibits many functions of $\alpha_v\beta_3$. In contrast, a mAb, 7G2, which recognized the β_3 integrin polypeptide, did inhibit the stimulation of phagocytosis by RGD for both PMN and monocytes (Brown et al., 1990). Thus, we hypothesized that the leukocyte integrin which stimulates phagocytosis is a β_3 or β_3-like integrin, but is not $\alpha_v\beta_3$. Furthermore, similar immunologic evidence suggested that this integrin does not contain α_{IIb}, the only other α chain known to associate with β_3, and which is expressed only on platelets. Because of these data, we hypothesized that the activating integrin on leukocytes was previously undescribed, and we called it Leukocyte Response Integrin (LRI).

Peptide Specificity of LRI. Recently, we have obtained further evidence that LRI is a unique integrin, because it has a unique peptide specificity (Gresham, Adams, et al., 1992 [in press]). These data arose from investigation of the interaction of peptides containing the sequence KQAGDV with LRI. It has been known for some time that the platelet β_3 integrin $\alpha_{IIb}\beta_3$ recognizes this sequence in addition to RGD (Santoro and Lawing, 1987). However, $\alpha_v\beta_3$ has little, if any, affinity for this peptide sequence. Therefore, we set out to determine whether LRI would recognize KQAGDV, by using this peptide to inhibit enhancement of PMN phagocytosis by fibrinogen. Indeed, LRI recognized KQAGDV about as well as $\alpha_{IIb}\beta_3$, and much better than $\alpha_v\beta_3$ (Table 1). In fact, when KQAGDV- and RGD-

Table 1. Fine Peptide Specificity of Arg-Gly-Asp-binding Integrins

Peptide	ID_{50} (μM)[a]		
	LRI	**IIb-IIIa**	$\alpha_v\beta_3$
KQAGDV[b]	9	15	>100
KGAGDV	3	>250	30
KQRGDV	>50	5	0.025

[a] ID_{50} calculated for fibrinogen-stimulated phagocytosis enhancement (LRI) or platelet aggregation (IIb-IIIa). Peptide inhibition of a_vb_3 binding to Vn was measured in an ELISA.

[b] 15 amino acid peptide from fibrinogen γ chain including hexapeptide shown in table. For LRI, hexapeptides alone are ~3 fold less potent than 15 aa peptide (not shown).

containing peptides were compared for their ability to inhibit fibrinogen-stimulated phagocytosis, KQAGDV was superior. This contrasts even with $\alpha_{IIb}\beta_3$, for which RGD-containing peptides are generally superior. The difference in peptide specificity between $\alpha_{IIb}\beta_3$ and LRI was made even clearer by amino acid substitutions in the core KQAGDV peptide (Table 1). While the peptide KQRGDV was superior as aligand for both $\alpha_{IIb}\beta_3$ and $\alpha_v\beta_3$, it had little inhibitory activity for PMN. In contrast, the peptide KGAGDV, which had essentially no activity in the $\alpha_{IIb}\beta_3$ assay, and very little effect on $\alpha_v\beta_3$ was the best peptide that we have synthesized for inhibition of LRI function. To test whether KGAGDV was a complete ligand for LRI, i.e., could stimulate ingestion without any other amino acids, we synthesized multivalent KGAGDV (Figure 2). Previous work had shown that for RGD

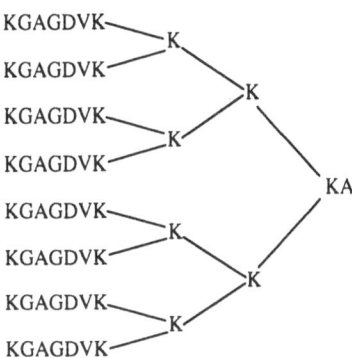

Figure 2. Branched KGAGDV. The structure of the multimeric, branched KGAGDV-containing peptide is shown. The peptide was synthesized as described (Brown et al., 1990) using fmoc chemistry. The branched peptide synthesis was patterned after the concept of Posnett et al. (1988).

peptides, while monovalent peptides inhibited stimulation of ingestion by ECM ligands, multivalent peptides were able to activate the cells in the absence of other signals (Brown et al., 1990; Gresham et al., 1989). Multivalent KGAGDV was indeed an excellent stimulatory ligand. Moreover, unlike multivalent RGD, stimulation by the KGAGDV ligand was unaffected by anti-$\beta 1$ antibodies, and the binding of multivalent KGAGDV to cells was unaffected by anti-$\beta 1$ or anti-$\beta 2$ antibodies, suggesting that it was quite specific for LRI (Figure 3). This ligand has been valuable not only to allow initiation of studies on the biochemical basis of signal transduction through LRI, but also as a ligand for affinity chromatography for the purification of LRI.

LRI in PMN Activation. Recently, using especially the antibodies which inhibit LRI function, we have investigated the participation of LRI in other leukocyte functions, such as chemotaxis and adhesion (Senior, R.M., Gresham, H.D., Griffin, G.L., Brown, E.J., and Chung, A.E., submitted for publication [1992]). We have found that LRI on PMN can recognize the basement membrane protein entactin. Entactin is a normal constituent of all basement membranes, which has domains which bind both laminin and type IV collagen, and which contains an RGD sequence that can function as a cell-binding sequence (Chakravarti et al., 1990; Chung and Durkin, 1990; Durkin et al., 1988). Entactin can be chemotactic for human PMN and can mediate PMN adhesion to protein coated surfaces. Both these functions are mediated by LRI, as judged by inhibition with 7G2 and antibody to IAP (see below) (Figure 4). A very important part of LRI interaction with entactin is that it is apparently β_2-independent. Both PMN from patients genetically deficient in the expression of β_2 and normal PMN treated with anti-β_2 monoclonal antibodies adheres to entactin-coated surfaces normally. Moreover, PMN adhesion to entactin, unlike adhesion to laminin, fibronectin, or fibrinogen (Bohnsack, Akiyama, et al., 1990; Vercellotti et al., 1983; Wright, Weitz, et al., 1988), does not require prior activation of the cells. Rather, adhesion to entactin apparently primes PMN for generation of the activated phenotype: highly phagocytic, with easily triggered degranulation and respiratory burst generation. These data have led us to hypothesize that adhesion to entactin is an early and critical event in the emigration of PMN from the bloodstream into sites of inflammation.

Biochemical Characterization of LRI. Most recently, we have made a polyclonal antibody to RGD-binding proteins from the cell line HL60, which had been induced to differentiate along a neutrophilic pathway by culture with DMSO (Carreno et al., 1991a; Carreno et al., 1992a). This polyclonal antibody apparently recognizes LRI, since it immunoprecipitates a heterodimer with appropriate Mr and inhibits LRI function on PMN and monocytes. Interestingly, this polyclonal antiserum recognizes lymphocytes as well as phagocytes, suggesting that LRI may be present on all leukocytes. Using this antiserum and the RGD-purified proteins from HL60, we have compared LRI with $\alpha_v\beta_3$. These experiments have shown that the RGD-binding proteins from both placenta and HL60 are recognized by the mAb 7G2, but polyclonal anti-$\alpha_v\beta_3$ does not bind appreciably to the HL60 proteins. Moreover, the antiserum made against the HL60 protein does not bind appreciably to $\alpha_v\beta_3$. These data suggest that there is minimal antigenic homology between $\alpha_v\beta_3$ and LRI and that the crossreactivity of 7G2 is quite fortuitous. Sequential immunoprecipitations with anti-LRI and anti-α_4 and anti-β_7 also demonstrate that LRI does not contain either of these integrin chains. Thus, all data to date suggest that LRI is quite unique; our most recent data suggest that it may not be a β_3 integrin at all, but an antigenically crossreactive receptor. Only better protein purification and cloning will define precisely the nature of LRI.

INTEGRIN ASSOCIATED PROTEIN (IAP)

Discovery of Anti-IAP Antibody B6H12. Even when the nature of LRI and its contribution to signal transduction are completely understood, the problem of how ligand-binding achieves cell activation remains. Perhaps our most important contribution to this problem has come from the discovery of a monoclonal antibody (B6H12) which inhibits LRI-mediated enhancement of phagocytosis, adhesion, and chemotaxis (Brown et al., 1990; Gresham, Adams, et al., 1992; Gresham and Brown, 1989a; Gresham and Brown, 1990a). Initially, we believed that this mAb recognized LRI (Gresham, Goodwin, et al., 1989), but subsequent studies proved that, rather than an integrin, it recognized a 50 kD membrane protein. To understand the nature of this protein and its role in signal transduction from LRI has become a major project of our laboratory. Antibody to the 50 kD protein inhibits several LRI functions, including: a) RGD-dependent enhancement of PMN IgG-mediated

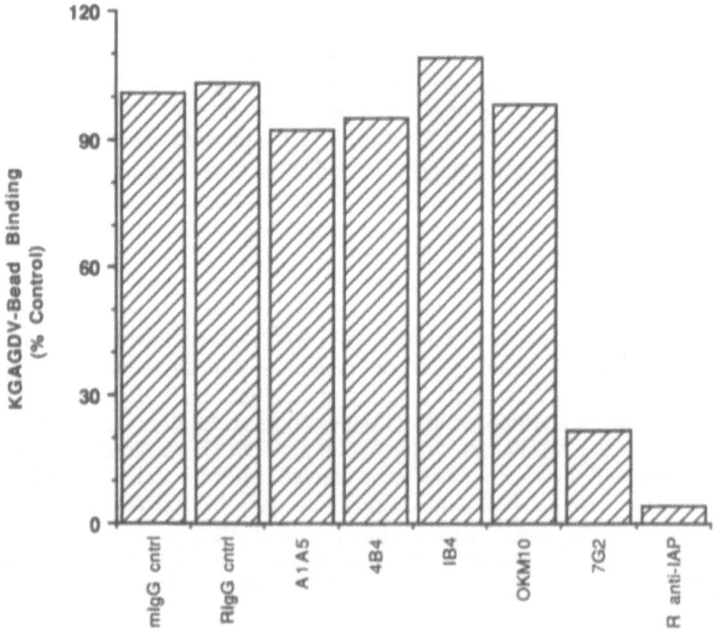

Figure 3. Binding of KGAGDV-coated Microspheres to PMN. PMN were incubated with 10 nM fMLP, and binding of KGAGDV-coated microspheres assessed as described (Gresham et al., 1992). Binding of albumin coated microspheres was <5% of the binding of the peptide-coated microspheres. Binding of microspheres is plotted as % of buffer treated cells, for irrelevant murine (mIgG) and rabbit (rIgG) antibodies, as well as for the anti-β_1 mAb A1A5 and 4B4; the anti-β_2 mAb IB4; the anti-α_M mAb OKM10; the anti-β_3 mAb 7G2, and rabbit polyclonal anti-IAP. Only anti-β_3 and anti-IAP inhibit microsphere binding, demonstrating that this ligand binds to LRI and not appreciably to other leukocyte integrins.

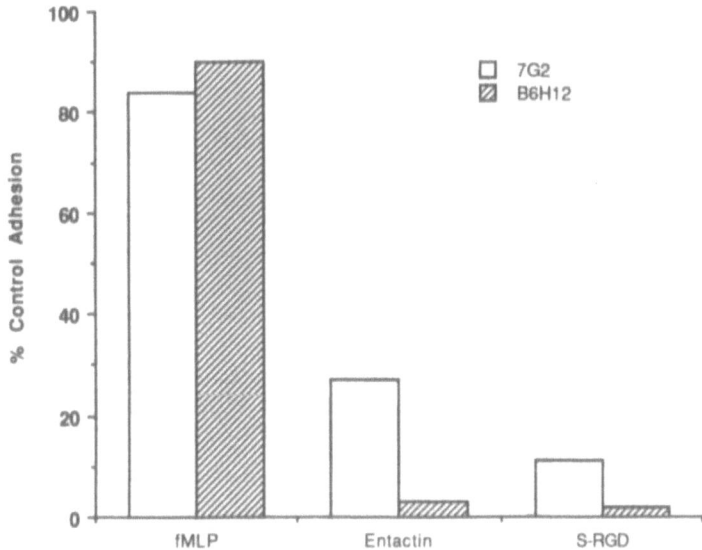

Figure 4. LRI-mediated Adhesion of PMN to Entactin. The binding of unstimulated PMN to surfaces coated with fMLP, entactin, or a synthetic dodecapeptide derived from the entactin sequence around the RGD site was measured in the presence or absence of the mAbs 7G2 (anti-β_3) or B6H12 (anti-IAP). The effect of the mAbs on adhesion is expressed as:

$$\frac{\text{(adherent cells [mAb treated])}}{\text{(adherent cells [control])}} \times 100$$

Both 7G2 and B6H12 inhibit entactin-mediated PMN adhesion, showing that this is an LRI-dependent event.

phagocytosis of opsonized erythrocytes or Group B streptococci; b) RGD-dependent enhancement of complement-mediated phagocytosis; c) PMN adhesion to entactin or RGD peptides; and d) PMN chemotaxis to entactin or RGD peptides. In fact, inhibition of a PMN function by both 7G2, the anti-$\beta3$ mAb which recognizes LRI and by antibody to the 50 kD protein strongly suggests that LRI participates in that phenomenon. In contrast, B6H12 has no effect on PMN activation, chemotaxis, or adhesion mediated by receptors other than LRI. Because of its close functional association with integrins, we have termed this 50 kD protein Integrin Associated Protein (IAP).

IAP is a Ubiquitous Protein. Our studies with IAP have demonstrated several unexpected phenomena. First, IAP is present on every cell that we have tested, including carcinomas, sarcomas, and fibroblasts in tissue culture, as well as on normal hematopoietic cells of all types (Brown et al., 1990). In all cases IAP is expressed on the plasma membrane exclusively (Figure 5). Recent Northern blot studies have shown message in liver, lung, brain, spleen, thymus, and heart, which reinforces the ubiquity of expression of the protein. This pattern of expression makes IAP very unlike LRI, which antibody studies show to be expressed on monocytes, PMN, and lymphocytes, but not on a wide variety of other cells, including EBV-transformed lymphocytes. Thus, whatever IAP's molecular function, it exists on cells independent of LRI. From this fact, we infer that it may have function(s) independent of LRI, perhaps in association with other integrins.

IAP cDNA and Gene. Recently, we have cloned IAP cDNA(s). Initial probes were obtained from the sequence of tryptic peptides from the purified molecule, and a combination of PCR and library screening performed to obtain cDNA encoding the entire coding sequence of IAP. The fact that the correct cDNA was cloned was confirmed by immunoprecipitation of *in vitro* translated products of the cDNA with several anti-IAP mAb

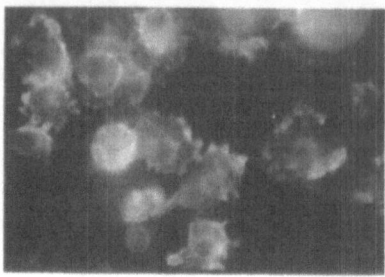 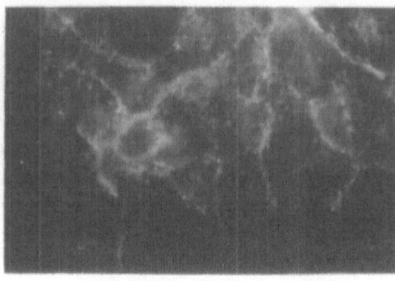

Figure 5. Immunofluorescent Localization of IAP. Human monocyte-derived macrophages (Bohnsack et al., 1985) (*Panel A*) and the A431 carcinoma cell line were stained with an anti-IAP mAb. 1F7 (Brown et al., 1990). Both cell lines show strong immunofluorescence on the plasma membrane and no significant intracellular pools.

and by transfection of the cDNA into CHO cells, with induction of a surface protein recognized by various anti-IAP mAb. An interesting and important feature of the *in vitro* translation was that successful generation of IAP protein absolutely required the presence of microsomes, suggesting that proper folding of IAP, and protection from proteolytic degradation, requires cotranslational interaction with membranes. This is a property of many integral membrane proteins, especially those which span the lipid bilayer several times. By Northern blot, IAP mRNA is 1.7 kb. Our longest clone is 1.4 kb, and is lacking the 5' end of the mRNA. This is because the 5' end of the mRNA contains 90% G+C, and we have been unable to get it sufficiently denatured to be copied by reverse transcriptase. This is often a feature of ubiquitously expressed, "housekeeping" genes (Kozak, 1989). The gene contains at least 5 exons and spans >21 kb of genomic DNA on chromosome 3. There is no evidence by Southern blotting for more than one gene.

IAP Protein Structure. The predicted protein structure for IAP would be expected to span the plasma membrane 6 times (Figure 6). There is no leader sequence, suggesting that both the NH_2 and the COOH termini face the cytoplasm. There is one major loop which is predicted to be extracellular, and it contains 5 of the 6 predicted N glycosylation sites. IAP is a highly glycosylated molecule, with 15-20 kD of apparent carbohydrate on a 32 kD core protein, so it is likely that these N-glycosylation sites are used. The fact that the extracellular face is apparently coated with carbohydrate may explain the difficulty of iodinating IAP in intact cells (Brown et al., 1990). There is no strong sequence homology to known ion channels, transporters, or receptors. The best homologies to proteins with known function are to *arah*, the high affinity transporter for arabinose in *E. coli* (Horazdovsky and Hogg, 1987; Horazdovsky and Hogg, 1989; Scripture et al., 1987), and to the rat brain Na^+ channel, type III (Noda et al., 1986), but these homologies are actually relatively weak (Table 2).

IAP Undergoes a Conformational Change. Recently, we have shown that there is about a 10-fold alteration in affinity of the mAb B6H12 for cells between 0° C and 37° C (Rosales et al., 1992). This also is true for the Fab' fragment of the mAb and for isolated IAP. Together, these data suggest that IAP undergoes a conformational change with temperature. While the physiologic significance of this change is not known, it is intriguing that a signal-transducing molecule is capable of such a conformational alteration. This suggests the hypothesis that the conformational states may represent "on" and "off" positions of the signal transducer. IAP expression does not change quantitatively during PMN activation, nor is there any evidence for constitutive recycling of IAP in resting or stimulated cells.

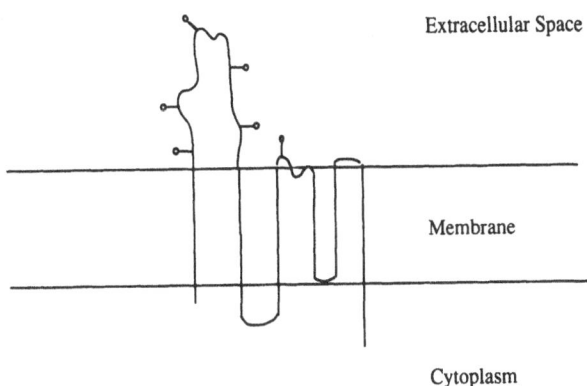

Extracellular Space

Membrane

Cytoplasm

Figure 6. A Model for IAP Structure. The relationship of IAP monomer to the plasma membrane is shown, based on the sequence of IAP cDNA. The structural model predicts 6 membrane spanning domains. There is one large extracellular loop which contains most of the potential N glycosylation sites. The longest stretch of intracytoplasmic sequence is predicted to be 15 amino acids at the COOH terminus of the molecule.

IAP Can Associate with Integrin Receptors. The data summarized above have led us to hypothesize that IAP is a membrane channel or transport system whose function is regulated, at least under some circumstances, by integrins. This hypothesis suggests that when ligand binds the integrin, a putative pore formed by multimeric IAP opens. What about the function of IAP on cells, such as erythrocytes, without integrins? Perhaps IAP can be gated by other receptor molecules as well. Alternatively, IAP on erythrocytes may be nonfunctional, a remnant from a period in the development of the cell when plasma membrane integrins were present (Patel et al., 1985; Patel and Lodish, 1984; Patel and Lodish, 1986).

Our data suggest that, at least under some circumstances, IAP can physically associate with certain integrin molecules (Brown et al., 1990). This has been shown in co-immunoprecipitation experiments, in which antibodies to IAP can co-precipitate integrin-like structures which are recognized by the anti-β3 mAb 7G2 described above. This physical association has led to the hypothesis that induction of IAP function (or its inhibition) is a very early step in signal transduction from ligand-binding by integrins (Figure 7).

At this point we can only speculate about the generality or importance of the association of IAP with LRI function. The fact that IAP is very widely expressed, on a

Table 2. Homologies of IAP to Other Proteins *

Protein	Overlap[§]	% Identity[#]	SD from mean[†]	Ref
AraH	101	26 (64)	39	(24,52)
Na channel	98	19 (65)	38	(35,36)

* Alignments made by the GCG program TFASTA (Devereux et al., 1984).

\# Percentage in parenthesis allows for conservative substitutions.

§ Amino acids in stretch of greatest homology.

† Calculated from an optimal alignment algorithm according to Pearson and Lipman (1988) and compared to the average score ± standard deviation for all GenBank sequences. Obviously, 37-41 S.D. from the mean are highly statistically significant homologies.

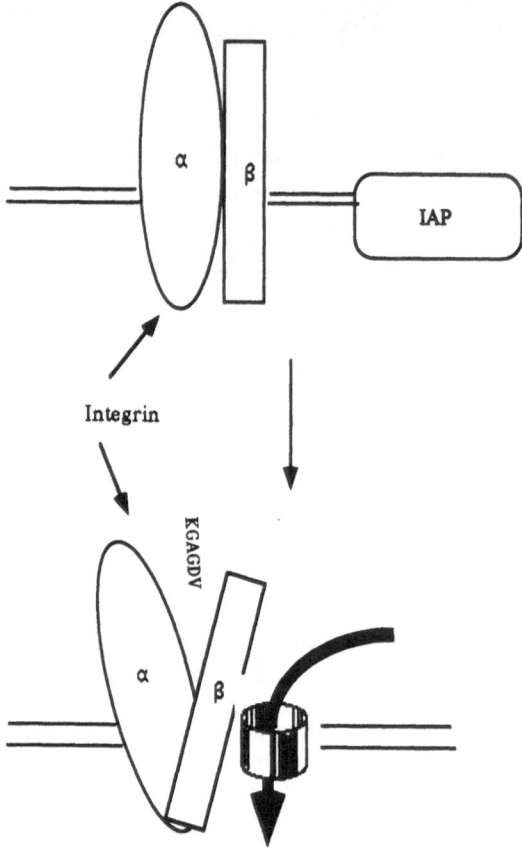

Figure 7. Hypothetical Role of IAP in LRI-mediated Signal Transduction. The data obtained so far suggest the possibility that both IAP-LRI association and IAP function are regulated by ligand binding to LRI. We propose that when ligand binds to LRI, the integrin undergoes a conformational rearrangement, as has been shown for other integrins (Frelinger et al., 1991; O'Toole et al., 1990), which leads to association with IAP. This association, in turn, causes a conformational rearrangement in IAP which activates a channel or a transporter and initiates a signal transduction cascade resulting in leukocyte activation.

variety of different cells, suggests the possibility that it is associated with multiple integrins. Certainly, it is not functionally coupled to all integrins. Data to date suggest a strong correlation with α_v or β_3 functions, as well as LRI. This makes the close relationship of LRI to the β_3 integrin family especially intriguing. Whether there are other channel-like proteins which may subserve a similar function to IAP for other integrins is not clear. There is no evidence at present for close homologues by either Northern or Southern blotting. When we understand the nature of the IAP membrane pore more completely, we will have made a great step in the understanding of signal transduction from ligand interaction with integrins.

REFERENCES

Bohnsack, J.F., 1992, CD11/CD18-independent neutrophil adherence to laminin is mediated by the integrin VLA-6. *Blood* 79:1545-1552.

Bohnsack, J.F., Akiyama, S.K., Damsky, C.H., Knape, W.A., and Zimmerman, G.A., 1990a, Human neutrophil adherence to laminin *in vitro*. Evidence for a distinct neutrophil integrin receptor for laminin. *J. Exp. Med.* 171:1221-1237.

Bohnsack, J.F., Kleinman, H., Takahashi, T., O'Shea, J.J., and Brown, E.J., 1990b, Connective tissue proteins and phagocytic cell function: laminin enhances complement and Fc-mediated phagocytosis by cultured human macrophages. *J. Exp. Med.* 161:912-923.

Bohnsack, J.F., O'Shea, J.J., Takahashi, T., and Brown, E.J., 1985, Fibronectin-enhanced phagocytosis of an alternative pathway activator by human culture-derived macrophages is mediated by the C4b/C3b complement receptor (CR1). *J. Immunol.* 135:2680-2686.

Brown, E.J., 1986, The interaction of connective tissue proteins with phagocytic cells. *J. Leuk. Biol.* 39:579-591.

Brown, E.J. and Goodwin, J.L., 1988, Fibronectin receptors of phagocytes: characterization of the arg-gly-asp binding proteins of human monocytes and polymorphonuclear leukocytes. *J. Exp. Med.* 167:777-793.

Brown, E.J., Hooper, L., Ho, T., and Gresham, H., 1990, Integrin-associated protein: A 50-kD plasma membrane antigen physically and functionally associated with integrins. *J. Cell Biol.* 111:2785-2794.

Brown, P. and Juliano, R.L., 1985, Selective inhibition of fibronectin-mediated cell adhesion by monoclonal antibodies to a cell-surface glycoprotein. *Science* 1448-1450.

Carreno, M.P., Gresham, H.D., and Brown, E.J., 1991, Characterization of a novel integrin involved in regulation of phagocytosis. *FASEB J.* 5:A549 (*abstract*).

Carreno, M.P., Gresham, H.D., and Brown, E.J., 1992, Isolation of a novel integrin involved in regulation of phagocytosis. *FASEB J.* 6:A1410 (*abstract*).

Chakravarti, S., Tam, M.F., and Chung, A.E., 1990, The basement membrane glycoprotein entactin promotes cell attachment and binds calcium ions. *J. Biol. Chem.* 265:10597-10603.

Chung, A.E. and Durkin, M.E., 1990, Entactin: Structure and function. *Am. J. Respir. Cell Mol. Biol.* 3:275-282.

Devereux, J., Haeberli, P., and Smithies, O., 1984, A comprehensive set of sequence analysis programs for the VAX. *Nucleic Acids Res.* 12:387-395.

Durkin, M.E., Chakravarti, S., Bartos, B.B., Liu, S.H., Friedman, R.L., and Chung, A.E., 1988, Amino acid sequence and domain structure of entactin. Homology with epidermal growth factor precursor and low density lipoprotein receptor. *J. Cell Biol.* 107:2749-2756.

Durum, S.K. and Oppenheim, J.J., 1989, Macrophage-derived mediators: Interleukin 1, tumor necrosis factor, interleukin 6, interferon, and related cytokines, in: "Fundamental Immunology," W.E. Paul, ed., Raven Press, New York, pp. 639-662.

Ferguson, T.A., Mizutani, H., and Kupper, T.S., 1991, Two integrin-binding peptides abrogate T cell-mediated immune responses *in vivo*. *Proc. Natl. Acad. Sci. USA* 88:8072-8076.

Frelinger, A.L.,III, Du, X., Plow, E.F., and Ginsberg, M.H., 1991, Monoclonal antibodies to ligand-occupied conformers of integrin $\alpha_{IIb}\beta_3$ (glycoprotein IIb-IIIa) alter receptor affinity, specificity, and function. *J. Biol. Chem.* 266:17106-17111.

Gresham, H.D., Adams, S.P., and Brown, E.J., 1992, Ligand binding specificity of the leukocyte response integrin expressed by human neutrophils. *J. Biol. Chem.* (*in press*).

Gresham, H.D. and Brown, E.J., 1989, Molecular mechanism of fibronectin-stimulated neutrophil phagocytosis. *J. Leuk. Biol.* 46:310 (*abstract*).

Gresham, H.D. and Brown, E.J., 1990, Fibrinogen gamma chain peptide KQAGDV stimulates neutrophil phagocytosis. *Clin. Res.* 38:480a (*abstract*).

Gresham, H.D., Goodwin, J.L., Anderson, D.C., and Brown, E.J., 1989, A novel member of the integrin receptor family mediates arg-gly-asp-stimulated neutrophil phagocytosis. *J. Cell Biol.* 108:1935-1943.

Hemler, M.E., Huang, C., and Schwarz, L., 1987, The VLA protein family: Characterization of five distinct cell surface heterodimers each with a common 130,000 molecular weight beta subunit. *J. Biol. Chem.* 262:3300-3309.

Horazdovsky, B.F. and Hogg, R.W., 1987, High-affinity L-arabinose transport operon. Gene product expression and mRNAs. *J. Mol. Biol.* 197:27-35.

Horazdovsky, B.F. and Hogg, R.W., 1989, Genetic reconstitution of the high-affinity L-arabinose transport system. *J. Bacteriol.* 171:3053-3059.

Hynes, R.O., 1992, Integrins: Versatility, modulation, and signaling in cell adhesion. *Cell* 69:11-25.

Kozak, M., 1989, The scanning model for translation: An update. *J. Cell Biol.* 108:229-241.

Krissansen, G.W., Elliott, M.J., Lucas, C.M., et al., 1990, Identification of a novel integrin β subunit expressed on cultured monocytes (macrophages): Evidence that one α subunit can associate with multiple β subunits. *J. Biol. Chem.* 265:823-830.

Loike, J.D., Sodeik, B., Cao, L., et al., 1991, CD11c/CD18 on neutrophils recognizes a domain at the N terminus of the Aα chain of fibrinogen. *Proc. Natl. Acad. Sci. USA* 88:1044-1048.

Maxfield, S.R., Moulder, K., Koning, F., et al., 1989, Murine T cells express a cell surface receptor for multiple extracellular matrix proteins. *J. Exp. Med.* 169:2173-2190.

Mecham, R.P. Laminin receptors. *Annu. Rev. Cell Biol.* 7:71-91.

Moulder, K., Roberts, K., Shevach, E.M., and Coligan, J.E., 1991, The mouse vitronectin receptor is a T cell activation antigen. *J. Exp. Med.* 173:343-347.

Nathan, C. and Sanchez, E., 1990, Tumor necrosis factor and CD11/CD18 (2) integrins act synergistically to lower cAMP in human neutrophils. *J. Cell Biol.* 111:2171-2181.

Nathan, C., Srimal, S., Farber, C., et al., 1989, Cytokine-induced respiratory burst of human neutrophils: dependence on extracellular matrix proteins and CD11/CD18 integrins. *J. Cell Biol.* 109:1341-1349.

Neff, N.T., Lowrey, C., Decker, C., et al., 1982, A monoclonal antibody detaches embryonic skeletal muscle from extracellular matrices. *J. Cell Biol.* 95:654-666.

Noda, M., Ikeda, T., Suzuki, H., et al., 1986, Expression of functional sodium channels from cloned cDNA. *Nature* 322:826-828.

Numa, S. and Noda, M., 1986, Molecular structure of sodium channels. *Ann. NY Acad. Sci.* 479:338-355.

O'Toole, T.E., Loftus, J.C., Du, X., et al., 1990, Affinity modulation of the $\alpha_{IIb}\beta_3$ integrin (platelet GPIIb-IIIa) is an intrinsic property of the receptor. *Cell Regulation* 1:883-893.

Patel, V.P., Ciechanover, A., Platt, O., and Lodish, H.F., 1985, Mammalian reticulocytes lose adhesion to fibronectin during maturation to erythrocytes. *Proc. Natl. Acad. Sci. USA* 82:440-444.

Patel, V.P. and Lodish, H.F., 1984, Loss of adhesion of murine erythroleukemia cells to fibronectin during erythroid differentiation. *Science* 224:996-998.

Patel, V.P. and Lodish, H.F., 1986, The fibronectin receptor on mammalian erythroid precursor cells: characterization and developmental regulation. *J. Cell Biol.* 102:449-456.

Pearson, W.R. and Lipman, D.J., 1988, Improved tools for biological sequence comparison. *Proc. Natl. Acad. Sci. USA* 85:2444-2448.

Pierschbacher, M.D. and Ruoslahti, E., 1984, Cell attachment activity of fibronectin can be duplicated by small synthetic fragments of the molecule. *Nature* 309:30.

Pommier, C.G., Inada, S., Fries, L.F., Takahashi, T., Frank, M.M., and Brown, E.J., 1983, Plasma fibronectin enhances phagocytosis of opsonized particles by human peripheral blood monocytes. *J. Exp. Med.* 157:1844-1854.

Pommier, C.G., et al., 1984, Differentiation stimuli induce receptors for plasma fibronectin on the human myelomonocytic cell line HL-60. *Blood* 64:858-866.

Pommier, C.G., O'Shea, J.J., et al., 1984, Studies of the fibronectin receptors of human peripheral blood leukocytes: Morphologic and functional characterization. *J. Exp. Med.* 159:137-151.

Posnett, D.N., McGrath, H., and Tam, J.P., 1988, A novel method for producing anti-peptide antibodies: production of site-specific antibodies to the T cell antigen receptor β chain. *J. Biol. Chem.* 263:1719-1725.

Pytela, R., Pierschbacher, M.D., and Ruoslahti, E., 1985, Identification and isolation of a 140kd cell surface glycoprotein with properties expected of a fibronectin receptor. *Cell* 40:191-196.

Rosales, C., Gresham, H.D., and Brown, E.J.., 1992, Expression of the 50kD integrin associated protein on myeloid cells and erythrocytes. *J.Immunol.(in press)*.

Rosen, H., Milon, G., and Gordon, S., 1989, Antibody to the murine type 3 complement receptor inhibits T lymphocyte-dependent recruitment of myelomonocytic cells *in vivo*. *J. Exp. Med.* 169:535-548.

Santoro, S.A. and Lawing, W.J., Jr., 1987, Competition for related but nonidentical binding sites on the glycoprotein IIb-IIIa complex by peptides derived from platelet adhesive proteins. *Cell* 48:867-873.

Savill, J., Dransfield, I., Hogg, N., and Haslett, C., 1990, Vitronectin receptor-mediated phagocytosis of cells undergoing apoptosis. *Nature* 343:170-173.

Scripture, J.B., Voelker, C., Miller, S., et al., 1987, High-affinity L-arabinose transport operon. Nucleotide sequence and analysis of gene products. *J. Mol. Biol.* 197:37-46.

Silverstein, S.C., Greenberg, S., Di Virgilio, F., and Steinberg, T.H., 1989, Phagocytosis, in: "Fundamental Immunology," W.E. Paul, ed.., Raven Press Ltd., New York, pp. 703-720.

Singer, I.I., Scott, S., Kawka, D.W., and Kazazis, D.M., 1989, Adhesomes: Specific granules containing receptors for laminin, C3bi/fibrinogen, fibronectin, and vitronectin in human polymorphonuclear leukocytes and monocytes. *J. Cell Biol.* 109:3169-3182.

Tamkun, J.W., Desimone, D.W., Fonda, D., et al., 1986, Structure of integrin, a glycoprotein involved in the transmembrane linkage between fibronectin and actin. *Cell* 46:271-282.

Unanue, E.R. and Allen, P.M., 1987, The basis for the immunoregulatory role of macrophages and other accessory cells. *Science* 236:551-557.

Vedder, N.B., Winn, R.K., Rice, C.L., Chi, E.Y., Arfors, K.-E., and Harlan, J.M. A monoclonal antibody to the adherence-promoting leukocyte glycoprotein, CD18, reduces organ injury and improves survival from hemorrhagic shock and resuscitation in rabbits. *J. Clin. Invest.* 81:939-944, 1988.

Vedder, N.B., Winn, R.K., Rice, C.L., Chi, E.Y., Arfors, K.-E., and Harlan, J.M., 1990, Inhibition of leukocyte adherence by anti-CD18 monoclonal antibody attenuates reperfusion injury in the rabbit ear. *Proc. Natl. Acad. Sci. USA* 87:2643-2646.

Vercellotti, G.M., McCarthy, J., Furcht, L.T., Jacob, H.S., and Moldow, C.F., 1983, Inflamed fibronectin: An altered fibronectin enhances neutrophil adhesion. *Blood* 62:1063-1069.

Wright, S.D., Craigmyle, L.S., and Silverstein, S.C., 1983, Fibronectin and serum amyloid P-component stimulate C3b-and C3bi-mediated phagocytosis in cultured human monocytes. *J. Exp. Med.* 158:1338-1342.

Wright, S.D., Licht, M.R., Craigmyle, L.S., and Silverstein, S.C., 1984, Communication between receptors for different ligands on a single cell: Ligation of fibronectin receptors induces a reversible alteration in the function of complement receptors on cultured human monocytes. *J. Cell Biol.* 99:336-339.

Wright, S.D. and Meyer, B.C., 1985, Fibronectin receptor of human macrophages recognize the sequence arg-gly-asp-ser. *J. Exp. Med.* 162:762-767.

Wright, S.D., Weitz, J.I., Huang, A.J., Levin, S.M., Silverstein, S.C., and Loike, J.D., 1988, Complement receptor type three (CD11b/CD18) of human polymorphonuclear leukocytes recognizes fibrinogen. *Proc. Natl. Acad. Sci. USA* 85:7734-7738.

DISCUSSION

C. FIGDOR

What would you like to do in the near future to characterize this IAP molecule?

E. BROWN

I might speculate about CA^{+2}, or Na^+, but we have no good evidence. Part of our problem has been the absence of a cell which does not express the IAP. If we use a polyclonal antibody CHO cells stain strongly and so we are in the process of using a variety of genetic approaches to try to make a negative cell. We have done some injection of *xenopus* oocytes but we have not come up with anything yet. Possibly this is a stretch-activated channel. What the cell may sense as integrins binding site is a stretch of membrane. Very little is known about stretch receptors but they are relatively non-specific.

C. FIGDOR

You mentioned that the conditions to immunoprecipitate the 50kD protein are critical. Could you comment on that?

E. BROWN

The conditions under which we see co-immunoprecipitation use very low concentrations of Triton. It turns out that platelets out of all the cells that we have looked at so far have the most of this 50 kD protein which can be solubilized by low concentrations of triton. Platelets also contain ligands for β_3 integrins, so you may be getting ligand binding at the same time as you are doing your immunoprecipitation. We have a little data from other cells indicating that when we put in the KGAGDV peptide we get better co-immunoprecipitation under those same sorts of conditions. Again, the limiting thing is how much IAP can actually be solubilized under these conditions.

D. LIVINGSTON

Would you expect that IAP must exist in an intact state for phagocytosis to take place?

E. BROWN

For this kind of stimulated phagocytosis.

D. LIVINGSTON

Can one make trans-dominant negative mutants of IAP in order to study its specific biochemical functions, *in vivo*?

E. BROWN

I think that is an experiment worth trying.

E. MIHICH

Are these pores required also for macrophage killer function, or only for phagocytosis? The other question refers to the first part of your presentation where you mentioned that activation of PMN is required for continuation of adhesion. Some time ago we found that indomethacin perhaps by inhibiting part of the prostaglandins metabolism, increases macrophage killer function; do you know whether indomethacin has an effect on the function you study?

E. BROWN

Let me answer the first question first. Remember that this pore is very specifically related to one particular integrin which we have called LRI and is present on macrophages. However, we do not know how to engage it during macrophage killer function. You would have to specifically engage it to expect this antibody to have an effect. In terms of the phospholipase A2 the rest of the data which I have not told you about are that we can use a variety of more or less specific inhibitors of activation of phospholipase A2 and inhibit not only phagocytosis but also macrophage adhesion to a variety of different substrates. Neither indomethacin nor blockers of lipoxygenases inhibit in a way that inhibitors of phospholipase A2 do. If you inhibit the phospholipase A2 and add arachidonic acid even in the presence of these inhibitors, you restore function. So our interpretation of those data is that this is not a function of one of the typical metabolic pathways of arachidonic acid and instead is either arachidonic acid itself or some other less well characterized metabolic product of the arachadonic acid. Arachidonic acid itself has been shown to be important in a variety of different kinds of membrane fusion events working through proteins called synexins. We have some very preliminary electron micrographs which show that if you block the phospholipase A2 which we believe is activated during phagocytosis and adhesion, by electron microscope you see the particle attached to the plasma membrane and just beneath the particle are lots of vesicles as if the cell were unable to insert membrane into that site.

G. LENAZ

You called this protein a pore but the evidence you have is the motif of a transmembrane helios. Have you looked a little more deeply at the structure of this protein?

E. BROWN

There are charged amino acids in the predicted transmembrane domains which could fit that model. I am not enough of a molecular modeler to say with any confidence that that would actually be that way. But yes, there are there potentials for hydrophilic interactions.

G. LENAZ

Is there any resemblance with known calcium channels? I ask this because calcium is needed for phospholipase A2.

E. BROWN

One of the reasons I showed the calcium data was that we do not think inhibition occurs at that step. Calcium is entering the cell even in the presence of inhibition. For IAP, although there is a general structural motif similarity, when you just compare amino acid sequences the homologies to known channels is pretty minimal, so that would suggest that this is a new sub-family of that motif.

G. TARONE

You showed very nicely that your antibody is interfering with phagocytosis and you also showed that this 50 K protein is ubiquitous. Did you see any effect of your antibodies on other integrin mediated phenomena which are not related to phagocytosis, such as cell adhesion?

E. BROWN

This antibody inhibits the ability of neutrophils to bind to entactin-coated surfaces just as well as it inhibits the ability of anything to stimulate phagocytosis. So in that sense, adhesion to surfaces coated with the ligand and stimulation of phagocytosis are identical.

M. HEMLER

Does it also inhibit adhesion of fibroblasts and other cell types?

E. BROWN

We have looked much less carefully at any other cell types. Preliminary data from other groups indicate that there are some other cell types which are inhibited by anti-IAP, particularly looking at substrates which you would expect to bind to α_V- or β_3-containing receptors.

M. HEMLER

During the first part of your talk you mentioned that your β_3 integrin on neutrophils must be something different from the $\alpha_V\beta_3$ or $\alpha_{IIb}\beta_3$ integrins, so are there any unique properties of this integrin in neutrophils and which are specifically interfered by this IAP protein?

E. BROWN

I am not sure that this is more specially interfered with than other $\alpha_V\beta_3$ functions, it is just that this is what we have studied. We cannot get antibodies to V to recognize neutrophils under any circumstance; we cannot get any immunoprecipitation with antibody to α_V and we have used a variety of different monoclonals which recognize α_V. We have used a few antibodies against $_{IIb}$ but not so many, mainly because nobody thinks that $_{IIb}$ is really expressed in these cells. So we have looked hard for α_V thinking there must be $\alpha_V\beta_3$ and we cannot find it.

M. HEMLER

Maybe you said this and I missed it. Do you know the actual affinity of the peptides for neutrophils?

E. BROWN

We cannot measure an affinity for the monovalent peptides. We can see specific binding of multivalent KGAGDV.

M. HEMLER

Does the V6H12 antibody inhibit that monovalent peptide interaction with the integrin?

E. BROWN

The multivalent peptide interaction -- if you do a conventional binding assay with monovalent peptides, the non-specific binding is sufficiently high that we cannot see specific binding because, at least in our hands, we cannot measure micromolar affinities in that way. We are in the process of resorting to other kinds of assays, trying to cross-link the peptide to the receptor and look at inhibition of that.

M. HEMLER

I usually think of cell adhesion, and especially signaling involving phagocytosis, as post-ligand binding events but what you are trying to say is that this other structure which is not an α or a β subunit is actually intimately involved with ligand binding to the integrin.

E. BROWN

My hypothesis would be that that has to do with receptor clustering, if we can ever do it we will find that binding of a monovalent peptide is not inhibited by the antibody.

M. HEMLER

So the final question is if you put your model together and you have the neutrophil stretched out on fibrinogen or some surface that requires this particular IAP structure, if you then consider what happens on laminin would you argue that the β_1 integrin would have to have comparable structure associated with it?

E. BROWN

It would have to have a comparable function associated with it. That would be the argument.

G. MARGUERIE

You mentioned that this molecule is expressed in all the cells that you tested, so it is expressed at the surface of platelets. Is it involved with platelet activation?

E. BROWN

I am embarrassed to say that I do not know the answer to that question.

G. MARGUERIE

The second thing is that apparently it is not expressed in CHO cells.

E. BROWN

It is expressed in CHO cells, but our monoclonal antibodies do not cross species. If you take a polyclonal antibody the CHO cells are well stained by the polyclonal antibody.

G. MARGUERIE

I see, because you take CHO cells and you transfect with β_3 cDNA and α subunit cDNA, you get a receptor at the surface of the cell which is not functional. So, I was thinking about an experiment to see the effects of transfecting with your molecule.

E. BROWN

I have no idea what the species specificity of interaction of this protein with $\alpha_V\beta_3$ is and we actually are doing that experiment that you have implied now.

R. JULIANO

I noticed that some of your initial curves of stimulation of phagocytosis were biphasic. Any explanation for this?

E. BROWN

I think that it is because RGD-containing proteins and peptides are interacting with several receptors on the surface of the cell which have different effects on phagocytosis. If you take the KGAGDV peptide which is the most specific one we have, you do not see the high-dose inhibition.

S. SHAW

I am very impressed by the wide tissue distribution of the molecule. I believe you mentioned that it is on erythrocytes. How many molecules are there estimated to be per erythrocyte, and how does that compare with other cell types?

E. BROWN

On erythrocytes there are about 12,000 to 20,000 copies. Compared to known channels on erythrocytes that is about 1/10 to 1/50 of the number that you find for band 3 or a water channel. On U937 cells and HL60 cells, there are about 200,000 copies per cell. On a normal monocyte or normal neutrophil there are about 20,000 to 50,000 copies.

S. SHAW

Has its identity with other molecules been tested, in particularly the CD molecules classified by the International Workshops of Leukocyte Differentiation Antigens?

E. BROWN

No, the other point about numbers that I am giving you is, that it is defined by FAB binding. This really exists as a tetromer in the membrane. The number of channels is even smaller.

S. SHAW

Of the classified CD molecules, a limited number have broad distributions generally consistent with your molecule -- for example, CD46, CD47, CD52, CD53. Have you compared these in detail with yours?

E. BROWN

I have looked through the CD lists for things that have similar properties to ours and looked up the papers when there was a possible match. There are a number of properties of this which I have not gone into, which distinguishes it from the things that I have found so far. One of the questions that we had is, since this is a heavily glycosylated molecule, might it be a blood group antigen, which again, we are in the process of trying to figure out.

S. SHAW

The implication would be that this molecule is serving a similar kind of function on those other cells on which it is expressed. It might be triggering cell responses to engagement of other cell surface receptors.

E. BROWN

Potentially yes. It is also possible that on red cells this is a remnant of differentiation from an earlier stage. The erythroblast has integrins.

J. McCARTHY

Just because there are multiple functions, one should see whether a specific protein is involved.

E. BROWN

Yes, there is a G-protein involved in the signal transduction for enhancing phagocytosis. We do not think it is directly associated with this protein because if you look at the serpentine receptors which are directly associated with G proteins, they have actually a big intracytoplasmic loop which is completely missing here.

P. MARCHISIO

Did you measure the intracellular pH?

E. BROWN

We have not done it yet.

V. QUARANTA

Have you seen bands of proteins other than integrins in the immunoprecipitate?

E. BROWN

I do not think we have seen anything that is specific that we can co-immunoprecipitate.

V. QUARANTA

In going back to what Shaw was saying, the widespread distribution suggests that perhaps there is a family of such proteins.

E. BROWN

Absolutely. I think that there must be something that we are missing because I think that just by analogy to other kinds of membrane pores there has to be some regulatory component to this channel that we have not found yet.

V. QUARANTA

How about structural variability? All your antibodies react with one particular loop of the molecule -- do you have any evidence for structural heterogeneity, perhaps cell-type specific, in other regions of the molecule.

E. BROWN

Actually we have looked very hard for that. We have purified the protein from 4 or 5 different cells and done tryptic peptide maps which are all very similar. The evidence at the moment is that there is a single gene.

R. BANKERT

Is it possible that your molecule is related to the heat stable antigen J11D?

E. BROWN

I have not looked directly at its relationship.

G. DOUGHERTY

Dr. Robert Kay in the Terry Fox Laboratory has cloned both the human and mouse heat stable antigen. It is a small GPI linked protein quite different from the molecule Dr. Brown has described.

S. HASKILL

I am curious about this association with different integrins. We are trying to understand why engagement of β_1 integrins fails to induce inflammatory genes in monocytes whereas engagement of β_1 does. It would be very interesting to know if there is a difference in the association of β_1 or β_2 with this protein.

E. BROWN

For this protein at the moment we have no evidence for association of β_1. The only evidence we have for association with β_2 is the following, and I think these experiments are very complicated to interpret. If you take a fibrinogen-coated surface, neutrophils bind to fibrinogen-coated surfaces, anti-β_2 antibodies block that. That has been interpreted to mean that fibrinogen is a ligand for $\alpha_M\beta_2$ and there are actually a little bit more data for that hypothesis. The antibodies to this 50 kD protein also inhibit neutrophil binding to fibrinogen. However, there may be some functional link between activation of this receptor and β_2 integrins, so I do not know that the inhibition by BGH12 is direct.

T. CAREY

That was a beautiful presentation. I have several questions. The first one relates to the picture you showed of A431 cells. Do they have like an order magnitude more of this receptor or are they in the same range as phagocytes?

E. BROWN

I do not know. Tissue culture cells tend to have more than normal cells to the extent that we have looked.

T. CAREY

My second question is one that may be obvious to everyone else; why did you pick this particular peptide or this set of peptides?

E. BROWN

We were struggling with the issue of what β_3 integrin could this be. One of the things that distinguishes α_{IIb} from $\alpha_V\beta_3$ is the ability to recognize this KQAGDV peptide. I know that it is now in the literature that $\alpha_V\beta_3$ recognizes KQAGDV but it is much less avid than $\alpha_2\beta_3$, so we wanted to test that peptide. The KGAGDV was originally included as a control; it was supposed to be a negative control and it was not negative and we were very upset. We resynthesized it and it did the same thing so we sequenced both of the peptides that we synthesized and they really were KGAGDV, so we began to believe our data.

T. CAREY

The final question has to do with your comment about receptor clustering which apparently came from inhibition of the β_3 binding. Is that there evidence that it is clustering?

E. BROWN

This is a hypothesis. I have no evidence that it really involves a receptor clustering.

S. GOODMAN

I am a bit bothered by the relationdship between phagocytosis and this molecule. So first to extrapolate Martin's question -- there is β_1 integrin-mediated phagocytosis on the macrophages. β_1-mediated phagocytosis does not involve this molecule, as far as you know. So, in your opinion, does that disassociate phagocytosis from adhesion? And the second question is, is there any relationship between the expression of this molecule in other cells and phagocytosis? Neutrophils are not quite as phagocytic as macrophages, so is there any relationship between phagocytosis and this molecule?

E. BROWN

Let me try to clearly distinguish what we are testing from phagocytosis. Phagocytosis as a process is not dependent on this molecule, it is dependent on β_2 integrin and that was the first bit that I talked about. It is the ability of clotting or matrix proteins to stimulate phagocytosis that is inhibited by BGH12. If you imagine yourself to be a monocyte or a neutrophil calmly wandering around the blood and suddenly somehow because of inflammatory peptides you sense that there is something going on. How do you tell when you are there? One way of telling that you are there might be that you can interact with extracellular matrix proteins. It is the ability of those proteins to stimulate this function which is dependent on IAP. Perhaps I should say just forget about phagocytosis; the ability of these peptides to activate neutrophils is dependent on this protein and so it is not directly related to phagocytosis. If you do the same experiment in monocytes or macrophages, and ask whether these peptides can stimulate phagocytosis in monocytes and macrophages and is that dependent on the 50 kD protein, the answer is yes.

D. LIVINGSTON

In that regard then if you were to extract membranes from red cells, monocytes and polys and do a blot with your antibody, how many proteins would you see?

E. BROWN

One.

D. LIVINGSTON

Always one? There are no cross-reacting bands? It is not that there is a family of unrelated proteins carrying a common epitope?

E. BROWN

The best evidence for that is that we have purified from platelets, placenta, monocytes and red cells the protein recognized by the antibody. Now, first of all, all of the antibodies recognize all of those proteins. Second of all, with tryptic digests you see essentially the same pattern, I mean I an not saying they are quantitatively identical but they are essentially the same pattern so that the best protein data that we have is that they are the same molecule. The genetic data are that we can only find one gene even at moderate stringency.

G. MARGUERIE

I know that this peptide and this KQAGDV is quite new. You do not find the sequence anywhere else and from I what I know about the many peptides that you test the values that you show in your table are quite good; would it be your guess that this is a specific receptor for fibrinogen?

E. BROWN

That has been our hypothesis on occasion, that this is essentially a fibrinogen receptor. The trouble with saying that in public is there are so many molecules on neutrophils which have been alleged to recognize fibrinogen that you just get into a huge fight when you say that. So we have preferred to shy away and just say it recognizes fibrinogen. Now if you do a KGAGDV search through the data base you do not find anything, there are no proteins known which contain that amino acid sequence so fibrinogen is an excellent candidate for an LRI ligand. Now the other thing which actually binds the receptor relatively well is entactin. One of the interesting things about entactin binding to this receptor is that, as opposed to almost everything else, a neutrophil binds to where activation of the cell is required, to see decent adhesion, neutrophils will bind to entactin via this receptor without prior activation. So our hypothesis, and whether its fibrinogen or entactin, it does not really make much difference, is that this is a very early step in the activation of the neutrophil as it emigrates into a site of inflammation. Binding to entactin via this receptor will actually activate the β_2 integrins so again we are putting it into a cascade of events which are involved in migration of neutrophils.

COORDINATE ROLE FOR PROTEOGLYCANS AND INTEGRINS IN CELL ADHESION

James B. McCarthy*[1,2], Amy P.N. Skubitz[1,2], Leo T. Furcht[1,2], Elizabeth A. Wayner[1,2], and Joji Iida[1]

University of Minnesota
*[1]Department of Laboratory Medicine and Pathology
[2]Biomedical Engineering Center
Box 609 UMHC
420 Delaware St., S.E.
Minneapolis, Minnesota, 55455 U.S.A.

ABBREVIATIONS USED

CS = Chondroitin Sulfate; **ECM** = Extracellular Matrix; **FN** = Fibronectin; **FN-C/H-I**, **FN-C/H-II, FN-C/H-III** = Fibronectin-derived Cell Adhesion/Heparin Binding Peptides I, II, III; **GAG** = Glycosaminoglycan; **MAb** = Monoclonal Antibody; **OVA** = Ovalbumin; α**DX** = p-nitrophenyl-α-D-xylopyranoside; β**DX** = p-nitrophenyl-β-D-xylopyranoside

INTRODUCTION

Cellular recognition of the extracellular matrix (ECM) is fundamentally important for development, the maintenance of normal adult tissue architecture and function, and for the pathogenesis of disease. The ECM exerts its effects in large part, by serving as an exoskeleton for cells within tissues, thereby providing structural, and other, information to cells that serve to regulate normal cell function. Understanding subtle changes in the cellular recognition of the ECM that occur as a result of malignant cell transformation can provide important information regarding the altered biology of tumor cells and could lead to novel therapeutic and/or diagnostic strategies for the treatment of tumor invasion and metastasis, two features of many malignancies that contribute to the mortality of cancer patients.

Regardless of their tissue of origin, ECM contain three basic types of macromolecules (collagens, proteoglycans (PG), and accessory glycoproteins) that interact with each other and assemble into a repeating array of biologically active structures that can be recognized by cells (reviewed in Faassen, 1992a). Due to this structural complexity, it has been necessary to use a reductionist approach for identifying the active sites within the ECM that can serve as cellular recognition sites. Early on, this approach included isolating individual ECM components from their more complex structures *in vivo*, followed by a characterization of the structure and function of these components *in vitro*. These studies were followed with proteolytic fragmentation of the isolated components, and further evaluation of the biological properties of the resulting fragments. Most recently, protein components of the ECM have been further dissected by using a synthetic peptide approach (Humphries, 1990; Faassen, 1992a). Collectively, the results demonstrate that cellular recognition of ECM components has a complex molecular basis, involving multiple sites within the ECM that can interact with distinct cell surface receptors.

Rapid advances in the area of cell surface receptors for the ECM have demonstrated that several different families of macromolecules serve to transmit information encoded within the structure of the ECM to the interior of cells (reviewed in Hynes, 1987, 1992; Ruoslahti, 1991; Albelda and Buck, 1990; Humphries, 1990; Springer, 1990; Faassen, 1992a). In particular, integrins and cell surface PG have emerged as two major types of receptors for the ECM. Integrins represent heterodimeric, transmembrane cell surface glycoproteins that interact with peptide determinants within ECM components, or with counterreceptors on the surface of adjacent cells. By contrast, cell surface PG represent a structurally diverse group of proteins to which highly anionic glycosaminoglycans (GAG) are covalently attached by an O-linkage to serine residues on the core proteins. As with integrins, cell surface PG can be expressed as transmembrane proteins that can function in the recognition of ECM-related ligands or potentially as counterreceptors for cell adhesion molecules on other cells (reviewed in Höök, 1984; Couchman and Höök, 1988; Gallagher, 1989; Faassen, 1992a), and earlier studies have demonstrated that a large melanoma cell chondroitin sulfate proteoglycan (CSPG) can be localized to microspikes on adherent cells (Garrigues, 1986), consistent with a role for cell surface CSPG in mediating the initial recognition of the ECM. Cell surface PG have also been identified that are covalently linked to plasma membrane lipids such as phosphatidylinositol, or bound noncovalently to cell surface receptors for PG, suggesting that these complex macromolecules may influence cell adhesion in very diverse ways. The anionic properties of GAG on cell surface PG are important for the binding of PG to their respective ligands. Understanding the molecular mechanisms by which these two types of receptors, as well as others mediate their effects can lead to significant new information regarding the mechanism(s) by which the ECM can modulate the behavior of normal and transformed cells.

Fibronectin (FN) has long served as a prototype ECM glycoprotein for understanding the molecular basis for cellular recognition of the ECM. Pioneering studies on FN identified a four amino sequence, arginyl-glycyl-aspartyl-serine (RGDS), that duplicates some of the cell adhesion promoting activities of the protein. Additional studies identified α5β1 integrin as the cell surface receptor for this sequence within the context of FN (reviewed in Hynes, 1987, 1992; Ruoslahti, 1991; Albelda and Buck, 1990; Humphries, 1990). However, other studies have clearly demonstrated that ligands other than RGDS within FN can support cell adhesion in an RGDS-independent fashion (McCarthy, 1986, 1988a, 1990; Rogers, 1987; Haugen, 1990; Wait, 1987; Visser, 1989; Liao, 1989; Wayner, 1989). One of these sites, defined by the synthetic peptide termed CS1 (Humphries, 1987), supports the adhesion of several cell types by an α4β1 integrin-dependent mechanism (Wayner, 1989; Mould, 1990; Guan, 1990). Additionally, we have identified several novel cell adhesion promoting synthetic peptides from the carboxyl terminal heparin binding domain of FN that can also support cell adhesion (McCarthy, 1986, 1988a, 1990; Haugen, 1990; Drake, 1992; Iida, 1992; Mooradian, 1992). These peptides, termed FN C/H I, II and III, differ structurally from either RGD or CS1, and were chosen on the basis of their potential to interact with cell surface GAG or PG, since these peptides represent cationic and relatively hydrophilic sequences.

In the current studies, we have further examined the molecular basis for the adhesion of highly metastatic human melanoma cells, termed A375SM cells (Fidler, 1986), to a 33 kD proteolytic fragment of FN A-chains that contains the carboxyl terminal heparin binding domain of FN a well as portion of the type IIIcs insert that includes the α4β1 integrin-binding sequence CS1 (McCarthy, 1986, 1988a; Iida, 1992). In addition to promoting the RGD-independent cell adhesion of many cell types, this fragment also inhibits the experimental metastasis of different tumor types in a tail vein lung colonization model (McCarthy, 1988b), implicating receptors for this fragment in the arrest, adhesion, and/or extravasation of tumor cells. Partial characterization studies of the cell surface PG on A375SM cells indicate that they express the overwhelming majority of their PG as CSPG. By employing specific monoclonal antibodies that inhibit α4β1 integrin function (Wayner, 1989), or by using strategies to interfere with CSPG expression and/or function (Faassen, 1992b), we demonstrate that the adhesion of these melanoma cells to this FN fragment involves contributions from both α4β1 integrin and cell surface CSPG.

Additionally, we have evaluated the ability of the various synthetic peptides from this fragment to support cell adhesion in the presence of these various receptor antagonists. One of these synthetic peptides, termed FN C/H III, is located near the amino terminal end of the

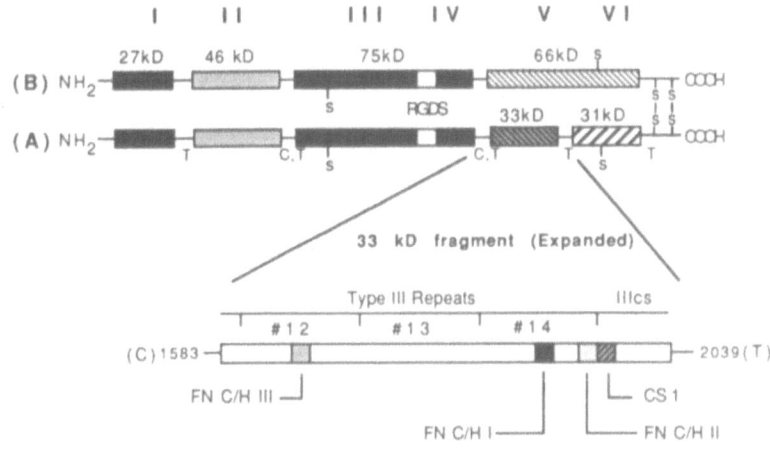

Name	Primary Sequence	Hydropathy Index	Net Charge
FN C/H I	YEKPGSPPREVVPRPRPGV (Residues #1906-1924)	-24.3	+2
FN C/H II	KNNQKSEPLIGRKKT (Residues #1946-1960)	-29.3	+4
FN C/H III	YRVRVTPKEKTGPMKE (Residues #1721-1736)	-23.7	+3
CS 1	DELPQLVTLPHPNLHGPEILDVPST (Residues #1961-1985, A chain only)	-9.9	- 4

Figure 1. Location and Primary Sequences of Cell Adhesion Promoting Synthetic Peptides within Intact FN. Shown is a schematic diagram depicting the location of the 33 kD carboxy terminal heparin binding fragment within FN and the four synthetic cell adhesion promoting peptides (FN C/H I, FN C/H II, FN C/H III, and CS1). FN C/H I, FN C/H II, and FN C/H III all bind heparin and are present within the 90 residue type III homologies common to all isoforms of FN. Peptide CS1, that does not bind heparin but does bind $\alpha4\beta1$ integrin, is restricted to isoforms of human plasma FN which contain the type IIIcs region (A-chains). The amino terminal end and the carboxy terminal limit of the 33 kD fragment is based on previous protein sequence data. Approximate locations of tryptic (**T**) and cathepsin D (**C**) sites on intact FN also are shown. Selected biological domains are indicated by the Roman numerals at the top of the figure. **I** = weak heparin-binding, **II** = collagen-binding (noncovalent), **III** = free sulfhydryl and known to contain site(s) which enhance cell adhesion to RGD, **IV** = RGD-mediated cell adhesion, **V** = carboxy terminal strong heparin-binding and cell adhesion, and **VI** = free sulfhydryl, carboxy-terminus of FN.

Peptides from FN were synthesized at the Microchemical Facility of the University of Minnesota using a Beckman System 990 peptide synthesizer. The procedures used were based on the Merrifield solid phase system as described previously (Stewart and Young, 1984). Lyophilized crude peptides were purified by preparative reverse-phase high performance liquid chromatography (HPLC) on a C-18 column, using an elution gradient of 0-60% acetonitrile with 0.1% trifluoroacetic acid in water. The purity and composition of the peptides were verified by HPLC analysis of hydrolysates prepared by treating the peptides under nitrogen in 6 N HCl overnight at 110° C (McCarthy et al., 1990; Haugen et al., 1990; Drake, 1992). The peptide sequences use the single letter amino acid code (K = lysine, R = arginine, H = histidine, E = glutamic acid, D = aspartic acid, Q = glutamine, N = asparagine, P = proline, G = glycine, S = serine, T = threonine, V = valine, I = isoleucine, L = eucine, Y = tyrosine). All peptides were synthesized with a carboxyl-terminal tyrosine to facilitate iodination when necessary. The net hydropathy indices for each synthetic peptide are calculated based on the values of Kyte and Doolittle (1982). According to this method, the more negative values reflect increasingly hydrophilic character. Net charge is calculated assumming a (+1) charge for lysine and arginine, and a (-1) charge for aspartic acid and glutamic acid at pH 7.0. Histidine is assumed as uncharged at this pH.

fragment. FN C/H III supports melanoma cell adhesion in a CSPG-dependent and α4β1 integrin-independent fashion. In contrast, FN C/H I, FN C/H II, and CS1, which are all clustered in close proximity toward the carboxyl terminal end of the fragment, appear to have a more complex molecular basis for recognition by the A375SM cells, in the sense that adhesion of the cells to these three synthetic peptides involves the action of both CSPG and α4 integrin subunits. The results are discussed in terms of a working model that depicts these various cell adhesion promoting sites within this fragment acting together to bind a melanoma cell surface receptor complex, consisting of CSPG and α4β1 integrin, thereby influencing early cellular recognition events that lead to A375SM human melanoma cell adhesion on the fragment.

RESULTS AND DISCUSSION

The 33 kD Fragment of FN Contains Several Cell Adhesion Promoting Sites

The location of the 33 kD fragment of FN, as well as the sequences and selected properties of the synthetic peptides used these studies, are shown in Figure 1. Cell adhesion to the region of FN encompassed by this fragment has been shown to involve α4β1 integrin (Wayner, 1989; Mould, 1990; Guan, 1990) or cell surface PG (Saunders, 1988; Drake, 1992). Peptide CS1, which is present only in certain isoforms of FN (Kornblihht, 1985), supports cell adhesion by an α4β1 integrin dependent mechanism (Wayner, 1989; Mould, 1990; Guan, 1990). This synthetic peptide does not bind heparin (McCarthy, 1990), is relatively hydrophobic, and has a net negative charge. In contrast to CS1, the three other synthetic peptides FN C/H I, FN C/H II and FN C/H III, are expressed in all isoforms of FN (Kornblihtt, 1985). These latter three peptides are characterized by a net positive charge, relatively hydrophilic hydropathy indices, and the ability to bind the GAG-heparin and to support cell adhesion.

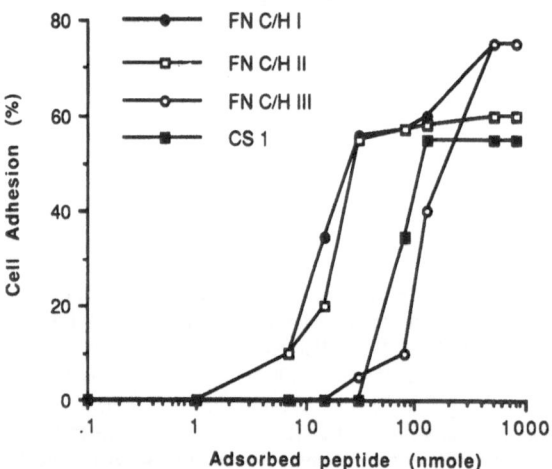

Figure 2. Comparison of the Cell Adhesion Promoting Activities of the Synthetic Peptides from the 33 kD Fragment. Microtiter wells were coated with various concentrations of peptides FN-C/H-I (*solid circles*), FN-C/H-II (*open squares*), FN-C/H-III (*open circles*) or CS1 (*solid squares*) that had been coupled to OVA using EDC. In accompanying experiments using radioactive tracers, the stoichiometry of the coupling reaction was determined and used to calculate the actual amount of the peptides bound to the wells as shown in the figure. Following adsorption of the peptide/OVA conjugates, the wells were then blocked with PBS/BSA. OVA/OVA, coated at the same concentrations used for peptide/OVA conjugates, did not promote adhesion to levels greater than 8% of input cells. Cell adhesion assays were carried out for 30 min. The data represent the means of triplicate determinations. SEM were less than 10% of mean values in each case.

Table 1

Anti-Integrin Monoclonal Antibody (Subunit Specificity)	Inhibition of Cell Adhesion to 33 kD fragment?
P4C2, P4G9 (α4 subunit)	Yes (P4G9 partial; 50-60%)
P1D6 (α5 subunit)	No
P4C10 (β1 subunit)	Yes
Normal mouse ascites	No

A comparison of the relative cell adhesion promoting activities of these four synthetic peptides is shown in Figure 2. Prior to coating onto the microtiters wells, the peptides were covalently coupled to ovalbumin (OVA) using 1-ethyl-3(3-dimethylaminopropyl)-carbodimide hydrochloride (EDC) as described (McCarthy, 1990; Iida, 1992) in order to facilitate the examination of their biological activities. By the use of trace amounts of radioactively labeled synthetic peptides in the coupling reaction, the stoichiometry of coupling was determined (approximately 4-5 peptides/OVA) and used to quantitate the actual amount of synthetic peptide bound to the surface of the wells. As a control for this approach, OVA was covalently coupled to itself and used in the various experiments. While all four synthetic peptides support cell adhesion to equal extents at high coating concentrations, they differ in their relative activities at lower coating concentrations. Specifically, FN C/H I and FN C/H II are approximately an order of magnitude more effective at promoting half maximal cell adhesion compared to FN C/H III and CS1 (Figure 2), suggesting that the peptides differ in their relative affinity for melanoma cell surfaces.

Importantly, these synthetic peptides are much less effective on a molar basis than the fragment, which promotes half maximal cell adhesion at coating levels of approximately 4.2 pmoles per well (not shown). Additionally, while the A375SM cells exhibit a fully spread morphology on the 33 kD fragment, none of the four synthetic peptides supported extensive spreading of the cells (not shown). Finally, when these synthetic peptides, or specific polyclonal antibodies generated against these synthetic peptides, are used to compete for cell adhesion to the fragment, only partial inhibition (40-60%) of adhesion is observed, consistent with previous results using other cell types to adhere to this fragment (McCarthy, 1990; Haugen, 1990). Collectively, the results suggest that these cells recognize and adhere to the 33 kD fragment by interacting with multiple sites on the fragment, and that each one of these sites provides only partial information for cellular recognition.

Adhesion of A375SM Cells to the 33 kD Fragment of FN Involves both α4β1 Integrin and CSPG

In order to partially characterize the nature of the PG expressed by these melanoma cells, subconfluent cultures of the cells were metabolically labeled overnight in the presence of ^{35}S-sulfate, and the cell layer was detergent-extracted in the presence of a cocktail of protease inhibitors as described (Faassen, 1992b; Drake, 1992; Iida, 1992). PG in the detergent extract were then purified by DEAE anion-exchange chromatography, and eluted from these columns as a single peak of radioactivity at approximately 0.4 M NaCl (not shown). Size exclusion analysis of the DEAE purified ^{35}S-labeled PG in this extract demonstrated a large hydrodynamic size of intact PG (eluting in the Vo of a dissociative Sepharose CL-4B column), and relatively small GAG chains (eluting at a K_{av} of approximately 0.6 of a Sepharose CL-6B column). Further analysis (not shown) of the ^{35}S GAG demonstrated that they were resistant to degradation by nitrous acid, and 90-95% sensitive to digestion with chondroitinase ABC, consistent with their identity as CS. Therefore, as with many other human melanoma cells characterized to date (Bumol and Reisfeld, 1982; Ross, 1983; reviewed in Gallagher, 1989), these cells express the overwhelming majority of their PG as CS, with an estimated GAG size of 10-12 kD based on the standards established by Wasteson (1971).

In order to initially verify a role for α4β1 integrin in mediating cell adhesion to the 33 kD fragment, specific monoclonal antibodies that neutralize the activity of α4, α5, or β1 integrin subunits were employed in cell adhesion assays. The assays used in these studies are relatively short term (20-30 min) and use coating concentrations of the cell adhesion promoting ligands that support half-maximal levels of cell adhesion under these conditions. These results, summarized in Table 1, demonstrate that one of the anti-α4 integrin monoclonal antibodies, as well as the anti-β1 integrin monoclonal antibody, could completely inhibit melanoma cell adhesion to the fragment, consistent with an important involvement of α4β1 integrin in mediating the cellular recognition of the 33 kD fragment. As controls, anti-α5 integrin monoclonal antibody and normal mouse ascities had no effect on melanoma cell adhesion to the fragment.

Since these cells express predominantly CSPG, we next evaluated the ability of specific antagonists of CSPG function and/or expression for their ability to interfere with melanoma cell adhesion to the 33 kD fragment. Approaches for interfering with CSPG included removal of cell surface CS by using short-term digestion with proteinase-free chondroitinase ABC, or overnight treatment of cultures with 1 mM, para-nitrophenyl β-D-xylopyranoside (βDX) prior to the cell adhesion assays (Faassen, 1992b). βDX is a compound that competes with galactosyltransferase I and prevents the covalent attachment of CS to xylosylated core proteins (Schwartz, 1977; Kato, 1978; Lohmander, 1979; Robinson and Gospadarowicz, 1984; Faassen, 1992b). βDX was included in the cell adhesion assays, since the effects of this compound are readily reversible and removal of the drug can lead to the rapid restoration of intact CSPG synthesis. As a control for βDX, the inactive analogue αDX was employed (Faassen, 1992b). The results indicated that treatment with chondroitinase ABC, or βDX, caused a 50-60% inhibition in cell adhesion to the fragment (not shown), while heparitinase or αDX had no effect on adhesion; consistent with an important involvement for CSPG in mediating cell adhesion to the fragment. This inhibition in cell adhesion was not due to a decrease in the expresssion of cell surface α4β1 integrin (as measured by flow cytometry using anti-α4 or anti-β1 integrin monoclonal antibodies), indicating that cell surface CSPG plays a direct role in mediating A375SM cell adhesion to the fragment. However, in contrast to the previous results using anti-integrin monoclonal antibodies, however, the removal of CS from the cell surface never inhibited cell adhesion to the fragment by more than 50-60%.

Table 2

Pretreatment[a]	Relative Cell Adhesion (%)[b]
P4C2 (anti-α4 subunit)	61
Chondroitinase ABC	57
P4C2 + Chondroitinase ABC	17
P1D6 (anti α5 subunit)	98
P1D6 + Chondroitinase ABC	56

[a] Microtiter wells were coated with 4.2 pmoles of the 33 kD fragment, which promoted half-maximal cell adhesion, and then blocked with PBS/BSA. Cells were pretreated with chondroitinase ABC and/or a 1:625 dilution of anti-α4 integrin (P4C2) or anti-α5 integrin (P1D6) monoclonal antibodies. In each experiment, the cell adhesion assays were carried out for 30 min in the continued presence of anti-integrin subunit monoclonal antbodies or the enzyme.

[b] Data are expressed as the percent relative cell adhesion compared to that observed in the absence of any pretreatment. Data represent the means of triplicate determinations. The standard errors of means were less than 5% of mean values in each case.

132

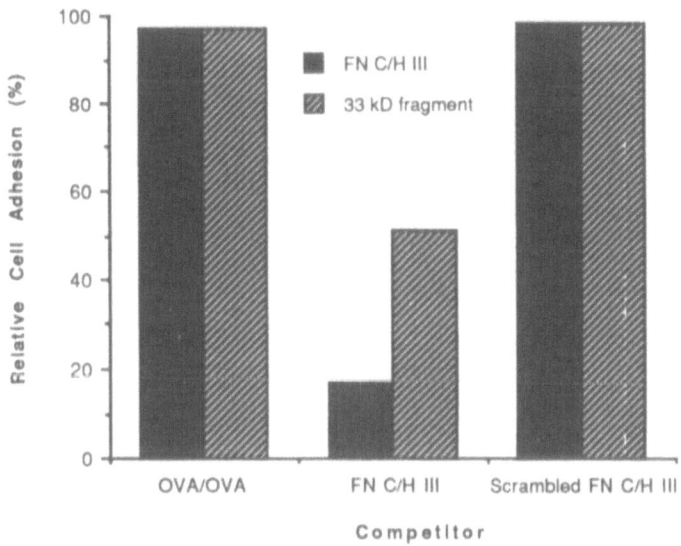

Figure 3. FN C/H III Represents a Novel Cell Adhesion Promoting Synthetic Peptide from within the 33 kD Fragment of FN. Microtiter wells were coated with 100 nmole of FN C/H III/OVA (*solid bars*) or 4.2 pmole of purified 33 kD fragment (*hatched bars*), and evaluated for their ability to support melanoma cell adhesion in the presence of 200 µg/ml of OVA/OVA, FN C/H III/OVA, or scrambled FN C/H III/OVA. SEMs were less than 10% of mean values in each case, and each experimental point represents the mean of triplicate values.

In an initial attempt to determine whether there could be a coordinated role for both cell surface CSPG and integrin in mediating melanoma cell adhesion to this FN fragment, the ability of chondroitinase ABC or βDX to inhibit cell adhesion in the presence of suboptimal levels of anti-α4 integrin monoclonal antibodies were examined. The results (Table 2) demonstrated that suboptimal levels of anti-α4 integrin monoclonal antibodies could potentiate the partial inhibition of cell adhesion caused by the enzymatic removal of cell surface CS. The specificity of these results was indicated by the failure of anti-α5 monoclonal antibody to inhibit melanoma cell adhesion to the fragment, and by the failure of anti-α5 integrin monoclonal antibody to potentiate the effects of chondroitinase ABC (Table 2). Virtually identical results were obtained when βDX was used in place of chondroitinase ABC (not shown). Collectively, the results support the notion that cell surface CSPG and α 4β1 integrin act in a closely coordinated fashion to mediate initial recognition and adhesion of melanoma cells to the fragment.

FN C/H III is a CS-dependent, α4β1 Integrin-independent Cellular Recognition Domain within the 33 kD Fragment

As shown above, FN C/H III, which is located at the amino terminal end of the 33 kD fragment, can support A375SM melanoma cell adhesion in a concentration dependent fashion. When this synthetic peptide was used in solution as a competitive inhibitor, it completely blocked cell adhesion to substrata coated with peptide FN C/H III, and partially competed for cell adhesion to the substrata coated with the 33 kD fragment (Figure 3). As controls, OVA coupled to OVA, or scrambled FN C/H III OVA (MKGKVTREVRYPTKPE), which has the same composition, charge, and net hydropathy as the parent synthetic peptide, had no effect on cell adhesion to either cell adhesion promoting ligand (Figure 3). In addition to supporting a partial role for active sites within this synthetic peptide in promoting melanoma cell adhesion to the fragment, the results demonstrate that the activity of FN C/H III is depdendent upon the primary sequence of the peptide, and that it is not due solely to the positive charge of the peptide.

In order to further define the nature of the receptor involved in melanoma cell recognition of this synthetic peptide, the ability of FN C/H III-coated substrata to support cell adhesion in the presence of antagonists of CSPG production or in the presence of inhibitory monoclonal antibodies to integrins were evaluated (summarized in Table 3). Removal of cell surface CS either by chondroitinase ABC or βDX inhibited melanoma cell adhesion to FN C/H III-coated substrata, whereas αDX and heparitinase had no effect. In contrast, neutralizing anti-α4 integrin monoclonal antibodies had no effect on melanoma cell adhesion to FN C/H III, even when used at very high (1:50 dilution of ascites) concentrations. These results demonstrate that cell surface CSPG is required for cell adhesion to FN C/H III, and that α4 integrin subunits would appear to play little or no role in cellular recognition of this ligand.

Table 3

Antagonist[a]	Relative Cell Adhesion[b]
βDX	15
αDX	98
Chondroitinase ABC	11
Heparitinase	96
P4G9 (α4 integrin)	97
P1D6 (α5 integrin)	99

[a] Microtiter wells were coated with 100 nmoles of FN C/H III/OVA and evaluated for the ability to support melanoma cell adhesion following pretreatment of the cells with 1 mM βDX, 1 mM αDX, chondroitinase ABC, heparitinase. Alternatively, microtiter wells adsorbed with 100 nmoles of FN C/H III/OVA were evaluated for their ability to support melanoma cell adhesion in the presence of a 1:50 dilution of anti-α4 integrin monoclonal antibody (P4G9), or anti-α5 integrin monoclonal antibody (P1D6).

[b] Data are expressed as the percent relative cell adhesion compared to that observed in the absence of any pretreatment. Data represent the means of triplicate determinations. The standard errors of means were less than 10% of mean values in each case.

Affinity columns consisting of peptide FN C/H III could bind purified CSPG, and this CSPG eluted at moderate ionic strength (approximately 0.4 M NaCl, not shown). Furthermore, alkaline borohydride released CS also bound to the column, but eluted at somewhat lower ionic strength from these columns (0.25 M NaCl), consistent with a role for the CS portion of the CSPG in binding to FN C/H III. Soluble FN C/H III was also used as a competitor for CSPG binding to affinity columns containing the 33 kD fragment (Figure 4). Pretreatment of approximately 7000 dpm of purified ^{35}S-CSPG with FN C/H III OVA could effectively inhibit CSPG binding to the fragment. In contrast, preincubation of the CSPG with either OVA coupled to OVA, or scrambled FN C/H III coupled to OVA, had no effect on the binding of the CSPG by the fragment. These results further support a direct CSPG-binding role for FN C/H III within the context of the fragment, and further indicate that the primary sequence of the synthetic peptide is important for its activities.

Melanoma Cell Adhesion to FN C/H I, FN C/H II and CS1 Involves a Complex Action of Integrin Subunits and Cell Surface CSPG

For the purposes of comparison to the cell adhesion data obtained using FN C/H III, we also evaluated the ability of peptides FN C/H I, FN C/H II, and CS1 to support melanoma cell adhesion in the presence of various antagonists of CSPG and integrins. An

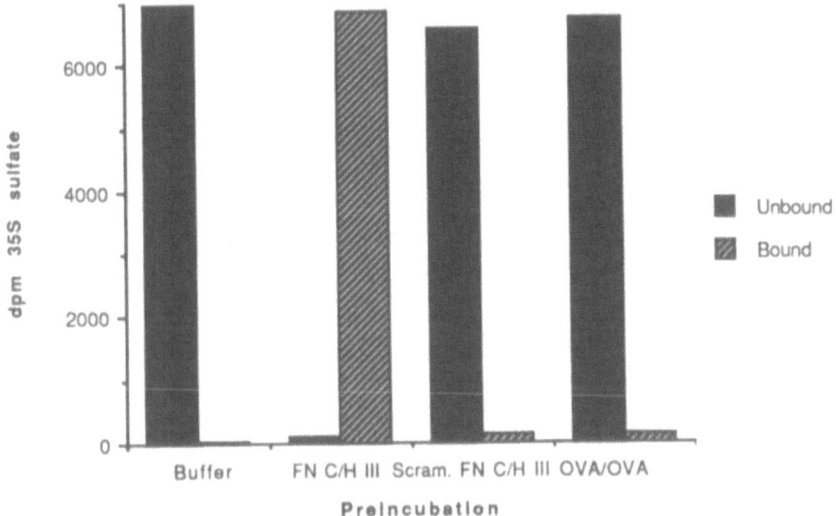

Figure 4. Binding of Purified CSPG to a 33 kD Fragment Affinity Column in the Presence of FN-C/H-III or Scrambled FN C/H III. Purified ^{35}S-CSPG (~7000 dpm) was pre-incubated with or without the indicated inhibitors (final concentrations of 500 µg/ml) for 2 h at 4° C and then applied to the 33 kD fragment affinity column that had been equilibrated with the buffer containing the corresponding inhibitors at concentrations of 500 µg/ml. Columns were washed with 6 ml of the buffer (*unbound, solid bars*) and then bound material (*hatched bars*) was eluted with 6 ml of buffer containing 0.2 M NaCl.

examination of the ability of affinity columns consisting of FN C/H I or FN C/H II to bind purified CSPG demonstrated that both synthetic peptides can bind CSPG, although the amount of NaCl required to elute the CSPG from either one of these column (0.25-0.3 M NaCl) was less than that required to elute CSPG from an FN C/H III column (not shown). As expected, affinity columns of CS1 had no ability to bind purified CSPG, even under very permissive binding conditions (i.e. applied to the column under 50 mM NaCl).

In contrast to FN C/H III, cell adhesion to FN C/H II was only partially (55-60%) inhibited by removal of cell surface CS by chondroitinase ABC or βDX (summarized in Table 4). Cell adhesion to FN C/H I was even less sensitive to this treatment, with only 15% inhibition of cell adhesion observed in the absence of cell surface CS. Importantly, cell adhesion to peptide CS1 also was partially sensitive to antagonists of CSPG function or expression (45-50%), despite the fact that peptide CS1 does not bind CSPG directly. When the ability of the three synthetic peptides to support adhesion in the presence of anti-integrin monoclonal antibodies was evaluated, it was observed that cell adhesion to all three synthetic peptides was sensitive to inhibition by anti-α4 integrin subunit monoclonal antibodies (Table 4). That this inhibition was specific was demonstrated by the failure of anti-α5 integrin monoclonal antibodies (Table 4) or normal mouse ascites (not shown) to inhibit melanoma cell adhesion to these three synthetic peptides.

A Working Model for the Complex Molecular Basis for Cellular Recognition of the 33 kD FN Fragment

As stated in the introduction, the ECM within tissues and basement membranes can be thought of as an array of closely spaced cellular recognition sites that act to support the recognition and adhesion of cells. The current studies have utilized a synthetic peptide approach for further defining the activities of several of these sites within the carboxyl terminal heparin-binding domain of FN A chains in supporting melanoma cell adhesion. By

Table 4

Antagonist[a]	Relative Cell Adhesion[b]		
	FN C/H I	**FN C/H II**	**CS1**
βDX	86	45	50
Chondroitinase ABC	85	43	53
Heparitinase	98	96	97
P4G9 (α4 integrin)	3	0.5	0.2
P1D6 (α5 integrin)	72	80	78

[a] Microtiter wells were coated with CS1/OVA, FN C/H I/OVA or FN C/H II/OVA at concentrations that promoted half-maximal cell adhesion, and then blocked with PBS/BSA. Cells were pretreated with 1 mM βDX overnight, or with chondroitinase ABC or heparitinase in PBS containing 1.25 mM $CaCl_2$ for 20 min. Alternatively, the cells were preincubated with a 1:50 dilution of anti-α4 integrin monoclonal antibody (P4G9), or anti-α5 integrin monoclonal antibody (P1D6) for 30 min at 37° C, and then evaluated for their ability to adhere to the various peptide coated substrata. Cell adhesion assays were carried out for 30 minutes in the continued presence of the various inhibitors.

[b] Data are expressed as the percent relative cell adhesion compared to that observed in the absence of any pretreatment. Each data point represents the mean of triplicate determinations. The SEM were less than 10% of mean values in each case.

the use of specific inhibitors (i.e. inhibitory monoclonal antibodies) of α4 integrin function and CSPG expression and/or function (i.e. chondroitinase ABC or βDX), A375SM cell adhesion to the 33 kD heparin binding fragment was demonstrated to involve both types of cell surface receptors. This was supported, in part, by the ability of either type of treatment to interfere with melanoma cell adhesion to the fragment, although removal of cell surface CS was never completely effective at inhibiting melanoma cell adhesion to the fragment. That suboptimal levels of anti-α4 integrin antibodies could potentiate the inhibitory effects of the removal of cell surface CS is consistent with the notion that these two types of receptors somehow cooperate with each other to support melanoma cell adhesion to this fragment.

Since certain monoclonal antibodies against α4 (or β1) integrin subunits could completely inhibit melanoma cell adhesion, whereas complete removal of cell surface CS was only partially effective in this regard, it would appear that a functional α4β1 integrin is of crucial importance in mediating melanoma cell adhesion to the fragment. In contrast, a functional CSPG, in the absence of a functional α4β1 integrin, was not able to support cell adhesion to the fragment under these conditions. While the precise reasons for this difference are not clear, potential explanations could relate to differences in the relative affinities of each receptor type for their respective ligand(s) within the fragment, or may be related to differences in the intracellular signals transmitted by each receptor. In support of the former possibility was the observation that FN C/H III, a ligand within the fragment that supports melanoma cell adhesion through a CS-dependent, integrin independent mechanism, is several orders of magnitude less effective than the intact fragment at supporting melanoma cell adhesion.

Each of these peptides is similar in the sense that each can only partially inhibit cell adhesion to the fragment when used individually as soluble competitors (McCarthy, 1990; Haugen, 1990; Mooradian, 1992; Iida, 1992). Similarly, specific immune IgG generated against each synthetic peptide, each of which can completely neutralize melanoma cell adhesion to their respective peptides, can only partially inhibit melanoma cell adhesion to the fragment (McCarthy, 1990; Haugen, 1990; Mooradian, 1992; Iida, 1992). Additionally, while these A375SM cells spread and migrate in a haptotactic fashion on this FN fragment, none of the synthetic peptides could individually duplicate these activities (not shown). These results are consistent with the notion that each one of these peptides represents a distinct cellular recognition site on the fragment, and that each peptide contains only partial mechanistic information for the cell.

An analysis of the activities of the four synthetic peptides in the presence of specific antagonists of CSPG or $\alpha4\beta1$ integrin function demonstrated some important differences regarding their respective mechanisms of action. FN C/H III represented a CSPG-dependent, $\alpha4\beta1$ integrin-independent cell adhesion site, based on its senstivity to antagonists of CSPG and its resistance to inhibitory monoclonal antibodies against $\alpha4$ integrin subunits. Furthermore, this peptide represents a major CSPG binding site on the fragment, based on its ability to completely inhibit CSPG binding to the fragment when used in solution. Importantly, a scrambled variant of FN C/H III could not inhibit cell adhesion or CSPG binding, indicating that its primary sequence (and not just net positive charge) is important for its activities.

Melanoma cell adhesion to the other three synthetic peptides, FN C/H I, FN C/H II, and CS1 was distinct from, and more complex than, that observed to FN C/H III. First, cell adhesion to all three synthetic peptides was completely sensitive to inhibitory monoclonal antibodies generated against $\alpha4$ integrin subunits. Cell adhesion to FN C/H I was only slightly (15%) sensitive to inhibitors of CSPG production, and FN C/H II was inhibited only by 50-60%, despite the fact that both synthetic peptides can bind the CSPG directly. Surprisingly, adhesion to the $\alpha4\beta1$ integrin binding peptide CS1 was also partially inhibited by chondroitinase ABC or βDX treatment (50%), despite the fact that this peptide does not bind CSPG directly. This inhibition of adhesion on CS1 was not due to nonspecific side effects of these treatments, as assessed by trypan blue exclusion data, ^{3}H-TdR or ^{35}S-methionine incorporation, or the relative inability of these treatments to interfere with adhesion to FN C/H I. While the exact mechanism by which removal of cell surface CS could interfere with $\alpha4\beta1$ integrin function remains to be elucidated, it is possible that CSPG may somehow modulate the affinity of the integrin for the active site within CS1 (Wayner, 1992), or that CSPG may modulate intracellular signals that are important for $\alpha4\beta1$ integrin to mediate cell adhesion.

Of additional importance in the immediate future will be to determine definitive ligand-receptor relationships for FN C/H I and FN C/H II for this cell type. Although both synthetic peptides can bind purified CSPG, this binding is somewhat lower affinity than FN C/H III, based on the amount of salt required to elute the CSPG from the respective affinity columns. Futhermore, interference with CSPG synthesis only partially inhibits cell adhesion to these two synthetic peptides, yet anti-$\alpha4$ integrin antibodies can completely inhibit adhesion to the peptides. These results suggest that these two FN C/H synthetic peptides might interact directly with integrin subunits, or alternatively that the peptides may be capable of somehow binding both CSPG and $\alpha4\beta1$ integrin. Both of these synthetic peptides are more effective on a molar basis at supporting melanoma cell adhesion compared to FN C/H III or CS1, consistent with the possibility that they may interact with more than one cell surface receptor. We are currently using an overlapping synthetic peptide approach, in combination with specific substitution of individual residues within the peptides, to help understand structural/functional relationships within these two synthetic peptides. Such an approach, coupled with direct affinity purification of cell lysates and the use of receptor/ligand cross-linking strategies, will all be important for addressing this important question.

Regardless of the exact explanation for some of these questions, the current results suggest a working model for the recognition of the 33 kD fragment of FN by A375SM melanoma cells (Figure 5). This model emphasizes the notion that cellular recognition of the fragment includes multiple sites throughout the fragment, and that these sites can be grouped loosely into two overlapping elements. One of these elements is CSPG dependent, in the sense that the four (and possibly other, see Bober-Barkalow and Schwarzbauer, 1992) sites fragment. Such ligand-induced clustering of ECM receptors is analogous to the way in within this element could either bind CSPG directly (FN C/H I, II, and III) and/or be influenced by the presence of cell surface CSPG (CS1). The second element, which is $\alpha4$-integrin-dependent, contains at least three active sites (FN C/H I, II and CS1), and possibly others (Mould and Humphries, 1991). The potential significance of the overlapping nature of these elements is that each could act in concert to efficiently bring melanoma cell surface CSPG and $\alpha4\beta1$ integrin into close proximity during initial cellular recogntion of the which lymphocytes utilize clusters of cell surface molecules to recognize specific target cells

(Springer,1990; Hynes, 1992). With respect to cellular recognition of the ECM, such clusters could have a profound effect on altering the conformation and binding specificity of the individual receptors sequestered within the clusters. Alternatively, the formation of such clusters could contribute to the generation of specific intracellular signals (Kornberg, 1991; Guan, 1991) that follow the initial cellular recognition of the fragment, thereby affecting cell adhesion, cytoskeletal reorganization, cellular motility, and other diverse aspects of cell phenotype.

Implicit in this "receptor cluster" model is the prediction that specific information transmitted into the cell is closely related to the specific combinations of cell surface receptors that may come into proximity as a result of their interaction with closely spaced cellular recognition domains with the ECM. Cell type-specific differences in the composition of such clusters could cause profound differences in the cell type specific signals transmitted to the interior of the cell. As an example, our recent studies suggest that a cell surface PI-linked heparan sulfate proteoglycan, MPIHP-63, is important in the recognition of FN C/H II by a metastatic mouse melanoma cell line, implicating changes in PI hydrolysis in mediating the adhesion of these cells to this fragment (Drake, 1992). Furthermore, analysis of corneal epithelial cell adhesion to the 33 kD fragment demonstrates that while these cells adhere to the fragment, they do not adhere to CS1 or to FN C/H II, but they do adhere to FN C/H I and to FN C/H III, suggesting that these cells utilize a different mechanism than melanoma cells in binding to this fragment (Mooradian, 1992). With respect to tumor biology, it will be interesting to compare normal cells with transformed counterparts at various stages of malignancy for their ability to adhere to the various synthetic peptides as well as to the fragment. Such studies may help in understanding the subtle changes in cellular recognition of the ECM that occurs coincident with tumor progression, and could help to provide mechanistic explanations as to how such changes contribute to ECM-mediated tumor cell adhesion, invasion, and metastasis formation.

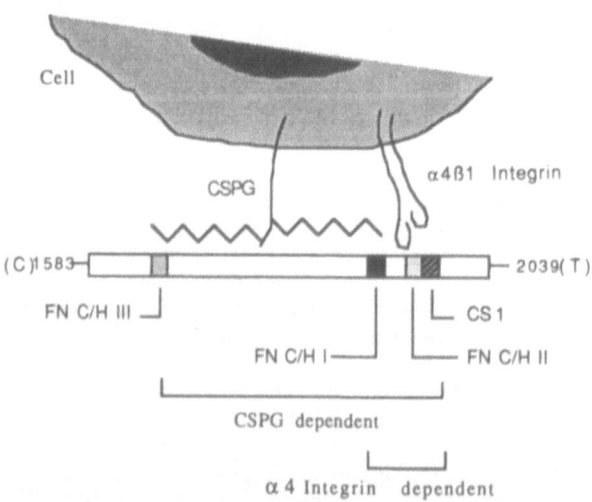

Figure 5. Diagram Illustrating a Coordinated Role for CSPG-Dependent and α4β1 Integrin-Dependent Cellular Recognition Sites in Mediating Melanoma Cell Adhesion to the 33 kD Fragment. Schematically depicted is a human melanoma cell initially recognizing substrata coated with the 33 kD fragment of human plasma FN-A-chains. The relative locations of the four synthetic peptides used in this study are shown. According to this model, the proximity of the CSPG-dependent and α4β1-integrin-dependent elements, within the context of a larger cellular recognition domain within this fragment, would act to focus cell surface CSPG and α4β1 integrin on the ventral surface of the plasma membrane. This clustering of cell surface receptors would act to modulate early cellular events that contribute to melanoma cell adhesion on this fragment.

ACKNOWLEDGEMENTS

We would like to acknowledge the excellent assistance of Judith Kahm for the production, purification, and characterization of the anti-peptide IgG, and of Daniel Mickelson and Truc Nguyen for the extraction, purification, and characterization of the ^{35}S-GAG. The advice of Drs. Theodore Oegema and David Klein for proteoglycan purification and characterization, and the thoughtful advice of Dr. Daniel Mooradian, is also gratefully acknowledged. This work was supported in part by NIH grants CA 43924 (J.B.M.), CA 21463 (L.T.F.), grants from the Juvenile Diabetes Foundation, American Cancer Society, and Minnesota Medical Foundation (A.P.N.S.) and grants from the Leukemia Task Force (J.B.M. and L.T.F.). L.T.F. is a recipient of the Allen-Pardee Professorship for Cancer Biology.

REFERENCES

Albelda, S.M. and Buck, C.A., 1990, Integrins and other cell adhesion molecules. *FASEB J.* 4:2868-2880.

Brennan, M.J., Oldberg, A., Hayman, E.G., and Ruoslahti, E., 1983, Effect of a proteoglycan produced by rat tumor cells on their adhesion to fibronectin-collagen substrata. *Cancer Res.* 43:4302-4307.

Bober-Barkalow, F.J. and Schwarzbauer, J.E., 1991, Localization of the major heparin-binding site in fibronectin. *J. Biol. Chem.* 266:7812-7818.

Bumol, T.F. and Reisfeld, R.A., 1982, Unique glycoprotein-proteoglycan complex defined by monoclonal antibody on human melanoma cells. *Proc. Natl. Acad. Sci. USA* 79:1245-1249.

Buck, C.A. and Horwitz, A.F., 1987, Cell surface receptors for extracellular matrix molecules *Annu. Rev. Cell Biol.* 3:179-205.

Couchman, J.R. and Höök, M., 1988, Proteoglycans and wound repair, in "The Molecular and Cellular Biology of Wound Repair," R.A.F. Clark and P. Henson, eds., Plenum Press, New York, pp 437-470.

Drake, S.L., Klein, D.J., Mickelson, D.J., Oegema, T.R., Furcht, L.T., and McCarthy, J.B., 1992, Cell surface phosphatidylinositol-anchored heparan sulfate proteoglycan initiates mouse melanoma cell adhesion to a fibronectin-derived, heparin-binding synthetic peptide. *J. Cell Biol.* 117:1331-1342.

Faassen, A.E., Drake, S.L., Iida, J., Knutson J.R., and McCarthy, J.B, 1992a, Mechanisms of normal cell adhesion to the extracellular matrix and alterations associated with tumor invasion and metastasis. *Adv. Pathol. Lab. Med.* 5:229-260.

Faassen, A.E., Schrager, J.A., Klein, D.J., Oegema, T.R., Couchman J.R., and McCarthy, J.B., 1992b, A cell surface chondroitin sulfate proteoglycan, immunologically related to CD44, is involved in type I collagen-mediated melanoma cell motility and invasion. *J. Cell Biol.* 116:521-532.

Fidler, I.J., 1986, Rational and methods for the use of nude mice to study the biology and therapy of human cancer metastasis. *Cancer Metast. Rev.* 5:29-49.

Gallagher, J.T., 1989, The extended family of proteoglycans: Social residents of the pericellular zone. *Curr. Opin. Cell Biol.* 1:1201-1218.

Garrigues, H.J., Lark, M.W., Lara, S., Hellström, I., Hellström K.E., and Wight, T.N., 1986, The melanoma proteoglycan: restricted expression on microspikes, a specific microdomain of the cell surface. *J. Cell Biol.* 103:1699-1710.

Guan, J-L. and Hynes, R.O., 1990, Lymphoid cells recognize an alternatively spliced segment of fibronectin via the integrin receptor $\alpha4\beta1$. *Cell* 60:51-63.

Guan, J-L., Trevithick, J.E., and Hynes, R.O., 1991, Fibronectin/integrin interaction induces tyrosine phosphorylation of a 120-kDa protein. *Cell Reg.* 2:951-964.

Haugen, P.K., McCarthy, J.B., Skubitz, A.P.N., Furcht, L.T., and Letourneau, P.C., 1990, Recognition of the A chain carboxy-terminal heparin binding region of fibronectin involves multiple sites: Two contiguous sequences act independently to promote neural cell adhesion. *J. Cell Biol.* 111:2733-2745.

Höök, M., Kjellen, L., Johansson S., and Robinson, J., 1984, Cell surface proteoglycans. *Annu. Rev. Biochem.* 53:847-869.

Humphries, M.J., Komoriya, A., Akiyama, S.K., Olden K., and Yamada, K.M., 1987,

Identification of two distinct regions of the type IIICS connecting segment of human plasma fibronectin that promote cell type-specific adhesion. *J. Biol. Chem.* 262:6886-6892.

Humphries, M.J., 1990, The molecular basis and specificity of integrin-ligand interactions. *J. Cell Sci.* 97:585-592.

Hynes, R.O., 1987, Integrins: A family of cell surface receptors. *Cell* 48:549-554.

Hynes, R.O., 1992, Integrins: Versatility, modulation, and signalling in cell adhesion. *Cell* 69:11-25.

Iida, J., Skubitz, A.P.N., Furcht, L.T., Wayner E.A., and McCarthy, J.B., 1992, Coordinate role for cell surface chondroitin sulfate proteoglycan and $\alpha 4\beta 1$ integrin in mediating melanoma cell adhesion to fibronectin. *J. Cell Biol.* 118:431-444.

Kato,Y., Kimata, K., Ito, K., Karasawa K., and Suzuki, S., 1978, Effect of β-D-xyloside and cycloheximide on the synthesis of two types of proteochondroitin sulfate in chick embryo cartilage. *J. Biol. Chem.* 253:2784-2789.

Kornblihtt, A.R., Umezawa, K., Vibe-Pedersen, K., and Baralle, F.E., 1985, Primary structure of human fibronectin: Differential splicing may generate at least 10 polypeptides from a single gene. *EMBO J.* 4:1755-1759.

Kornberg, L.T., Earp, H.S., Turner, C.E., Prockop, C,. and Juliano, R.L., 1991, Signal transduction by integrins: Increased protein tyrosine phosphorylation caused by clustering of $\beta 1$ integrins. *Proc. Natl. Acad. Sci. USA* 88:8392-8396.

Kyte, J. and Doolittle, R.F., 1982, A simple method for displaying the hydropathic character of a protein. *J. Mol. Biol.* 157:105-132.

Lark, M.W., Laterra, J., and Culp, L.A., 1985, Close and focal contact adhesions of fibroblasts to a fibronectin-containing matrix. *Fed. Proc.* 44:394-403.

Liao, N.S., St. John, J., McCarthy, J.B., Furcht, L.T., and Cheung, H.T., 1989, Adhesion of lymphoid cells to the carboxyl-terminal heparin-binding domains of fibronectin. *Exp. Cell Res.* 181:348-361.

Lohmander, S., Madsen, K., and Hinek, A., 1979, Secretion of proteoglycans by chondrocytes: Influence of colchicin, cytochalasin β and β-D-xyloside. *Arch. Biochem. Biophys.* 192:148-157.

McCarthy, J.B., Hagen S.T., and Furcht, L.T., 1986, Human fibronectin contains distinct adhesion and motility promoting domains for metastatic melanoma cells. *J. Cell Biol.* 102:179-188.

McCarthy, J.B., Chelberg, M.K., Mickelson D.J., and Furcht, L.T., 1988a, Localization and chemical synthesis of fibronectin peptides with melanoma adhesion and heparin binding activities. *Biochemistry* 27:1380-1388.

McCarthy, J.B., Skubitz, A.P.N., Palm, S.L., and Furcht, L.T., 1988b, Metastasis inhibition of different tumor types by purified laminin fragments and a heparin binding fragments of fibronectin. *J. Natl. Cancer Inst.* 80:108-116.

McCarthy, J.B., Skubitz, A.P.N., Zhao, Q., Yi, X-Y., Mickelson, D.J. and Furcht, L.T., 1990, RGD-independent cell adhesion to the carboxy-terminal heparin binding fragment of fibronectin involves heparin-dependent and -independent activities. *J. Cell Biol.* 110:777-787.

Mooradian, D.L., McCarthy, J.B., Cameron, D.J., and Furcht, L.T., 1992, Rabbit corneal epithelial cell adhere to two distinct heparin-binding synthetic peptides derived from fibronectin. *Invest Opthalmol. Vis. Sci. (in press)*.

Mould, A.P., Wheldon, L.A., Komoriyama, A., Wayner, E.A., Yamada, K.M., and Humphries, M.J., 1990, Affinity chromatographic isolation of the melanoma cell adhesion receptor for the IIICS region of fibronectin and its identification as the integrin $\alpha 4\beta 1$. *J. Biol. Chem.* 265:4020-4024.

Mould, A.P. and Humphries, M.J., 1991, Identification of a novel recognition sequence for the integrin a4b1 in the COOH-terminal heparin-binding domain of fibronectin. *EMBO J.* 10:4089-4095.

Robinson, J. and Gospodarowicz, D., 1984, Effect of p-nitrophenyl-β-D-xyloside on proteoglycan synthesis and extracellular matrix formation by bovine corneal endothelial cell cultures. *J. Biol. Chem.* 259:3818-3824.

Rogers, S.L., Letourneau, P.C., Peterson, B.A., Furcht, L.T., and McCarthy, J.B., 1987, Selective interaction of peripheral and central nervous system cells with two distinct cell-binding domains of fibronectin. *J. Cell Biol.* 105:1435-1442.

Ross, A.H., Cossu, G., Herlyn, M., Bell, J.R., Steplewski, Z., and Koprowski, H., 1983,

Isolation and chemical characterization of a melanoma-associated proteoglycan antigen. *Arch. Biochem. Biophy.* 225:370-383.

Ruoslahti, E., 1991, Integrins. *J. Clin. Invest.* 87:1-5.

Saunders, S. and Bernfield, M., 1988, Cell surface proteoglycan binds mouse mammary epithelial cells to fibronectin and behaves as a receptor for interstitial matrix. *J. Cell Biol.* 106:423-430.

Schwartz, N.B., 1977, Regulation of chondroitin sulfate synthesis. *J. Biol. Chem.* 252:6316-6321.

Springer, T.A., 1990, Adhesion receptors of the immune system. *Nature* 346:425-434.

Stewart, J.M. and Young, J.D., 1984, "Solid Phase Peptide Synthesis," 2nd ed. Pierce Chemical Co., Rockford, IL.

Visser, M.R., Vercellotti, G.M., McCarthy, J.B., Goodman, J.L., Herbst, T.J., Furcht, L.T., and Jacob, H.S., 1989, Herpes simplex virus inhibits endothelial cell attachment and migration to extracellular matrix proteins. *Am. J. Pathol.* 134:223-230.

Wait, K.A., Mugnai, G., and Culp, L.A., 1987, A second cell binding domain on fibronectin (RGDS-independent) for neurite extension of neuroblastoma cells. *Exp. Cell Res.* 169:311-327.

Wasteson, A., 1971, A method for the determination of the molecular weight and molecular-weight distribution of chondroitin sulfate. *J. Chromatogr.* 59:87-97.

Wayner, E.A., Garcia-Pardo, A., Humphries, M.J., McDonald, J.A., and Carter, W.G., 1989, Identification and characterization of the T lymphocyte adhesion receptor for an alternative cell attachment domain (CS-1) in plasma fibronectin. *J. Cell Biol.* 109:1321-1330.

Wayner, E.A. and Kovach, N.L., 1992, Activation-dependent recognition by hematopoietic cells for the LDV sequence in the V region of fibronectin. *J. Cell Biol.* 116:489-497.

DISCUSSION

E. BROWN

I have actually two questions. In the original descriptions of α_4-mediated adhesions, one of the important things was that it was heparin-independent, that is heparin did not inhibit cell adhesion. I mean this was a heparin-binding fragment of fibronectin, cell adhesion to it was not inhibited by heparin. In your case is cell adhesion inhibited by heparin?

J. McCARTHY

That seems to be a cell type-specific phenomenon, since we have other cells (for example, mouse melanomas) that are resistant to soluble heparin and I do not fully understand that. However, heparin will inhibit the adhesion of these cells to fragment.

E. BROWN

And then I was just wondering about the relationship between your work and the data from years ago of Lloyd Culp, which showed glycosamine glycan involvement in focal adhesions on fibronectin.

J. McCARTHY

Oh sure, that inspired a lot of our own research. What they and others have demonstrated is that focal-adhesion formation on certain adherent cells requires a cell surface heparin sulfate, the RGD-containing domain of fibronectin, and a heparin-binding ligand on the plate. Our results are similar in that the effects of the PG on cell adhesion are subtle in nature, and likely to involve modifying the activity of the integrin.

E. BROWN

Did he ever identify what the proteoglycan was?

J. McCARTHY

No, to my knowledge he never did. His work focused on the metabolism of the glycosaminoglycans within focal adhesions and he demonstrated that these are very dynamic structures. At that time the core proteins were not known, and Lloyd has since gone on to do some other things. I know that John Couchman and Anne Woods are continuing that sort of work, and I think that with their fibroblasts they feel that there is a plasma membrane-linked heparin sulphate in their focal adhesion.

R. JULIANO

Do you know that these peptides are actually accessible and intact?

J. McCARTHY

We do, but only on the basis of either inhibition studies and on the basis of antibodies against the peptides to recognize the fragment by a lyser, coated under conditions that we would coat them for a cell adhesion assay, so to that extent we know, the answer is yes. How they all work when they are in the context of the fragment is another question, but, I would say yes, they are exposed on the surface.

R. JULIANO

Way back at the beginning you show inhibition by the anti-$\alpha_4\beta_1$ antibody to the intact fragment. Could you ever achieve a hundred percent inhibition using just the antibody?

J. McCARTHY

Just anti-α_4? With one of the antibodies we can, with two of them the effect is only 50-60%, and we do not. Again, understand the reason for that. Maybe it speaks to the problem of not always knowing why a monoclonal antibody blocks adhesion; does it block ligand binding or does it do something else? But all three anti-α_4s have at least some and certainly significant effect.

E. MIHICH

You said at one point that the final goal would be to try to correlate the function of these fragments to progression. Have you started already in that direction with the facts you have, looking, for instance, at nodule growth, or perhaps metastatic causes, and if you did start, what did you find out?

J. McCARTHY

What I had in mind was trying to use the fragment in combination with the peptides to evaluate possible progression-related differences in cellular recognition. We have looked at the ability of the fragment to inhibit nodule growth of mouse cells, and it is very effective in the tail vein experimental metastasis assay. We are currently evaluating these peptides for their ability to do the same.

E. MIHICH

Are they effective in the cell detachment or attachment phase of these metastatic models?

J. McCARTHY

We take a bolus of tumor cells, put them into the tail veins of mice. So we are only looking at attachment, arrest, attachment and extravasation. Some of these peptides are quite effective at inhibiting metastasis, and some of them are not so effective. And in particular, in our hands, peptides 1 and 3 seem to be fairly effective at doing so. And interestingly if you compare different tumor types, melanomas and lymphomas (these are all collaborative studies that we are doing with a fellow in Japan, Dr. Saiki), the peptides appear to differ in their ability to block metastasis, so there may be some sort of cell-type specific differences there. But our interest was in just trying to use these peptides to understand very subtle differences. When we got into this business to begin with, we thought we would find that the cells of differring metastatic or invasive potential would exhibit dramatic differences in the ability to adhere to fibronectin, however, that was not the case. When we got to smaller fragments we still could not see much of a difference, but as we get it down to peptides you can start to see very subtle differences in the ways these cells recognize ECM components, so that is what we are focusing on now, what our major goal is.

V. QUARANTA

The question is whether the proteoglycan originally described in melanoma cells can still be considered a "tumor-specific antigen?"

J. McCARTHY

I believe that is not true. Since it has been cloned and examined more closely, it would appear that the protein is more widely distributed that originally thought. It is important to keep in mind, however, that even if the core is there, it is expressed as a part-time proteoglycan. These cells express the MPG as a chondroitin sulphate, but it is a very low percentage of the total protein that is on the surface, lower compared to even some other melanoma cells, and why that is so is not understood. But it is not tumor-specific, as such, it just happened to be very immunogenic on melanomas.

S. SHAW

To what extent have you surveyed diverse cell types in terms of their differential capacity to bind via their proteoglycans in a specific fashion to these several peptides?

J. McCARTHY

We have surveyed a lot of different cell types to stick to these peptides, and they all behaved quite differently. Mouse melanoma cells for example have an apparently different repertoire of proteoglycans; they have a lipid linked heparin sulphate, and a CD44-related chondroitin sulphate. Those cells will stick spread and move on every one of these peptides. So somehow they are getting different information from the peptides than this human melanoma that we are working with. We have also looked at corneal epithelial that will stick to certain of these peptides, while some of the peptides will support motility, the rest of them will not. So we know bits and pieces of information relative to cell type-specific behavior on the peptides. It is clear that there are differences either in the nature of the receptor, or the state of the signals that are modulated by the receptor. The important point is that peptides represent a potentially powerful approach for studying such differences at the molecular level.

S. SHAW

That comprises a powerful combinatorial element in determining the specificity of binding.

J. McCARTHY

Oh, absolutely.

S. SHAW

My other question has to do with shedding or removal of proteoglycans, as has been reported for syndecan. For your cells, is there shedding of proteoglycans which serves to regulate the adhesion?

J. McCARTHY

Not as a way of regulating the adhesion process. They are shed, and they are shed as both GAGS as pieces of proteoglycan and that is all we know.

S. SHAW

But that shedding, is it induced by the adhesion *per se*?

J. McCARTHY

Not obviously, no, but we have not evaluated that in any systematic, detailed fashion. It is an interesting possibility.

G. TARONE

Can you see focal adhesion when you plate your cells on the 30K peptides, in other words, are the proteoglycan binding sites required to make focal adhesion?

J. McCARTHY

We have not studied that like it should be studied. It should be evaluated under very stringent conditions (for example, in the absence of protein synthesis) and we have not done that yet. We do know that if the cells are adherent on the fragment for a long time they do exhibit stress fibers (and we assume focal adhesions) but we really have not looked at it in the short term like the focal adhesion people do.

G. TARONE

You also mentioned that your melanoma cells express $\alpha_5\beta_1$ that recognize the RGD fragment. So these cells interact with fibronectin via $\alpha_5\beta_1$, β_1, and proteoglycans. Can you give us an idea of what is the relative importance of these three types of interaction when cells adhere to the intact fibronectin molecule.

J. McCARTHY

There are a couple of variables here that have to be looked at; first of all, how do you plate fibronectin. If you plate fibronectin at different pHs you can change its conformation and hence change the active sites exposed on the molecule as it is plated. There are papers that have appeared years ago that clearly demonstrate that if you take fibronectin and put it in different environments you can show that heparin, for example, is more or less effective at blocking a fibronectin-coated bead-binding to cell surfaces. With respect to how the material is coated, we coat all this at a neutral pH, and that seems to expose the heparin-binding a little more; when coated using rollers buffer it seems to be less exposed. So how one can influence how things happen using those coating conditions for fibronectin (that is, neutral pH); we can take that fragment, in solution, and we can partially block adhesion to fibronectin, not completely, so it has some role to play.

A. SONNENBERG

Are these adhesions assays done with trypsin- or EDTA-treated cells?

J. McCARTHY

They are done with EDTA.

P. MARCHISIO

Normal human melanocytes do express $\alpha_V\beta_3$ and $\alpha_3\beta_1$. And the expression of $\alpha_V\beta_3$, as well as that of $\alpha_3\beta_1$, is highly enhanced by the treatment with phorbol esters like PMA *in vitro.* But do your cells express $\alpha_V\beta_3$, and in this case is there a possibility of a role in recognizing the fragment of fibronectin?

J. McCARTHY

That is an interesting question. I do not know the answer to that. They are not my cells. These studies are being performed in collaboration with Dr. Glynis Scott in Rochester, New York. I know they do not express $\alpha_4\beta_1$, I know they do adhere to that fragment, I know they do not adhere very well at all to peptides 1, 2, and CS1, which are clustered on the fragment. They also adhere actually quite well to peptide 3, as well as to two other peptides, 4 and 5, that I did not talk about. So, there are at least three sites in that fragment that we know will work for a melanocyte. But we do not yet understand the reason for the difference between melanocytes and melanoma.

S. GOODMAN

Going back to Guido's question, there is always this problem about the differential role of these various adhesion mechanisms, so you can have very nice data that the proteoglycans are involved in attachment upon this type of fibronectin fragments. But there is still this question. Can you know what are they doing? Yesterday, I asked the same question, about taking epithelia off an intact basement membrane, when other mechanisms must exist. So is there any situation where you have more or less the naked cell which just has proteoglycans? Do proteoglycans on their own mediate active adhesion and spreading in any system that you know of?

J. McCARTHY

The problem is that we are starting to get into some real vague terminology in our own laboratory. For example, we have PI linked heparin sulphate. We just published a paper which said it initiates cell adhesion to fibronectin peptide-2. What we meant to emphasize by this word was that we could not understand how a PI linked thing could directly mediate cell adhesion as people generally think about it. How is it going to bridge? How is the membrane to support adhesion? In that particular case, the GAGs are important, but so is the core. We have antibodies against the core protein, those antibodies will block adhesion to that peptide even if those antibodies do not block binding of the proteoglycan to the peptide. So, it is a complicated situation.

S. GOODMAN

Yes, I admit it is really a lousy word -- in the absence of integrins.

J. McCARTHY

Is there a cell expressing no integrins? The answer is no.

S. GOODMAN

Is there a way that you can block out, have you tried to block out the early effects of integrins with the β_1-specific antibody, and demonstrate that your cells got an adhesion non-integrin mediated. I mean like we have, although we took another site from you. I am not too happy with cells going down as round balls on a substrate; I am talking about cells going down and spreading.

J. McCARTHY

Yes, I think again you have to consider what the nature of the proteoglycan would be. The linked HSPG seems to initiate spreading on its own. If we put the cells onto the fragment, these mouse melanoma cells, and treat the mouse melanoma cells with heparitinase, they still stick on the fragment, but their spreading is retarded, or their morphology is a little bit different. The effects that these molecules have are subtle, how they do what they do? They could feed information in through the tail, they could also feed in perhaps laterally from one receptor to another. I think the diversity of PG expression of different cell types could help to explain some of the recognition events that occur. Such differences may help to explain some of the cell type-specific differences seen by Dr. Hemler when he transfects his integrin constructs into cells of different backgrounds. So, we would like to consider recognition as a receptor cluster and that is, if you will, a bar code for how cells recognize the actual site of a matrix component. It obviously could involve more than just proteoglycans and integrin. How these clusters differ could be complex, ranging from modifying receptor affinities for ECM components to changing the composition or activities of specific signal-transducting pathways.

G. TARONE

This is a speculation rather than a question. We were hearing yesterday about antibodies which can increase the affinity of the integrins, and I was discussing this morning with Francisco Sanchez-Madrid whether this can be a physiological phenomenon or simply an artifact induced by antibodies. Your presentation suggested to me that proteoglycans can perhaps function as physiological regulators of integrin affinity.

J. McCARTHY

I would strongly support that speculation. I think the difference here is that this model that we have says that the matrix assembles these things together, that it brings them into proximity. And at the same time, proteoglycan could associate with various integrin subunits, depending on what the integrin-binding sites are on the specific ECM molecule in question.

G. TARONE

That could be experimentally tested probably by means of chromatography or something like that.

J. McCARTHY

Yes.

V. QUARANTA

In this last round of discussions, I got a little bit confused. So you are saying that the specificity of the binding to the fibronectin fragments is actually in the sugar portions.

J. McCARTHY

We are saying that the sugar portions bind at peptide-3, and the protein core portion may or may not bind to another site.

V. QUARANTA

My point is, are there additional post-translational modifications, like sulphation, phophorylation, etc.

J. McCARTHY

Well sulphation on the GAGs, there is phosphorylation on that xylose that seems to be important for GAG extension. We have not evaluated the sulphation patterns in any great detail as of yet.

V. QUARANTA

And also the length of carbohydrates may vary. So what do you think this specificity is coming from?

J. McCARTHY

I think GAG recognition of that fragment involves multiple sites along the fragment, and that is where specificity comes in. And each one of them has a contribution to make relative to that.

V. QUARANTA

These are also very stiff chains, right?

J. McCARTHY

That is correct.

V. QUARANTA

So, is there any mechanical role that you think of this? You seem to have some lubricant properties.

J. McCARTHY

If I understand the question, I think you are asking if there are lubricant properties attributed to the GAGs. We have not evaluated that but it is certainly an interesting possibility.

ADHESION MOLECULES AT ENDOTHELIAL CELL
TO CELL JUNCTIONS

Maria Grazia Lampugnani, Massimo Resnati, Marco Raiteri, Marco Pittiglio,
Luigi Ruco, and Elisabetta Dejana*

* Istituto di Ricerche Farmacologiche Mario Negri
Via Eritrea 62
20157 Milano, ITALY

ABBREVIATIONS USED

TNF = Tumor Necrosis Factor; **γ-IFN** = γ-interferon; **PECAM** = Platelet Endothelial
Cell Adhesion Molecule; **Vn** = Vitronectin; **Fg** = Fibrinogen; **vWf** = vonWillebrand factor;
Fn = Fibronectin; **Lm** = Laminin; **Tsp** = Thrombospondin; **Thr** = Thrombin; **Coll** =
Collagen

SUMMARY

Endothelial cells express a variety of adhesive receptors that regulate their adhesion to
extracellular matrix and the organization of cell-cell junctions. Most of the endothelial cell
receptors for matrix proteins belong to the integrin superfamily. Endothelial cells, as many
other cell types, have many different integrins on their surface. This indicates that the
interaction with matrix proteins is a complex phenomenon that requires multiple recognition
mechanisms. We have only a very limited knowledge of the molecules present at endothelial
cell-cell junctions. Integrins have been found to be localized in these structures. This
suggests that they can play a role also in this homotypic type of cell interaction. Other
molecules, however, structurally and functionally distinct from integrins (including cadherins
and immunoglobulins) have been found in endothelial junctions. The reciprocal role of these
proteins remains to be fully defined.

INTRODUCTION

The endothelium constitutes the inner lining of the entire cardiovascular system. As
other types of epithelia, endothelial cells are polarized. The luminal front facing the blood is
continuously exposed to circulating cells and plasma components. The opposite abluminal
front is in contact with the extracellular matrix and the surrounding tissues.

Endothelial cell monolayer integrity is maintained by specific adhesive receptors that
regulate cell to cell and cell to matrix interaction (Dejana, 1991).

Endothelial cell-cell junctions play an important role in regulating vascular permeability
and leukocyte extravasation. These structures have been divided for morphological and
functional characteristics in: i) communicating (gap) junctions; ii) adherens junctions; and iii)
occluding (tight) junctions. The type and the number of endothelial cell junctions varies
along the vascular tree, they are well organized and numerous in the large vessels where the
control of permeability is strict while they almost disappear in the postcapillary venules

where cell extravasation and exchange of plasma constituents need to be particularly efficient (for review, see Simionescu and Simionescu, 1991).

Endothelial cell interaction with matrix proteins modulates their migration and growth. The subendothelial matrix is, in general, a thrombogenic surface that promotes platelet adhesion and the activation of the coagulation system. In normal conditions the presence of the endothelium represents a good protection against thrombotic phenomena and plasma protein infiltration in the vascular media. The capacity of endothelial cells to remain attached to the vascular surface and to migrate and proliferate to cover exposed subendothelial structures is an important defense mechanism against the development of vascular injury.

In addition, the development of endothelial cell differentiated characteristics during embryogenesis and the maintenance of polarity are regulated, as for other cell types, by the interaction with matrix components. The extracellular matrix exerts therefore a more complex role than just providing a substrate for cell attachment.

INTEGRIN RECEPTORS IN ENDOTHELIAL CELLS

As mentioned above most of the endothelial cell receptors for extracellular matrix proteins belong to the integrin family. In general, the integrin family consists of a large number of heterodimers that regulate a variety of cell adhesion functions (for review see: Albelda and Buck, 1990; Hemler, 1990; Hynes, 1992; Ruoslahti and Giancotti, 1989; Ruoslahti and Pierschbacher, 1987; van Mourik at al., 1989). Almost all cell types express these structures and integrins are well preserved along the phylogenetic tree. All integrins are formed by two non-covalently linked subunits: the larger termed α chain and the smaller β chain. Both subunits are integral membrane proteins that present a small C-terminal cytoplasmic domain and a large N-terminal extracellular domain and a transmembrane segment.

Table 1 lists the principal members of the integrin family identified in endothelial cells.

The $\alpha^1\beta_1$ complex is a receptor for laminin and it was found to be present in endothelial cells of different type of vessels but not of human aorta, umbilical and femoral vein (Albelda and Buck, 1990; Defilippi et al., 1991).

Table 1. Integrins in Endothelial Cells

Subunits	Ligand/function
$\alpha^1\beta_1$	lm, coll
$\alpha^2\beta_1$	lm, coll, fn, cell-cell junctions
$\alpha^3\beta_1$	fn, lm, coll
$\alpha^5\beta_1$	fn, cell-cell junctions
$\alpha^6\beta_1$	lm
$\alpha^v\beta_3$	vn, fg, vWf, tsp, fn, lm, thr
$\alpha^v\beta_5$	vn

vn = vitronectin; fg = fibrinogen; vWf = von Willebrand factor; fn = fibronectin; lm = laminin; tsp = thrombospondin; thr = thrombin; coll = collagen.

The $\alpha^2\beta_1$ is the receptor for collagen in platelets and in other cells of hematopoietic and non-hematopoietic origin. In endothelial cells it behaves differently acting as a receptor for laminin and binding less efficiently to fibronectin and collagen (Languino et al., 1989).

The $\alpha^3\beta_1$ receptor is, in general, a multifunctional integrin: it binds to fibronectin, collagen and laminin (Wayner and Carter, 1987).

The $\alpha^5\beta_1$ complex is abundant in cultured endothelial cells but is poorly expressed *in vivo* (Albelda and Buck, 1990). $\alpha^5\beta_1$ in endothelial cells (as well as in other cell types) preferentially binds fibronectin (Conforti et al., 1989).

The $\alpha^6\beta_1$ integrin is moderately expressed in cultured endothelial cells. This molecule is the laminin receptor in platelets (Sonnemberg et al, 1989) and can play an identical role in endothelial cells.

Endothelial cells from umbilical vein express high levels of $\alpha^v\beta_3$. This integrin has the same β chain as the platelet complex GpIIb-IIIa but a distinct α chain. These two integrins share the characteristic of having a low substratum specificity. Both molecules bind to different matrix and plasma proteins. The $\alpha^v\beta_3$ integrin in endothelial cells besides vitronectin also recognizes fibrinogen (Cheresh, 1987; Dejana et al., 1990), vonWillebrand factor (Dejana et al., 1989), fibronectin (Conforti et al., 1990), thrombospondin (Lawler et al., 1988), laminin (Kramer et al., 1989) and thrombin (Bar-Shavit et al., 1991). The capacity of this molecule to bind thrombin, fibrinogen and vitronectin suggests that it can play a role in the activation of the coagulation and immune system. This integrin is not specifically expressed in the endothelium but is quite widespread. Table 2 compares the distribution of $\alpha^v\beta_3$ in different tissues and in endothelial cells. Only the endothelium of the microvasculature expresses low amount of this molecule while the endothelium from muscular arteries, arterioles venules and veins shows high levels of this integrin on its membrane (Figure 1). Interestingly, other epithelia like that of the ileum or the colon are negative. The epidermis was negative for β_3 and $\alpha^v\beta_3$ but positive for α^v thus suggesting that α^v is linked to another β chain.

Endothelial cells also express the newly defined $\alpha_v\beta_5$. This receptor, in its purified form, recognizes vitronectin essentially (Smith et al., 1990).

Endothelial cells have at least four fibronectin and four laminin receptors. The reason for this redundancy is still obscure. Possibly multiple interactions with the same molecule are required for the stability of the adhesion process. Matrix components are large proteins formed by different functional domains. Different integrin receptors can bind to distinct sites on the same ligand and transmit distinct information from the extracellular to the intracellular compartment.

We have recently found that endothelial cells express integrins both on their basal and apical surface *in vitro* and *in vivo* (Conforti et al., 1992). This indicates that, in addition to their role in promoting endothelial cell attachment to extracellular matrix proteins, integrin receptors of endothelial cells can be exposed to blood stream and eventually be available for binding of plasma proteins and circulating cells.

INTEGRIN ACTIVITY IN ENDOTHELIAL CELLS

Possible mechanisms of constitutive regulation of integrin activity could include mRNA splicing, posttranslational modification of the receptor or association of the receptor with modifying components (gangliosides, glycosaminoglycans) (for review see Hynes, 1992). Alternative splicing for integrin subunits have been described also in the endothelium. Ionic concentration and the phospholipid composition of the membrane can dramatically modify integrin receptor affinity and specificity for different substrata (Conforti et al., 1990; Kirchhofer et al., 1991). Membrane phospholipid microenvironment might change not only in different cell types or in culture conditions but also during cell movement. This might be a very efficient and fast way the cell uses for regulating receptor affinity for matrix ligands.

In general, in circulating cells, adhesion to substrata or to plasma proteins seems to be more effectively regulated by modulation of integrin activity than by changes in their number. This has been particularly studied for platelet and leukocyte integrins (for review see Hynes, 1992). It is,however, still unknown whether the same type of "fast" modulation of these receptors acts in other non circulating cells such as the endothelium.

A way by which integrins may transfer information to the cell is through their interaction with cytoskeletal proteins. During the first hours of endothelial cell attachment to substrata the basal surface of the cells forms several types of contacts (called focal contacts or adhesion plaques [Burridge et al., 1988]) which are the area of closest interaction between the substratum, the cell membrane and the membrane insertion sites of actin microfilaments. During cell adhesion integrins have been found to be clustered in focal contacts.

It is still debated whether the cytoskeletal proteins regulate integrin clustering or whether integrins, after adhesion to their specific substratum, trigger cytoskeletal organization. Experimental evidence supports the latter possibility (Dejana et al., 1988).

Table 2. Distribution of $\alpha^v\beta_3$ in Different Tissues by Immunohistochemistry

Type of tissue	Integrin chain		
	$\beta3$	αv	$\alpha v\beta3$
Endothelium			
Capillaries	+	+	+
Veins	+++	+++	+++
Arteries	+++	+++	+++
Epithelial			
Keratinizing epithelia	-	++	-
Non keratinizing epithelia	-	++	-
Salivary glands	++	++	++
Thyroid	+++	+++	+++
Breast		++	++
Ileum	-	-	-
Colon	-	-	-
Liver hepatocytes	-	-	-
Liver sinusoids	+	+	+
Kidney	+++	+++	+++
Placenta		+++	+++
Mesenchimal			
Fibroblasts	++	++	++
Myofibroblasts	++	++	++
Adipocytes	-	-	-
Smooth muscle	+++	+++	+++
Cartilage	-	-	-
Nervous			
Neurons	-	-	-
Glial cells	-	-	-
Microglial cells	-	-	-
Schwann	+	++	+

Cryostat sections of different tissues were immunostained with monoclonal antibodies directed to αv, $\beta3$ or the complex $\alpha v\beta3$, using avidin-biotin-peroxidase method. The antigen was expressed in endothelial cells from muscular arteries, arterioles, venules and veins including human umbilical veins. Capillary endothelium was less intensely stained.

This idea also is indirectly supported by the observation that integrin organization is dependent, at least during the first hours of spreading, on the binding to their specific ligand. For instance, when endothelial cells are plated on vitronectin only $\alpha^v\beta_3$ clusters, while when they are plated on fibronectin only $\alpha^5\beta_1$ organizes in focal contacts.

Cytoskeletal organization and in particular actin microfilament assembly can regulate cell proliferation and motility. The exact mechanism by which this happens is still unknown but it is possible that the actin filament network transfer mechanical signals to the endothelial cell nucleus that promotes cell division (Ingberg, 1990).

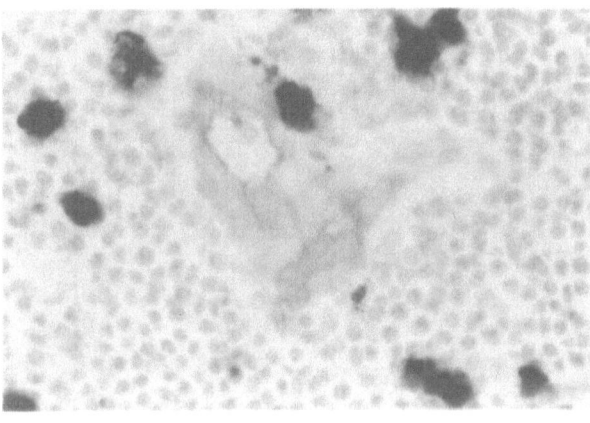

Figure 1. Cryostat Section of a Lymph Node Venule Immunostained with a Monoclonal Antibodies Directed to αvβ3 Using Avidin-biotin-peroxidase Method. The antigen is expressed in the cytoplasm and on the apical surface of endothelial cells. Magnification 1000x.

ENDOTHELIAL CELL TO CELL JUNCTIONS

Besides their location at the basal aspect of the membrane, in contact with matrix proteins, integrins (in particular $\alpha^2\beta_1$ and $\alpha^5\beta_1$) have been found at cell-cell junctions in endothelial cells (Lampugnani et al., 1991). All the other endothelial integrins remained diffuse on the cell membrane.

This integrin distributions seems to be specific of polarized cells. For instance in keratinocytes, two β_1 integrins are located at cell-cell contacts ($\alpha^2\beta_1$ and $\alpha^5\beta_1$) (Carter et al., 1990; De Luca et al., 1990) while no such distribution was found in smooth muscle cells or skin fibroblasts.

When antibodies directed to $\alpha^2\beta_1$ or $\alpha^5\beta_1$ were added to a confluent endothelial cell monolayer a detectable increase in endothelial permeability to high molecular weight proteins was observed (Lampugnani et al., 1991). This indicates that integrins play a functional role in regulating endothelial cell permeability properties. It is still obscure how integrins organize at cell to cell contacts. If a matrix ligand is indeed required for the intercellular role of integrins, this may be produced by endothelial cells during the relatively long time required for building up monolayers. Good candidates are matrix molecules such as fibronectin, laminin or collagen all major ligands for $\alpha^2\beta_1$ or $\alpha^5\beta_1$. We found that the latter matrix proteins are indeed produced by endothelial cells and are concentrated in strands corresponding to cell-cell interaction rims. However, antibodies directed to these matrix proteins were inactive in inhibiting cell-cell interaction while they were potent in blocking cell-substrate adhesion. When endothelial cell were detached from their matrix and maintained in suspension, they aggregated and this was inhibited by Arg-Gly-Asp containing peptides and antibodies to $\alpha^5\beta_1$ (Figure 2). This suggests that integrins might also promote cell to cell contacts by an homotypic (integrin-integrin) type of interaction.

In endothelial cells other molecules have been described to be localized at intercellular contacts: PECAM (or endoCAM or CD31), a recently sequenced integral protein belonging to the immunoglobulin family (Newman et al., 1990; Albelda et al., 1990), and a new endothelial specific cadherin (Lampugnani et al., 1992) (Figure 3).

The biological role of PECAM remains to be fully defined. Probably, endothelial cell-cell adhesion is mediated by an homotypic type of interaction between different PECAM molecules. However, since this protein is also present in platelets and leukocytes it is possible that it could play a more general role in promoting adhesion of different types of cells. Recent studies (Zehnder et al., 1992) showed that PECAM can be phosphorylated after cell activation. This might modulate its cellular adhesiveness.

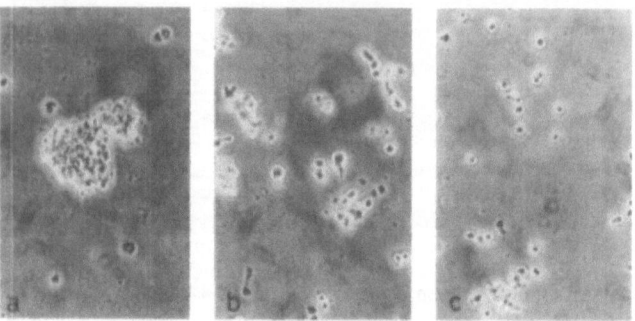

Figure 2. Homotypic aggregation of human umbilical vein endothelial cells after 4 hours incubation with: **a.** goat preimmune serum (1:10 dilution); **b.** goat serum to α5β1 (1:10 dilution); **c.** goat preimmune serum (1:10) and Gly-Arg-Gly-Asp-Ser-Pro (1mM) peptide.

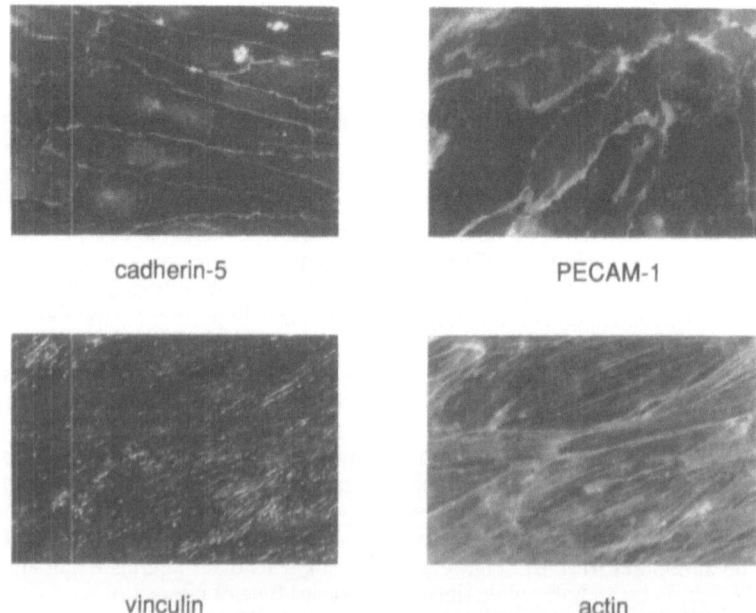

Figure 3. Indirect Immunofluorescence Staining of Confluent Human Umbilical Vein Endothelial Cells in Culture. Cells were decorated with a monoclonal antibody to cadherin-5, PECAM, vinculin and fluorescein labelled phalloidin for actin staining. Cadherin-5 and PECAM reactivity is restricted to areas of cell contacts. Vinculin is localized in dotted spots corresponding to focal contacts and along the cell rims, actin is organized in stress fibers and along cell junctions. Magnification 1000x.

The new endothelial cadherin (7B4 from the name of a monoclonal antibody able to bind to it or cadherin-5) is a Ca^{++}-dependent adhesion molecule strictly located at cell to cell junctions in endothelial cells both *in vitro* (Figure 3) and *in situ*. Cadherin-5 is bound at appositional surfaces of cultured endothelial cells only as they become confluent and is stably expressed at intercellular boundaries of confluent monolayers. This molecule is present in all types of endothelium (veins and arteries, capillary and large vessels). *In vitro*, addition of antibodies directed to this molecule increased permeation of macromolecules across monolayers even without any obvious change of cell morphology. When endothelial cell permeability was increased by agents such as thrombin, elastase and tumor necrosis factor (TNF) in combination with γ-interferon (IFN), its distribution pattern at intercellular contact rims was severely altered. The monoclonal antibody 7B4 immunoprecipitated a major protein of 140 kD from metabolically and surface labelled endothelial cells. NH_2-terminal sequencing of the antigen and sequencing of peptides from tryptic peptide maps revealed that the molecule is a new member of the cadherin family (Lampugnani et al., 1992).

The interrelationship between PECAM, cadherin-5 and integrins at cell-cell contacts is unclear. Integrins might directly bind these proteins in a heterotypic type of interaction or just integrate and support their role in maintaining endothelial cell boundaries.

It also is unclear which are the cytoskeletal proteins involved in PECAM, cadherin-5 and integrin organization. As reported for other cell types actin bundles and vinculin can be observed at cell-to-cell contacts (Figure 3).

Overall these data indicate that cell to cell junctions in endothelial cells are very complex structures where different molecules can be organized and play a coordinated role in maintaining the permeability and antithrombotic properties of the monolayer. Further studies are required to understand how these molecules interplay and which is their reciprocal relevance.

ACKNOWLEDGEMENTS

This work was supported by the Italian National Research Council (Progetto Finalizzato Biotecnologie e Biostrumentazione; Progetto Speciale: "Peptidi Bioattivi"; Progetto Finalizzato ACRO) and by the Commission of the European Communities (BRIDGE: BIOT-CT90-0195). Massimo Resnati's fellowship was provided by Banca Popolare di Milano.

REFERENCES

Albelda, S.M. and Buck, C.A., 1990, Integrins and other cell adhesion molecules. *FASEB J.* 4:2868-2880.

Albelda, S.M.,Oliver, P.D., Romer, L.H., and Buck, C.A. 1990, EndoCAM: A novel endothelial cell-cell adhesion molecule. *J. Cell Biol.* 110:1227-1237.

Bar-Shavit, R., Sabbah, V., Lampugnani, M.G., Marchisio, P.C., Fenton, II, J.W., Vlodavsky, I., and Dejana, E., 1991, An arg-gly-asp sequence within thrombin promotes endothelial cell adhesion. *J. Cell Biol.* 112:335-344.

Burridge, K., Fath, K., Kelly, T., Nuckolls, G., and Turner, C., 1988, Focal adhesions: transmemebrane junctions between the extracellular matrix and the cytoskeleton. *Ann. Rev. Cell Biol.* 4:487-525.

Carter, W.G., Wayner, E.A., Bouchard, T.S., and Kaur, P., 1990, The role of integrins $\alpha^2\beta_1$ and $\alpha^3\beta_1$ in cell-cell and cell-substrate adhesion of human epidermal cells. *J. Cell Biol.* 110:1378-1404.

Cheresh, D., 1987, Human endothelial cells synthesize and express an arg-gly-asp directed adhesion receptor involved in attachment to fibrinogen and von Willebrand factor. *Proc. Natl. Acad. Sci. USA* 84:6471-6475.

Conforti, G., Dominguez-Jimenez, C., Zanetti, A., Gimbrone, M.A., Cremona, O., Marchisio, P.C., and Dejana, E., 1992, Human endothelial cells express integrin receptors on the luminal aspect of their membrane. *Blood* 40:437-446.

Conforti, G., Zanetti, A., Colella, S., Abbadini, M., Marchisio, P.C., Pytela, R., Giancotti, F., Tarone, G., Languino, L.R., and Dejana, E., 1989, Interaction of fibronectin with cultured human endothelial cells: Characterization of the specific receptor. *Blood* 73:1576-1585.

Conforti, G., Zanetti, A., Pasquali-Ronchetti, I., Quaglino, Jr., D., Neyroz, P., and Dejana, E., 1990, Modulation of vitronectin receptor binding by membrane lipid composition. *J. Biol. Chem.* 265:4011-4019.

Defilippi, P., van Hinsberg, V.,Bertolotto, A., Rossino, P., Silengo, L., and Tarone, G., 1991, Differential distribution and modulation of expression of alpha 1/beta 1 integrin on human endothelial cells. *J. Cell Biol.* 114:855-863.

De Luca, M., Tamura, R.N., Kajiji, S., Bondanza, S., Rossino, P., Cancedda, R., Marchisio, P.C., and Quaranta, V., 1990, Polarized integrin mediates human keratynocyte adhesion to basal lamina. *Proc. Natl. Acad. Sci. USA* 87:6888-6892.

Dejana, E., 1991, Endothelial cell integrin receptors, in: "Vascular Endothelium: Interactions with Circulating Cells, " J.L. Gordon, ed., Elsevier, Amsterdam, pp 31-41.

Dejana, E., Colella, S., Conforti, G., Abbadini, M., Gaboli, M., Marchisio, P.C., 1988, Fibronectin and vitronectin regulate the organizaton of their respective arg-gly-asp adhesion receptors in cultured human endothelial cells. *J. Cell Biol.* 107:1215-23.

Dejana, E., Lampugnani, M.G., Giorgi, M., Gaboli, M., and Marchisio, P.C., 1990, Fibrinogen induces endothelial cell adhesion and spreading via the release of endogenous matrix proteins and the recruitment of more than one integrin receptor. *Blood* 75:1509-1517.

Dejana, E., Lampugnani, M.G., Giorgi, M., Gaboli, M.,Federici, A.B., Ruggeri, Z.M., and Marchisio, P.C., 1989, vonWillebrand factor promotes endothelial cell adhesion via an arg-gly-asp dependent mechanism. *J. Cell Biol.* 109:367-375.

Hemler, M.E., 1990, VLA proteins in the integrin family: structures,functions and their role on leukocytes. *Annu. Rev. Immunol.* 8:365-400.

Hynes, R.O., 1986, Integrins: Versatility, modulation and signaling in cell adhesion. *Cell* 48:549-554.

Ingberg, D.E., 1990, Fibronectin controls capillary endothelial cells growth by modulating cell shape. *Proc. Natl. Acad. Sci. USA* 87: 3579-3583.

Kirchhofer, D., Grzesiak, J., and Pierschbacher, M.D. Calcium as a potential physiological regulator of integrin-mediated call adhesion. *J. Biol. Chem.* 266:4471-4477.

Kramer, R.H., Mc Donald, K.A., and Vu, M.P., 1989, Human microvascular endothelial cells use $\beta 1$ and $\beta 3$ integrin receptor complexes to attach to laminin. *J. Biol. Chem.* 264:15642-15649.

Lampugnani, M.G., Resnati, M., Dejana, E., and Marchisio, P.C.,1991, The role of integrins in the maintenance of endothelial monolayer integrity. *J. Cell Biol.* 112:479-490.

Lampugnani, M.G., Resnati, M., Raiteri, M., Pigott, R., Pisacane, A., Houen, G., Ruco, L.P., and Dejana, E.,1992, A novel endothelial specific membrane protein is a marker of cell-cell contacts. *J. Cell Biol.* (*in press*).

Languino, L.R., Gehlsen, K.R., Wayner, E., Carter, W.G., Engvall, E., and Ruoslahti, E., 1989, Endothelial cells use $\alpha^2\beta_1$ integrin as a laminin receptor. *J. Cell Biol.* 109:2455-2462.

Lawler, J., Weinstein, R., and Hynes, R.O., 1988, Cell attachment to thrombospondin: the role of arg-gly-asp, calcium and integrin receptors. *J. Cell Biol.* 107:2351-2361.

Newman, P.J., Berndt,M.C., Gorski, J., White, II G.C., Lyman, S., Paddock, C., and Muller, W., 1990, A.PECAM-1(CD31) cloning and relation to adhesion molecules of the immunoglobulin gene superfamily. *Science* 247:1219-1222.

Ruoslahti, E. and Giancotti, F.G., 1989, Integrins and tumor cell dissemination. *Cancer Cells* 1:119-126.

Ruoslahti, E. and Pierschbacher, M.D., 1987, New perspectives in cell adhesion: RGD and integrins. *Science* 238:491-493.

Simionescu, M. and Simionescu, N.,1991, Endothelial transport macromolecules: transcytosis and endocytosis. *Cell Biol. Reviews* 25:5-80.

Smith, J.W., Vestal, D.J., Irwin, S.V., Burke, T.A., and Cheresh, D.A., 1990, Purification and functional characterization of integrin avß5. An adhesion receptor for vitronectin. *J. Biol. Chem.* 265:11008-11013.

Sonnenberg, A., Modderman, P.W., and Hogervost, F., 1988, Laminin receptor on platelets is the integrin VLA-6. *Nature* 336:487-489.

van Mourik, J.A., von dem Borne, A.E.G.Kr., and Giltay, J.G., 1989, Pathophysiological significance of integrin expression by vascular endothelial cells. *Biochem. Pharm.* 39:233-239.

Wayner, E.A. and Carter, W.G., 1987, Identification of multiple cell adhesion receptors for collagen and fibronectin in human fibrosarcoma cells possessing unique α and common β subunit. *J. Cell Biol.* 105:1873-1884.

Zehnder, J.L., Hirai, K., Shatsky, M., McGregor, J.L., Levitt, L.J., and Leung, L.L.K., 1992, The cell adhesion molecule CD31 is phosphorylated after cell activation. Down regulation of CD31 in activated T lymphocytes. *J. Biol. Chem.* 267:5243-5249.

DISCUSSION

C. FIGDOR

You showed that the combination of cytokine you used, interferon-γ and TNF, has no effect on the up-regulation of this molecule, but it had an effect on the distribution. When you compare it with CD31 do you see the same type of change in distribution or does CD31 remain distributed as it is?

E. DEJANA

We did not do as a complete study as for 7B4, but we are going to do it.

C. FIGDOR

And why did you use the combination of cytokines, do you need them both?

E. DEJANA

We used γ-interferon and TNF because that was the most effective treatment, but you can get similar data also with TNF alone.

C. FIGDOR

And with IL-β?

E. DEJANA

No, IL1 is less effective in our hands.

G. TARONE

In terms of the the relation of the localisation of integrins to the cell-cell contact, did you ever compare the localization of integrins and cadherin-5 at the cell-to-cell contact? Is it possible that this cadherin is a ligand for the integrins in the cell-cell contact?

E. DEJANA

We are doing this experiment actually. It is very difficult to do it in tissue section, because, as you know, integrins also stain other cell types so it is very difficult to get a specific staining. It might be, even if it would be against the dogma, that a cadherin is a ligand for an integrin. This molecule has a RGD sequence, which however does not seem to be exposed. For PECAM an homotypic type of interaction exists but there is also evidence that PECAM can bind something else. We do not know whether it is an integrin ligand.

A. SONNENBERG

One question, one comment. Your immunoprecipitation data indicate that also α6β4 is expressed on endothelial cells.

E. DEJANA

We never went into the detail of that. When we tried to immunoprecipitate with anti-β 4 antibodies we were unsuccessful, maybe we did not use the right conditions.

A. SONNENBERG

And the question is, whether there is any evidence that cadherin-5 is associated with catenins. I refer specifically to the fact that in one of your gels, you see some proteins precipitated in the area of 90-120 kD.

E. DEJANA

Indirect evidence suggests that the three catenins are linked to cadherin-5, but we are waiting for specific antibodies to confirm these results.

J. CALVETE

You showed a lane where cadherin was digested with trypsin both in the presence and the absence of calcium. However, cadherins seem to be quite resistant to enzymatic digestion in the presence of calcium. How can you explain this discrepancy, and how were your digestion experiments done?

E. DEJANA

We do not have an explanation so far. This cadherin seems to be particularly sensitive to enzymatic digestion. It might be an interesting characteristic because in endothelial cells it is located at the junctions and it might be that enzymes released by tumor cells or leukocytes digest the molecule and promote, in this way, cell extravasation.

J. CALVETE

Was it digested to small pieces or to large domains?

E. DEJANA

With 7B4 antibody we could not immunoprecipitate any low molecular band. We did not use different times of digestion, however. Certainly if you decrease the time of exposure to trypsin in the presence of calcium you partially protect the molecule.

J. CALVETE

So, these were immunoblots after precipitation. Then, the only thing you can say is that at least the epitope for your monoclonal was destroyed by trypsin.

E. DEJANA

Yes.

V. QUARANTA

What sort of morphological changes do you see in the endothelial monolayers when you add 7B4 or the anti-α_5? Are the effects tangible?

E. DEJANA

When you add the 7B4 antibody you do not see a dramatic change in the morphology of the cells but you see a detectable increase in the passage of peroxidase through the monolayer. In contrast, when we used α_2 and α_5 integrin antibodies we observed the appearance of discontinuities in the monolayers.

J. CALVETE

Could it be that cadherins may be acting at an earlier stage?

E. DEJANA

Yes, we are working on this. We are adding the antibodies to the cells while they are coming in contact to see whether you can change the organization of the junctions.

P. HERRLICH

In order for this change in functional permeability to be a natural phenomenon, we must expect that the initial interaction has to be on the apical side, and that there would be a signal transfer to the basolateral cell surface. So, can this be imitated? Do you have an interaction with a molecule (with an integrin, for instance) on the apical surface of the endothelial cell and then have redistribution in the junctions?

E. DEJANA

It may well be that a cell that is adhering on the apical surface of the endothelium then changes the junction structure. We do not have direct evidence for this. I believe that another hypothesis might be that when you have a leukocyte adhering on the surface of endothelial cells it can get activated and release lytic enzymes locally. These cadherins are really very sensitive to these enzymes, so you might have a small lytic digestion of these molecules. On the other hand cytokines, I mean TNF and γ-interferon, even if after a long time of incubation, change the organization of the junctions, this might be another way to favor cell extravasation.

R. JULIANO

There are some simple *in vitro* models of angiogenesis; have you looked to see if there are any changes in expression or distribution of the cadherins during vessel formation?

E. DEJANA

Yes, we are doing this now.

D. LIVINGSTON

How stable are those molecules? How stable is cadherin-5?

E. DEJANA

You mean when it is purified or when it is expressed on the cell membrane?

D. LIVINGSTON

No, you label it and look at its turnover.

E. DEJANA

We did not look at that.

D. LIVINGSTON

The reason I ask that is that if it is not particularly stable one could do an anti-sense experiment and ask questions about some of the biological implications of what you have.

E. DEJANA

Sure. This is another thing.

D. LIVINGSTON

One of the requirements is that the protein not be rock stable.

G. MARGUERIE

I would like to make some hypotheses. When you have formation of a granuloma, one of the first reaction is the leukocyte adhesion to the endothelium and then penetration of leukocytes into the subendothelium. How can this happen with your molecule? During the development of metastasis E-cadherin are up- and down-regulated and that regulates the level of adhesiveness of the metastatic cells. Probably the leukocytes send some message to the endothelial cells so that the gene of your cadherin can also be up- and down-regulated. But you mentioned in fact that when you stimulate with interferon or TNF, you rather see a relocation and no decrease of the molecule. What was the time difference between the stimulation and the immunoprecipitation and did you look at the messenger level?

E. DEJANA

Yes, this is going to be done. The time of incubation with the TNF and interferon is a long time of incubation, 72 hours, because this is the time required for the increase in permeability of the endothelial cell monolayer. I do not know what might happen before. We have to investigate different times of incubation. However, I think that you might also hypothesize that the activity of this molecule is changed just because it is relocated on the membrane. To me it is easier to think that for a fast regulation, for example, you might have a phsphorylation of the cytoskeletal proteins linked to this molecule and then you lose the specific localization of cadherin-5 in the junctions and it is diffused on the membrane.

P. MARCHISIO

But extravasation is a slow process, is it not?

E. DEJANA

No, I think that for example the passage of leukocyte may be very fast, for tumor cells it may be something different.

P. MARCHISIO

I have a very short comment. When you study a wounded monolayer of endothelial cells and treat it with hepatocyte growth-factor, scatter-factor, the cell to cell molecules, including integrins, and the antigen recognized by 7B4 and the CD31 antigens are down-regulated in the surface of the cells within minutes of addition of the scatter factor and long before you have a morphological change of the cells at the border.

E. DEJANA

Are they down-regulated by immunoprecipitation analysis or just because you see that at immunofluorescence?

P. MARCHISIO

By immunofluorescence you see the disappearance of the signal around cells.

E. DEJANA

It might be diffusion?

P. MARCHISIO

Yes, it might be a diffusion.

G. TARONE

Coming back to the question of Gerard, I must add one thing. We have shown that TNF and interferon-γ after a long treatment (48/72 hours) deeply change the expression of integrins on the surface of endothelial cells. The vitronectin receptor $\alpha V\beta 3$ and the laminin receptor $\alpha_6\beta_1$, are down-regulated, and $\alpha_1\beta_1$ is up-regulated. The other integrins are unaffected. Some of these effects are due to regulation of gene expression. Is it possible that in the case of cadherin 5 there is a similar mechanism of regulation?

E. DEJANA

You might have a fast way of regulation of the organization of this molecule probably through phosphorylation mechanisms. You might also have a slow way of regulation, that is through down-regulation of the synthesis of the molecule.

E. MIHICH

In relation to Dr. Juliano's question, when you reach that point of studying angiogenesis in systems such as the rabbit cornea or the chorioallantoid membrane of the chicken egg you may wish to use some new inhibitors of angiogenesis that are now available to Dr. Folkman in Boston and others; but I have a question. You said that this molecule was present only in endothelium and in some kind of endothelium related macrophages. I was wondering whether you had checked sinusoids of lymph nodes and whether you had extended your distribution studies to hepatic sinusoids epithelial cells, and also Kuppfer cells which are more frank macrophages.

E. DEJANA

To be honest, no. That is something missing. It might well be that the sinusoids do not express the molecule or it is organized in a different way.

E. MIHICH

Maybe it is so from a functional point of view, also from the immunological point of view. Now more attention is given to the so-called regional immunity which, in the liver, involves epithelial cells in addition to the Kupffer cells.

E. DEJANA

Well we are looking now at different types of hemangiomas. In different types of hemangiomas you have islands of endothelium express 7B4 and other areas that do not express it.

S. GOODMAN

I do not think you mentioned if the 7B4 antigen is expressed in microvascular endothelium as well.

E. DEJANA

In all the tissues that we have screened it is present in all types of vessels.

D. LIVINGSTON

Are these primary epithelial cultures, umbilical vein epithelial, endothelial cultures that you are using?

E. DEJANA

We studied both primary cultures and endothelial cells after several passages. The antigen is always present.

D. LIVINGSTON

Right. Have you ever looked in replicating endothelial cells? What happens to this protein during mitosis?

E. DEJANA

No, we have not looked at it.

D. LIVINGSTON

Because if a protein of this type played any role specifically in the dramatic changes in cell shape that occurs during M, you might expect to see a reorganization at that time.

E. DEJANA

Sure. There is another aspect of the problem that is interesting, concerning the lines of endothelium that we know do not express 7B4.

D. LIVINGSTON

Lines or strains?

E. DEJANA

Lines. Lines of endothelium. It might well be that there is a sort of negative growth signal to the endothelium.

D. LIVINGSTON

So, can I push this for a second? If you take a confluent monolayer in the presence of basic TGF or any of the other appropriate growth factors and attack it with your antibody, do the cells go through a round of replication?

E. DEJANA

We are working on this, as well as with the purified soluble antigen to see whether you can change endothelial cell growth capacity.

D. LIVINGSTON

Or perhaps coat latex spheres with cadherin-5 and ask for a round of replication.

T. CAREY

Simon Goodman's comment on expression in vascular endothelium was interesting and reminded me of something I was going to ask you. Do you find that small vessels in tumors or vessels in endometrium, we call these neo-vascular vessels, have decreased expression of cadherin or PECAM?

E. DEJANA

Luigi Ruco did some work on tumors, and the vessels were positive, but I am not sure that he really examined all the types of vessels and how they were organized. So this is certainly a thing that has to be done.

T. CAREY

And how about in endometrium, decidual endometrium?

E. DEJANA

He did not do that.

S. SHAW

You commented earlier on in your talk on the differences between large and small vessel endothelium in terms of their integrin profiles. To what extent is that plastic and can be changed for a given cell type, or source, and what are the presumably physiologic stimuli which influence that during the differentiation of the endothelial cell?

E. DEJANA

You mean in terms of expression of one integrin or the other?

S. SHAW

Yes.

E. DEJANA

I think that actually the best data are those of Guido Tarone. They really analyze different stimuli for the expression of the different integrins. Cytokines can down-regulate the $\alpha_V\beta_3$, and $\alpha_6\beta_1$, while $\alpha_1\beta_1$ is increased in umbilical vein endothelium. While in the microcirculation $\alpha_1\beta_1$ is always present, it is absent in large vessels. In terms of other stimuli that can up-regulate or down-regulate other integrins along the vascular tree, I think that very little is known. At least as far as I know.

D. LIVINGSTON

Henry Mihich's question is relevant. If you study regenerating liver do you detect this protein? Because that is a wonderful system for studying them.

E. DEJANA

We do not know. There is still a lot of work to be done and this is a very nice suggestion.

REGULATION OF T CELL ADHESION WITH T CELL DIFFERENTIATION AND WITH ACUTE ACTIVATION BY MIP-1β CYTOKINE IMMOBILIZED ON CD44 PROTEOGLYCAN

Yoshiya Tanaka, David H. Adams, Tamas Schweighoffer, and Stephen Shaw*

*Experimental Immunology Branch
National Cancer Institute
National Institutes of Health
Bethesda, Maryland 20892 U.S.A.

ABBREVIATIONS USED

CD44 = Cluster Determinant-44; **GAG** = Glycosaminoglycan; **HEV** = High Endothelial Venules; **IL8** = Interleukin-8; **mAb** = Monoclonal Antibody; **TGFβ** = Transforming Growth Factor-β; **VCAM-1** = Vascular Cell Adhesion Molecule-1

INTRODUCTION

Adhesion is essential to many aspects of the T cell physiology. Although initially it seemed plausible that this would be mediated by only a few molecules, it is now apparent that there are many adhesion molecules on T cells. Resting T cells have more than a dozen, activated cells more than two dozen, and more will undoubtedly be discovered. The challenge is to find simplifying paradigms to understand similarities and differences between these molecules and elucidate general principles by which they function. One such important concept which has emerged is that of adhesion *cascades*.

Adhesion is not a single event, but rather a carefully coordinated series of events in which different molecules play different roles. Since the details of an adhesion cascade between a T cell and an apposing cell are dictated by the ensemble of molecules present on both cell types, there will be different adhesion cascades for T cell interactions with different cell types. We subdivide these into two families of cascades which are differentiated by whether or not there is involvement of the T cell receptor.

We designate one family as T cell receptor-dependent (or antigen-specific). This includes the interactions of T cells with various antigen-bearing cells including macrophages, B cells or any target for cell-mediated cytotoxicity (Makgoba et al., 1989). This kind of cascade is obviously central to the processes of T cell activation and T cell effector function. The other family of adhesion cascades we refer to as T cell-receptor-independent (or antigen-independent). This includes the interactions involved in the migratory life of the T cell. Of such interactions, we have studied most intensely the process of T cell binding to vascular endothelial cells (Shimizu et al., 1992). Since this process is a necessary first step for T cell movement into tissue, it is a crucial one. Our presentation at this meeting explores T cell interactions with endothelial cells. However, fundamental similarities between the two

Cell Adhesion Molecules, Edited by M.E. Hemler
and E. Mihich, Plenum Press, New York, 1993

families make the same considerations relevant to other adhesion cascades (Schweighoffer and Shaw, 1992).

A model of the steps involved in T cell interaction with endothelium is represented in Figure 1. This is the T cell version of a "consensus model" which has evolved from studies in many laboratories. Findings with granulocytes have been particularly important in formulating this model; but studies of platelets, monocytes, and lymphocytes have confirmed and extended what appears to be a very general formulation of the processes involved in leukocyte-endothelial cell interactions.

It is easiest to describe the rationale for the cascade by working backwards. Strong adhesion is essential, and is mediated by members of the integrin family, molecules widely discussed at this meeting. However, the integrins on circulating resting T cells are in a functionally inactive state. Therefore, in order for them to play a role, there must be a preceding step which activates integrin function. We refer to that step as triggering. To accomplish this a ligand at or near the endothelial surface must engage a receptor on the T cell. Although soluble cytokines have been found to trigger integrin function on granulocytes, this step has been an enigma for T cells. We now present data which indicate that an inflammatory cytokine can be the trigger for T cells. More important, our data indicate that this cytokine acts not in solution, but as a form bound to the endothelial surface.

Triggering and integrin-mediated adhesion are not sufficient for T cell adhesion to endothelium. Since the trigger ligand is present at the endothelial surface and the relevant trigger receptor is present on the T cell, there needs to be at least a brief period of contact for triggering to occur. The first step, referred to as tethering, involves a loose transient adhesion to provide such (brief) contact. This step is mediated by the selectin family of molecules whose structure appears to be uniquely suited for this role of mediating initial contact (Siegelman, 1991).

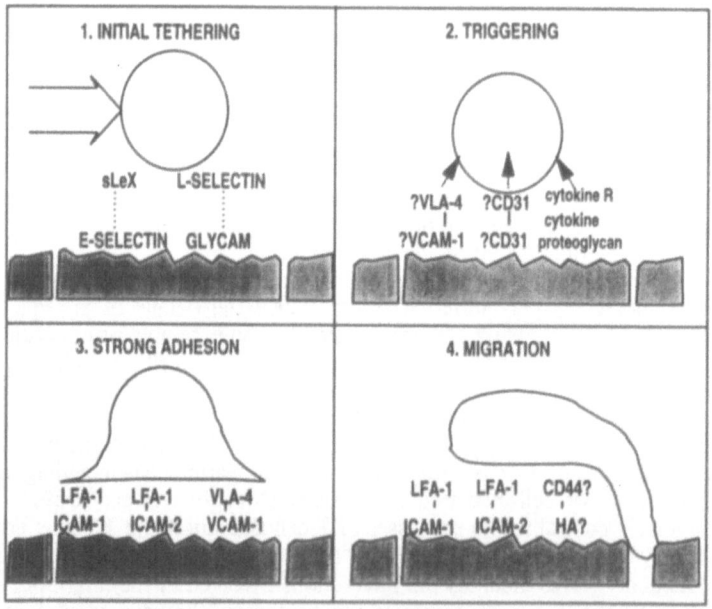

Figure 1. Model of TCR-Independent Adhesion to Endothelium. Proposed Sequence of Events in an Adhesion Cascade Mediating T Cell Adhesion to Endothelium (Shimizu et al., 1992). The cascade involves: (1) tenuous adhesion or tethering of a flowing T cell via selectin-mediated interactions; (2) delivery of a triggering signal to up-regulate integrin function; (3) strong integrin-mediated adhesion; and (4) migration of the T cell through the endothelium into the surrounding tissue. Potential receptor/ligand interactions are shown.

This brief review, which summarizes findings presented at the Pezcoller meeting, explores two aspects of regulation of T cell adhesion cascades. The first aspect relates to the integrins which mediate strong adhesion (step 3). We describe the regulation of these molecule during post-thymic T cell differentiation. The second aspect is the description of cytokine-mediated triggering of T cells, which relates to step 2. Because of the format of this review, citations relating to background information will be largely to other reviews rather than primary literature.

Two Concepts of Regulated Expression of Adhesion Molecules with T Cell Different-iation

We have sought to understand "peripheral" T cell differentiation, that is the progressive differentiation of the CD4 (or CD8) lineage of T cells after they have left the thymus. It involves many complex processes but we believe there are two important general principles which simplify considerably the understanding of this overall process. The first relates to fundamental differences between naive and memory cells and the second relates to specialization among memory cells. Adhesion is important to both (Horgan et al., 1992b).

The first principle is that the most fundamental subdivision among circulating CD4 cells, is the dichotomy between naive and memory cells. Naive cells have never undergone antigen-specific activation. When they do, they undergo complex changes and eventually (presumably a matter of days) they revert to a relatively quiescent memory cell. Acutely activated cells generally are *not* found in circulation and are not relevant to the present discussion.

Since many molecules contribute to adhesion, thorough understanding of the "personality" of T cells requires understanding of as many molecules as possible. Consequently, we have approached the analysis of peripheral T cell differentiation by exploring as wide a range as possible of surface molecules on the T cells. What we find is that many of the molecules which we know to be relevant to T cell adhesion show regulated expression during peripheral T cell differentiation. Systematic analysis demonstrates major differences between naive and memory cells. Memory cells express higher levels than naive cells of at least 8 adhesion molecules, including 5 integrins: LFA-1, VLA-3, VLA-4, VLA-5, and VLA-6 (Horgan et al., 1992b; Shimizu et al., 1990). Functional analysis demonstrates that memory cells bind more effectively than naive cells via these integrins to purified immobilized ligands including ICAM-1 (for LFA-1), fibronectin (for VLA-4, VLA-5), VCAM-1 (for VLA-4) and laminin (for VLA-6). Furthermore, when T cells are assayed for binding to cultured endothelium, memory cells bind much more effectively than naive cells via multiple pathways, including those mediated by LFA-1 and by VLA-4 (Shimizu et al., 1991).

Our findings integrate satisfyingly with findings from many investigators that memory cells but not naive cells are present in virtually all tissues which have been examined, including gut and skin (Mackay, 1991; Shimizu et al., 1992). Our findings of preferential expression and utilization of adhesion molecules by memory cells would account for such localization. But where do naive cells go? Elegant studies by Mackay (Mackay et al., 1990) demonstrate profuse preferential movement of naive cells into lymph node. Although the precise molecular mechanisms which account for this preferential movement remain to be defined, the physiologic importance of this strategy is profound. Naive cells *must* be activated in lymph node by antigen drained from tissue, often in migratory antigen-presenting cells (such as Langerhans cells). Only after naive cells become memory cells will they be eased from the lymph node into circulation and have the appropriate level of expression of A-1, VLA-4, and other adhesion molecules to facilitate their entry into non-lymphoid tissue.

We now consider the second simplifying principle, specialization of memory cells, which is also closely related to the issue of selective cell movement. The analysis began with more detailed characterization of the expression of all integrin chains on CD4+ T cells (Horgan et al., 1992a). The results demonstrate a simple bimodal expression of $\alpha3$, $\alpha5$, $\alpha6$, and $\beta1$. Furthermore, there is a simple correlation so that the same subset of cells which express high levels of $\beta1$ also expresses high levels of $\alpha3$, $\alpha5$, $\alpha6$; and the reciprocal subset expresses lower levels of each of these chains. However, a very different pattern emerges with $\alpha4$. Its expression is heterogenous rather than clearly bimodal. When the parameters $\alpha4$ and $\beta1$ are examined jointly, the pattern show at least 4 subsets, rather than simply 2.

When the β1 vs α4 complexity is dissected with markers for naive and memory cells, a simple conclusion emerges: naive cells have low homogeneous expression of α4 and β1 while memory cells show complex heterogeneity. Many memory cells express high levels of α4 and β1 but others show elevation of either one or the other.

Of particular interest is the subset of cells which express high levels of α4 but low levels of β1; we call these CD4 memory β1lo or simply 4Mβ1lo. Given the studies of Holzmann and Weissman demonstrating alternate pairing of α4 with β7 (Holzmann et al.., 1989; Holzmann and Weissman, 1989), we postulated that this subset might express β7. Extensive screening of pre-existing mAb identified one, designated Act-1 which is selectively expressed on the 4Mβ1lo cells (Schweighoffer et al., 1992); this mAb was originally described by Lazarovits and colleagues (Lazarovits et al., 1984). Act-1 mAb proves to be poor in immunoprecipitation, but success was achieved using the strategy of inclusion of Mn++ in the solubilizing buffer to stabilize the molecule. The results demonstrate that Act-1 immunoprecipitates the α4β7 integrin. Direct comparison with β7-specific antisera by immunoprecipitation, by sequential precipitation and by double antibody ELISA demonstrate that Act-1 binds to α4β7 but not to α4 alone (for example in α4β1) or to β7 (for example in $\alpha_{IEL}\beta7$). The heterogeneity of memory cell α4 vs β1 takes on new conceptual significance in light of the understanding of β7 expression. First, the results demonstrate that naive cells have low levels of β1 (and probably β7); however as they become memory cells some acquire higher expression of β1 and others acquire higher expression of β7 (the 4Mβ1lo cells). Second, since Holzmann and Weissman had described α4β7 as an integrin mediating specific binding to gut endothelium, this suggested that the 4Mβ1lo cells which express α4β7 are gut-homing cells.

Figure 2 shows a schematic analysis of the subdivision of CD4 cells and shows some of the additional phenotypic characterization of the subdivision within memory cells (Schweighoffer et al., 1992). The 4Mβ1lo cells are unique in two other respects. First, unlike all other CD4 memory cells they are low for α6. Thus their phenotype is precisely concordant with that described by Mackay and colleagues for cells selectively found in gut lymph: β1lo, α6lo memory cells (Mackay et al., 1992). Second, the 4Mβ1lo cells lack the CLA marker described by Picker, Butcher and co-workers as a skin homing receptor (Picker et al., 1990); that marker is found only on some of the cells of the 4Mβ1hi phenotype. Our data strongly suggest that the 4Mβ1lo subset of memory cells are those which selectively migrate to mucosal sites while the reciprocal 4Mβ1hi memory cell subset will migrate to skin and other tissue.

The finding that naive cells are homogenous in integrin phenotype is consistent with the interpretation that they are not specialized, but rather have a uniform capacity to enter all lymph nodes. In contrast, the diversified phenotype of memory cells is consistent with the interpretation that they acquire a distinct phenotype during their transition to memory cells. Those which differentiate in skin-draining lymph nodes will preferentially augment β1 expression and tend to enter sites in the skin while those which differentiate in mucosal-draining lymph nodes will preferentially augment β7 expression and tend to bind mucosal endothelium. Although these comments emphasize integrin phenotype, a number of other markers show heterogeneity in memory but not naive cells (Horgan et al., 1992b); we believe that there is a rather extensive specialization of memory T cell subsets.

Paradigm of Cytokine Immobilized on Proteoglycan as a Triggering Mechanism in T Cell Interactions with Endothelial Cells

As noted in the introductory comments on adhesion cascades, an important missing piece in our understanding of T cell binding to endothelium has been the lack of insight as to what provides the trigger of integrin function. Studies of granulocytes have convincingly demonstrated that cytokines and chemotactic factors present in solution can induce granulocyte adhesion (Butcher, 1991); some of these factors, such as IL8, are also chemotactic for granulocytes (Huber et al., 1991). Although it had not been demonstrated for T cells, we reasoned that cytokines might also be able to trigger their adhesion. We undertook an analysis of a variety of cytokines for their capacity to induce T cell chemotaxis and T cell adhesion. Among the cytokines studied, one cytokine stood out in its capacity to do both: MIP-1β (Tanaka et al., 1992). The chemotactic response was measured in a

early T cells

CD4+   CD8+

CD45RA+   CD45RO+
 naive   memory

CD31+   CD31-   ß1-low   ß1-high
                  ß7-high   ß7-low
                  a4-high   a4-high/low
                  a6-low   a6-high
                  CLA-low   CLA-high/low

home to gut   home to skin and other
            non lymphoid tissue

Figure 2. Schematic of Some of the Most Fundamental Subdivisions within Circulating T cells, with Emphasis on CD4 Memory Cells.

standard assay of movement through a polycarbonate filter in response to a gradient in MIP-1β concentration. The response occurred at low MIP-1β concentrations (1-10ng/ml) and showed a bell-shaped dose-response curve usually observed in such assays. Adhesion induction was measured by the ability of soluble MIP-1β to augment the binding of purified resting CD8 cells to VCAM-1 or fibronectin immobilized on plastic. This response was also observed at low MIP-1β concentrations.

MIP-1β is one of the rapidly expanding *chemokine* family of cytokines (Oppenheim et al., 1991; Schall, 1991). We tested a number of cytokines in that family because some of them were reported to induce granulocyte adhesion (especially IL8) and some T cell chemotaxis (IL8 and RANTES) (Larsen et al., 1989; Schall et al., 1990). In general this family of cytokines seems to have pro-inflammatory activity (Stoeckle and Barker, 1990). cDNA clones for MIP-1β have been isolated repeatedly in different labs as an abundant early product with activation of T cells and macrophages (Miller et al., 1989; Sherry et al., 1988; Wolpe and Cerami, 1989; Zipfel et al., 1989). The precise function of MIP-1β is not yet known. Most of the reported studies have been carried out with myeloid cells.

It seems to us unlikely that *soluble* cytokine is an effective trigger of integrin function in physiologic circumstances. Because of blood flow, soluble cytokine will be washed rapidly away. In contrast, we suggest that cytokine would be effective if *immobilized* at the endothelial surface. We postulated that cytokine could be immobilized by interaction with a proteoglycan on the endothelial surface (Tanaka et al., 1992). The concept that cytokines interact in functionally important ways with proteoglycans has been established in two contexts: 1) elegant studies of cytokines (such as TGFβ) binding to extracellular matrix (Lopez-Casillas et al., 1991; Nathan and Sporn, 1991; Ruoslahti and Yamaguchi, 1991); and 2) demonstration that proteoglycans on a cell act as critical low affinity receptors to facilitate subsequent interaction with higher affinity receptors on the same cell (Klagsbrun and Baird, 1991; Yayon et al., 1991). We are proposing a variation on these themes that proteoglycan on one cell can bind and "present" a cytokine to an apposing cell.

MIP-1β, like other chemokines, has a glycosaminoglycan (GAG) binding site. Since endothelial cells express GAGs, we reasoned that this could be a physiologic mechanism by which MIP-1β is retained at the endothelial surface. To test the concept that immobilized cytokine can trigger integrin adhesion, we developed the following simplified model system. We co-immobilized heparin-BSA (as an artificial proteoglycan) and VCAM-1 (as a VLA-4 ligand) on plastic wells (Tanaka et al., 1992). MIP-1β was then added, the unbound MIP-1β was washed out of the system, and then the adhesion of resting CD8 T cells was analyzed. The results demonstrate that MIP-1β can induce augmented adhesion, even though the soluble form has been washed out. The MIP-1β-mediated induction is observed only when the GAG is present to retain MIP-1β and the VCAM-1 is present as a ligand for integrin-induced adhesion.

The foregoing results established that immobilized MIP-1β could function as an inducer of T cell adhesion in the manner we postulated; but what might be a biologically relevant proteolgycan? CD44 has the desired attributes: 1) forms of CD44 carry GAG chains (Brown et al., 1991; Jalkanen and Jalkanen, 1992); 2) CD44 is expressed on endothelial cells (Pals et al., 1989); and 3) CD44 has been demonstrated to participate in T cell binding to endothelium in some assays (Berg et al.., 1989; Haynes et al.., 1989). Therefore we purified CD44 and substituted this natural material for the synthetic heparin-BSA. The results with the natural proteoglycan were analogous to those with the synthetic one. When MIP-1β was added and the free MIP-1β washed away, augmentation was observed only when both CD44 and VCAM-1 were present (Tanaka et al., 1992). The involvement of GAG chains on CD44 was supported by findings of inhibition by free heparin present during the MIP-1β preincubation and the involvement of VLA-4 by mAb blocking with relevant mAb. These data are consistent with concept that: 1) GAG chains on CD44 retain the MIP-1β so some is not washed away, and 2) retained MIP-1β induces T cell VLA-4 function, which then mediates binding to VCAM-1.

Our conclusion is that proteoglycans retain MIP-1β and thereby make it available for functional induction of T cell adhesion. Although we have not yet directly demonstrated binding of MIP-1β to proteoglycan, the conclusion seems justified since: 1) both CD44 and an artificial proteoglycan (heparin-BSA) function in analogous fashions; 2) the binding of MIP-1β to GAG is predicted from a structural motif in MIP-1β (Wolpe et al., 1988); and 3) other chemokines with similar structural motifs bind heparin (Oppenheim et al., 1991; Schall, 1991).

To test the hypothesis that MIP-1β might be immobilized on endothelium, immunohistochemical studies were performed on sections of lymphoid tissue using an antisera specific for a peptide sequence at the C-terminus of MIP-1β. Marked staining was observed in the vicinity of high endothelial venules (HEV) of reactive lymph node (Tanaka et al., 1992), which are the sites of lymphocyte entry into lymph nodes (Gowans and Knight, 1964; Mackay et al., 1990). Staining was luminal as well as basal. These data support the concept of endothelial retention of MIP-1β and presentation at the luminal surface of the endothelium. Many additional aspects remain to be explored including: how broadly MIP-1β is expressed at other endothelial sites; which other cytokines are similarly present at the endothelial surface; whether the endothelium *per se* is producing the MIP-1β; and which GAGs (or additional mechanisms) are responsible for cytokine retention at the luminal surface of endothelium.

At this juncture, we are using CD44 as a model proteoglycan. We have no evidence yet that CD44 on endothelial cells performs this function of cytokine retention. Indeed, the following lines of evidence are reasons that it *might* not: 1) the CD44 used so far in our experiments has been purified from PBML, rather than endothelial cells. It remains to be determined whether CD44 on any or all endothelium has the appropriate glycosylation to mediate this function; 2) there are a number of other proteoglycans present on endothelium which could account for GAG-mediated cytokine retention, for example Syndecan and Yyudocan (Jackson et al., 1991; Kojima et al., 1992); 3) the involvement of CD44 in T cell binding to endothelium has been attributed to cell CD44-binding to a hyaluronidase sensitive moiety on endothelium (Aruffo et al., 1990). Nevertheless, none of these are inconsistent with the concept of CD44 having a role on the endothelial cells in cytokine retention, as well as a role on the T cell in T cell binding to endothelium. A number of the proteoglycans, including the endothelial candidates CD44 and syndecan, show structural variation in both protein core and glycosylation. Such structural variations would be expected to influence cytokine retention and thereby introduce additional combinatorial possibilities into the adhesion cascades involved in specific binding of T cells and other circulating cell types to endothelium (Butcher, 1991; Shimizu et al., 1992).

ACKNOWLEDGEMENTS

We thank many generous colleagues for providing reagents; collaborators A. Lazarovits, D. Buck, H. Hirano, K. Horgan, S. Hubscher, G. Ginther-Luce, W. Newman, Y. Shimizu, and U. Siebenlist, for participation in the studies described in this review.

REFERENCES

Aruffo, A., Stamenkovic, I. Melnick, M., Underhill, C.B., and Seed, B., 1990, CD44 is the principal cell surface receptor for hyaluronate. *Cell* 61:1303.

Berg, E.L., Goldstein, L.A., Jutila, M.A., Nakache, M., Picker, L.J., Streeter, P.R., Wu, N.W., Zhou, D., and Butcher, E.C., 1989, Homing receptors and vascular addressins: Cell adhesion molecules that direct lymphocyte traffic. *Immunol. Rev.* 108:5.

Brown, T.A., Bouchard, T., St John, T., Wayner, E., and Carter, W.G., 1991, Human keratinocytes express a new CD44 core protein (CD44E) as a heparin-sulfate intrinsic membrane proteoglycan with additional exons. *J. Cell Biol.* 113:207.

Butcher, E.C., 1991, Leukocyte-endothelial cell recognition: Three (or more) steps to specificity and diversity. *Cell* 67:1033.

Gougos, A. and Letarte, M., 1990, Primary structure of endoglin, an RGD-containing glycoprotein of human endothelial cells. *J. Biol. Chem.* 265:8361.

Gowans, J.L. and Knight, E.J., 1964, The route of recirculation of lymphocytes in the rat. *Proc. Roy. Soc. B.* 159:257.

Haynes, B.F., Telen, M.J., Hale, L.P., and Denning, S.M., 1989, CD44 - A molecule involved in leukocyte adherence and T-cell activation. *Immunol. Today* 10:423.

Holzmann, B., McIntyre, B.W., and Weissman, I.L., 1989, Identification of a murine Peyer's patch-specific lymphocyte homing receptor as an integrin molecule with an alpha chain homologous to human VLA-4. *Cell* 56:37.

Holzmann, B. and Weissman, I.L., 1989, Peyer's patch-specific lymphocyte homing receptors consist of a VLA-4-like alpha chain associated with either of two integrin beta chains, one of which is novel. *EMBO J.* 8:1735.

Horgan, K.J., Luce, G.G., Tanaka, Y., Schweighoffer, T., Shimizu, Y., Sharrow, S.O., and Shaw, S., 1992a, Differential expression of VLA-α4 and VLA-β1 discriminates multiple subsets of CD4+ CD45R0+ "memory" T cells. *J. Immunol.* in press.

Horgan, K.J., Tanaka, Y., and Shaw, S., 1992b, Post-thymic differentiation of CD4 T lymphocytes: Naive vs memory subsets and further specialization among memory cells. *Prog. Chem. Immunol.* 54:72.

Huber, A.R., Kunkel, S.L., Todd, III, R.F., and Weiss, S.J., 1991, Regulation of transendothelial neutrophil migration by endogenous interleukin-8. *Science* 254:99.

Jackson, R.L., Busch, S.J. and Cardin, A.D., 1991, Glycosaminoglycans: molecular properties, protein interactions, and role in physiological processes. *Physiol. Rev.* 71:481.

Jalkanen, S. and Jalkanen, M., 1992, Lymphocyte CD44 binds the COOH-terminal heparin-binding domain of fibronectin. *J. Cell Biol.* 116:817.

Klagsbrun, M. and Baird, A., 1991, A dual receptor system is required for basic fibroblast growth factor activity. *Cell* 67:229.

Kojima, T., Leone, C.W., Marchildon, G.A., Marcum, J.A. and Rosenberg, R.D., 1992, Isolation and characterization of heparan sulfate proteoglycans produced by cloned rat microvacular endothelial cells. *J. Biol. Chem.* 267:4859.

Larsen, C.G., Anderson, A.O., Appella, E., Oppenheim, J.J., and Matsushima, K., 1989, The neutrophil-activating protein (NAP-1) is also chemotactic for T lymphocytes. *Science* 243:1464.

Lazarovits, A.I., Moscicki, R.A., Kurnick, J.T., Camerini, D., Bhan, A.K., Baird, L.G., Erikson, M., and Colvin, R.B., 1984, Lymphocyte activation antigens I. A monoclonal antibody, Act-I, defines a new late lymphocyte activation antigen. *J. Immunol.* 133:1857.

Lopez-Casillas, F., Cheifetz, S., Doody, J., Andres, J.L., Lane, W.S., and Massague, J., 1991, Structure and expression of the membrane proteoglycan Betaglycan, a component of the TGF-beta receptor system. *Cell* 67:785.

Mackay, C.R., 1991, T-cell memory: The connection between function, phenotype and migration pathway. *Immunol. Today* 12:189.

Mackay, C.R., Marston, W.L., and Dudler, L., 1990, Naive and memory T cells show distinct pathways of lymphocyte recirculation. *J. Exp. Med.* 171:801.

Mackay, C.R., Marston, W.L., Dudler, L., Spertini, O., Tedder, T.F., and Hein, W.R., 1992, Tissue-specific migration pathways by phenotypically distinct subpopulations of memory T cells. *Eur. J. Immunol.* 22:887.

Makgoba, M.W., Sanders, M.E., and Shaw, S., 1989, The CD2-LFA-3 and LFA-1-ICAM-1 pathways: Relevance to T-cell recognition. *Immunol. Today* 10:417.

Miller, M.D., Hata, S., Malefyt, R.D.W., and Krangel, M.S., 1989, A novel polypeptide secreted by activated human T lymphocytes. *J. Immunol.* 143:2907.

Nathan, C. and Sporn, M., 1991, Cytokines in context. *J. Cell Biol.* 113:981.

Oppenheim, J.J., Zachariae, C.O.C., Mukaida, N., and Matsushima, K., 1991, Properties of the novel proinflammatory supergene "intercrine" cytokine family. *Ann. Rev. Immunol.* 9:617.

Pals, S.T., Hogervorst, F., Keizer, G.D., Thepen, T., Horst, E., and Figdor, C.C., 1989, Identification of a widely distributed 90-kDa glycoprotein that is homologous to the Hermes-1 human lymphocyte homing receptor. *J. Immunol.* 143:851.

Picker, L.J., Terstappen, L.W.M.M., Rott, L.S., Streeter, P.R., Stein, H., and Butcher, E.C., 1990, Differential expression of homing-associated adhesion molecules by T cell subsets in man. *J. Immunol.* 145:3247.

Ruoslahti, E. and Yamaguchi, Y., 1991, Proteoglycans as modulators of growth factor activities. *Cell* 64:867.

Schall, T.J., 1991, Biology of the RANTES/SIS cytokine family. *Cytokine* 3:165.

Schall, T.J., Bacon, K., Toy, K.J., and Goeddel, D.V., 1990, Selective attraction of monocytes and T lymphocytes of the memory phenotype by cytokine RANTES. *Nature* 347:669.

Schweighoffer, T., Tanaka, Y., Horgan, K.J., Luce, G.E., Lazarovits, A.I., and Shaw, S., 1992, Selective expression of integrin α4β7 on a subset of human CD4+ memory T cells with hallmarks of gut-trophism. submitted

Schweighoffer, T. and Shaw, S., 1992, Adhesion cascades: Diversity through combinatorial strategies. *Curr. Opin. Cell Biol.* in press

Sherry, B., Tekamp-Olson, P., Gallegos, C., Bauer, D., Davatelis, G., Wolpe, S.D., Masiarz, F., Coit, D., and Cerami, A., 1988, Resolution of the two components of macrophage inflammatory protein 1, and cloning and characterization of one of those components, macrophage inflammatory protein 1b. *J. Exp. Med.* 168:2251.

Shimizu, Y., Newman, W., Gopal, T.V., Horgan, K.J., Graber, N., Beall, L.D., van Seventer, G.A., and Shaw, S., 1991, Four molecular pathways of T cell adhesion to endothelial cells: Roles of LFA-1, VCAM-1 and ELAM-1 and changes in pathway hierarchy under different activation conditions. *J. Cell Biol.* 113:1203.

Shimizu, Y., Newman, W., Tanaka, Y., and Shaw, S., 1992, Lymphocyte interactions with endothelial cells. *Immunol. Today* 13:106.

Shimizu, Y., van Seventer, G.A., Horgan, K.J., and Shaw,. S., 1990, Roles of adhesion molecules in T cell recognition: Fundamental similarities between four integrins on resting human T cells (LFA-1, VLA-4, VLA-5, VLA-6) in expression, binding, and costimulation. *Immunol. Rev.* 114:109.

Siegelman, M., 1991, Sweetening the selectin pot. *Curr. Biol.* 1:125.

Stoeckle, M.Y. and Barker, K.A., 1990, Two burgeoning families of platelet factor 4-related proteins: Mediators of the inflammatory response. *New Biologist* 2:313.

Tanaka, Y., Adams, D.H., Hubscher, S., Hirano, H., Siebenlist, U., and Shaw, S., 1992, Proteoglycan-immobilized MIP-1β induces adhesion of T cells. *Nature*, in press.

Wolpe, S.D., and Cerami. A. 1989. Macrophage inflammatory proteins 1 and 2: Members of a novel superfamily of cytokines. *FASEB J.* 3:2565.

Wolpe, S.D., Davatelis, G., Sherry, B., Beutler, B., Hesse, D.G., Hguyen, H.T., Moldawer, L.L., Nathan, C.F., Lowry, S.F., and Cerami, A., 1988, Macrophages secrete a novel heparin-binding protein with inflammatory and neutrophil chemokinetic properties. *J. Exp. Med.* 167:570.

Yayon, A., Klagsbrun, M., Esko, J.D., Leder, P., and Ornitz, D.M., 1991, Cell surface, heparin-like molecules are required for binding of basic growth factor to its high affinity receptor. *Cell* 64:841.

Zipfel, P.F., Balke, J., Irving, S.G., Kelly, K., and Siebenlist, U., 1989, Mitogenic activation of human T cells induces two closely related genes which share structural similarities with a new family of secreted factors. *J. Immunol.* 142:1582.

DISCUSSION

J. McCARTHY

This is a great story. Have you attempted to treat your dishes with chondroitinase? Maybe that will be a little sloppy, but if you treat the dishes with chondroitinase, you could make sure that chondroitin sulphate is in fact responsible for the binding.

S. SHAW

Those studies are currently in progress.

G. NICOLSON

Heparin can bind a number of molecules, of course, especially those released by endothelial cells. In your model for endothelial cells CD44 could be binding to a number of different cytokines and I find that to be a very interesting approach. We know that it binds growth factors, and so on. Do you think that this might be a kind of general mechanism for its presentation as CD44 in the various splice variants? You alluded to the fact that these splice variants may have different roles. Is that consistent with the side-chain binding results? I wonder if you could elaborate on this? In other words, are the heparan-sulfate side chains really binding these factors? And how would the splice variants come into play?

S. SHAW

First, there is evidence from many laboratories that proteoglycans are important in binding and presentation of cytokines in a variety of contexts. So, yes, we think that the mechanism we are proposing relative to cytokines and endothelial cells is part of a large picture of very general biological relevance. Second, with regard to CD44, it is clear that there are marked differences between cell types in the GAG side chains on CD44. This is likely controlled both by differential splicing of the CD44 protein core and by cell-types specific regulation of GAG addition. Graeme, do you want to comment?

G. DOUGHERTY

I am sure that Peter Herrlich is better positioned to answer this particular questions than me. However, it is evident from our studies that certain of the CD44 splice varaents that are generated, including CD44R1 and CD44R2, do contain serine-glycine motifs that could potentially act as sites of addition of chondroitin sulfate side-chains. Whether these sites are actually utilized or not, I do not think is clear at the moment.

S. SHAW

As one talks to people who know more than we do about structural specificity of GAGs, it is clear that there are other elements of specificity in GAG structure -- for example, details of sulfation. The protein core will determine whether a side chain can be added. Thereafter the exact structure of the GAG will be determined both by the protein core and the many enzymes in a given cell type which determine the details of the GAG composition/structure. I would speculate that the details of the GAG structure will be important in conferring specificity on its interaction with cytokines.

M. HEMLER

Is CD44 fairly widely expressed on all of these cell types?

S. SHAW

CD44 is widely expressed.

M. HEMLER

So you could envision that not only the VLA-4 function is increased, but many other integrin functions?

S. SHAW

Yes. There is a beautiful review by Nathan and Sporn called "Cytokines in Context." It describes how the effects of cytokines make no sense if you look at them in isolation. But as you look at them in the context of extracellular matrix in which they are presented then it makes quite good sense. Their point is the same one that I am making, namely, that every detail in the process is important and that every component in the process contributes an element of specificity to what happens subsequently. So that is why I really loved the presentation by Dr. McCarthy. He too was building a combinatorial model with GAG's playing an important role in determing the specificity of the response.

A. SANTONI

Did you study the combination between CD3 or CD2 and MIP-1β on T cell adhesion? Do you have any distinct feature between CD2 or CD2-induced T cell adhesion to PECAM-1 and this type?

S. SHAW

No, we have not looked at that yet. Let me explain why. We think of the "life" of T cells as having two components, antigen-specific and antigen-independent. So far, we think the "triggering" molecules sort out into one of the other. The primary trigger molecule for antigen-specific T cell responses is CD3; we tend to think of CD2 as a molecule so similar in triggering to CD3, that it must fit into that category. For our thinking about antigen-independent triggering, we think of CD31 as a prototype. So far these two processes seem relatively independent. For example, we have not seen synergism between CD3 and CD31 triggering of adhesion. In concept we put MIP-1β into the antigen-independent family of triggers. Therefore, our prediction is that MIP-1β might not be expected to synergize with CD2.

S. HASKILL

Steve, I wonder if you have thought about collagen in this story. It turns out that MIP-1β's unique in this family, in that monocytes go crazy on collagen to make MIP-1β. In comparison to activated lymphocytes there is probably almost 100-fold difference in the levels.

S. SHAW

You mean when monocytes meet collagen they produce huge amounts of MIP-1β?

S. HASKILL

Incredible amounts in comparison with fibronectin or endothelial cells.

S. SHAW

That is interesting. There is interplay between the specificity of the extracellular matrix and the responding cell type in determining production of MIP-1β which we are inferring is critical to T cell interaction with endothelium.

C. FIGDOR

Steve, are there any antibodies against CD44 which can prevent MIP-1β binding, or the adhesion of T cells. And a second question is -- you said that when you have memory cells that they choose either to be β_1-expressors or β_7-expressors. But what happens when you stimulate those cells? Do they retain their phenotype or do they change?

S. SHAW

Good questions. First, regarding anti CD44 antibodies. There are two parts to that question, one about the MIP-1β binding, and the other about the T cell binding. Let me start with the second. There is a problem with studies of CD44 inhibition of T cell binding because CD44 is generally present on both sides. So in most systems where we have tried blocking with CD44 it enhances rather than inhibits. As to the first part, we have just started trying to detect MIP-1β binding to CD44 and to inhibit such interaction with CD44 mAb. We have a request pending with Gene Butcher for Hermes-3, whch is the antibody which he described which inhibited T/endothelial interaction. That might be a prime candidate.

V. QUARANTA

Along these lines, you show that there is a heterogeneity of trigger molecules in naive T cells. Did you imply that there is some sort of relationship between this heterogeneity and what you then see in the memory cells; in other words, is there some sort of imprinting? And if there is, where is that coming from?

S. SHAW

I think there are two elements in that question. First let me deal with the issue of imprinting. The only way this model makes sense to me is if there is in the lymph node something which imprints into a T cell the decision, for example, to turn on $\alpha_6\beta_1$ and go back to the skin. The question is, how would that be conferred? I think that there are at least two candidate mechanisms. One is by the antigen presenting cell in the lymph node. For example, the Langerhans cell actually brings antigen from skin to the draining lymph node, and apparently becomes the APC which presents that antigen to the T cell in the lymph node. Since the Langerhans cell has a distinctive molecular phenotype, it may do the imprinting. The other possibility relates to the afferent lymph itself. There is a rich flow of "juices", for lack of a better word, coming from the tissue in the afferent lymph. If you cut off the afferent lymphatic, the lymph node withers up and becomes hypocellular. So it is clear that there is a very strong trophic effect of these factors coming in from tissue. I would bet that molecules in that lymph shapes the phenotype of cells in lymph node, which in turn would allow them to shape the phenotype of the memory cells which are generated there. So, I think that there is imprinting, and I would speculate that those may be the mechanisms for it. The other part of the question relates to the heterogeneity of the trigger molecules. I do not have a clever way to integrate what we know about heterogeneity of trigger molecules on T cell subsets with what we know about susequent memory cell specialization. For example, I do not know whether heterogeneity among naive cells in their CD31 expression will correlate with how they specialize as memory cells subsequently. However, one thing that is apparent is that trigger heterogeneity will contribute to the specificity of adhesion. Whether or not a T cell has the right trigger molecule (i.e., receptor) will determine whether or not it can bind, independent of whether it has the right integrin.

E. ROOS

You said that certain subsets are more susceptible to the effect of MIP-1β than others. Could you clarify that, CD8 or CD4 and in naive or memory cells?

S. SHAW

Yes. I am confident of our findings that CD8 cells respond better than CD4s to MIP-1 β. In a very nice recent review of Gene Butcher's, one of his tables shows unpublished data from Schall, who reports preferential chemotaxis of CD8 cells to MIP-1β. I have not seen the data, but that one line from the chart suggests that his and our findings are concordant. In terms of naive vs. memory I am less confident. In general MIP-1β appears to preferentially act on naive cells. However, although I am pretty confident that we can distinguish naive and memory CD4 cells by phenotype; I am not confident that I know what the phenotypic markers of CD8 naive cells are. So I would like to be a little cautious on that.

S. GOODMAN

I think the concept of synergy between the growth factors and the integrins is going to be really very, very important in the future, and you have a really great assay for looking at it there. I just wondered why, especially MIP-1β? In other words, have you looked at whether other growth factors are capable of promoting the same stimulation over CD44?

S. SHAW

We have screened or are screening a number of cytokines and none of the others have the effect of MIP-1β. We have probably looked at a dozen. Some of them might have effects, but if so they are very weak compared to MIP-1β. Thus we think there is something special about MIP-1β for T cells.

S. GOODMAN

Including ones that obviously bind to CD44, I mean?

S. SHAW

Unfortunately, we do not know which ones bind to CD44. However, we have tested others in the same chemokine family as MIP-1β which would be expected to bind to proteoglycan, for example, RANTES. Schall and co-workers have reported that RANTES is a memory-specific chemotactic agent. We do not find much adhesion induction by that. So I am surprised. It is not the result that I would have expected. One of the reasons we screened a fairly big panel of cytokines is that we thought we would find that quite a few induced adhesion, and did not expect that one would stand out.

E. MIHICH

I refer to the first part of your talk -- the different profiles between naive and memory cells. I was concerned about the kinetics of the changes between the two profiles. There are also changes in cytoskeleton, that accompany the memory cell modification from naive to memory cells. What happens first? Do you know whether the cytoskeleton changes happen first, or the surface changes that you are describing? And then when memory cells are reactivated by antigen in secondary reaction, what happens? What is the difference between primary response in the naive cells and secondary response of the memory cells, in relation to the profiles?

S. SHAW

In my simplified slide of T cell maturation, I distinguished three categories of cells: naive, activated and memory. For the subsequent analysis, we have specifically excluded activated cells, which generally are not found in circulation. So we are looking at naive and memory cells, and not the intervening differentiation which is quite complex, and probably required at least several days. With regard to your question on cytoskeleton, I do not know. I need some education as to what might be the differences between naive and memory cells in cytoskeleton. With regard to your question on activation requirements, the answer is simple. For most stimuli, it is much easier to activate a memory cell than a naive cell.

M. HEMLER

Just very briefly. What sort of time course does the cytokine have for the induction of adhesion?

S. SHAW

Quite rapid. At least as rapid as other reported triggers, like CD3 and CD31. It has got to be extremely rapid to explain the adhesion that we are interested in, namely flowing T cells binding to endothelium. Our experiments say that some induction is apparent within several minutes and the peak is at 10-15 minutes. The cumbersome nature of the assay makes it hard to say if some response occurs even earlier.

E. ROOS

You mentioned that you saw MIP-1β on endothelium in tissue sections. Was that in very heavily inflamed areas or did you see it in other tissues, in normal tissues as well?

S. SHAW

As I said, those studies were done by our collaborator, Steven Hubsher. He has not yet looked at a wide variety of tissues. The most striking results have been in the reactive lymph node which I showed you, where its expression is most prominent on HEV. This is satisfying with respect to the concept of naive cells entering lymph node through HEV, and of MIP-1β found there probably acting preferentially on naive cells.

D. LIVINGSTON

Do T cells respond to growth factors in the bound state to endothelium or other surfaces? Are such factors presented to them in the very elegant context such as you described?

S. SHAW

My prejudice is that the idea of proteoglycan immobilization is not as relevant to growth factors such as IL2 as it is to chemokines such as MIP-1β. For example, consider the issue of quantity of the soluble factor produced. A cytokine which acts in solution does not need to be present in high quantity. The classic growth factors seem to be produced in small quantities. In contrast, for something which is going to be immobilized, particularly on an abundant GAG, there has to be a lot of it, which is the case for MIP-1β. So I do not think low abundance growth factors will be immobilized in a widespread way like MIP-1β. However, it certainly is possible that proteoglycans could be involved in local presentation of growth factors, such as at the local contact site between cells. So, it would not surprise me if IL2 were "presented" to T cells, but I doubt it is retained and distributed on GAGs in the same way I am postulating for MIP-1β.

D. LIVINGSTON

The implications for the control of replication are serious and if indeed it occurs, if replication occurs in an attached state, this would have interesting implications for the mechanisms which lead to the development of certain T cell lymphomas.

S. SHAW

I agree. I do not believe that growth factors such as IL2 contain classic GAG-binding sites. So, the narrow hypothesis regarding presentation by GAG-binding is probably not relevant. However the broader hypothesis is that growth factors are presented by interaction with other molecules; I think that is quite likely to be right. But classic interaction sites other than GAG-binding sites have not been defined, such that for a growth factor like IL2. I can predict that it will interact with a presenting molecule X or Y.

D. LIVINGSTON

If you make an endothelial cell produce IL2, will T cells grow once bound to such an endothelial surface?

S. SHAW

I have no idea.

S. GOODMAN

On the original question, there is good evidence that many matrix components would actually immobilize growth factors, and actually, I sort of put your comment round the other way. I find it much more likely that one of the major functions of the extracellular matrix may be to localize and concentrate growth factors so that it has just those effects.

S. SHAW

But Dr. Livingston's question was for factors that are different from MIP-1-β that do not have a heparin-binding site and are viewed as growth factors, like IL2, IL4.

S. GOODMAN

Yes, TGF-β.

S. SHAW

Yes, but that also has a heparin-binding site.

D. LIVINGSTON

A good one, that certainly puts T cells into G-zero.

S. GOODMAN

There are binding sites demonstrated on matrix for TGF-β.

S. SHAW

I would quibble a little bit with your distinction between education factors and growth factors. I had the impression that such distinctions were beginning to break down, so maybe that is the wrong way to ask the question. The distinction which we thought we saw in approaching the field was chemotactic factor vs. growth factors vs. differentiation factors. Now, we see breakdown in those categories. Cytokines have unpredictable overlap from one of those areas into another. So we wonder how valid such distinctions are.

D. LIVINGSTON

I was speaking really only of factors that have the ability to overcome certain blocks in the cell cycle.

S. SHAW

OK, that is a more formal definition than I was thinking of.

G. DOUGHERTY

I wonder if you could discuss how this model fits in with the relatively slow appearance of T cells within sites of inflamation. How do you see these events happening kinetically?

S. SHAW

It is clear that the process can happen rapidly. I have been struck by data of Art Anderson. If he puts IL8 in a mouse ear, he can get an inflamatory infiltrate within minutes. So, presumably that means that IL8 has migrated to endothelium and interacted with it in a short timespan. I think the delays will be built into how long it takes to get the IL8 produced. In addition, one of the other funny features of these GAG-binding "intercrines" is that their diffusion must be slowed by the reservoir of GAG in the ECM. It is analogous to column chromatography. It may take a while for the cytokine to actually get from the site of production out to the endothelium which serves it.

G. NICOLSON

I think you have really hit upon it. If you think of the vascular system as a very dynamic system where elements will be carried away very rapidly, anything produced in a local environment and held there by molecular interactions are much more likely to be stimulatory, rather than if it travels some distance away and gets diluted out tremendously by the circulation. So I think what you are describing here is extremely important, probably for the whole process of inflamation. That is, that elements are displayed locally, immobilized, so that they stay in place, more or less, for a period of time and they are not just drifting away very rapidly in the circulation. This afternoon I will actually show that endothelial cells can release several types of molecules, some of which appear in immobilized forms and others which appear in soluble forms. Some of these are growth factors for tumor cells, some of these are motility factors and so on, so it is a very complex system. But I think you are exactly right in thinking that there are molecules that need to be immobilized, to stay in place for a period of time, to stimulate the processes.

S. SHAW

Actually, one thing I did not know until yesterday is that the GAGs are actually quite rigid rods. That feature of GAGs may make them extremely favorably inclined for this function, the fact that they are rods sticking out and accessible to the tips of T cells.

G. NICOLSON

Well, they are very hydrophylic, so they are bound to be expressed off the cell surface.

ACTIVATION OF LFA-1: THE L16 EPITOPE IS A CATION-BINDING REPORTER

Carl G. Figdor and Yvette van Kooyk

Division of Immunology
The Netherlands Cancer Institute, Antoni van Leeuwenhoek Huis
Plesmanlaan 121, 1066 CX
Amsterdam, THE NETHERLANDS

ABBREVIATIONS USED

ADP = Adenosine diphosphate; **CTL** = Cytotoxic T Lymphocyte(s); **dPBS** = Depleted Phosphate-buffered saline; **EC** = Endothelial Cells; **EDTA** = Ethylenedi-aminetetraacetic Acid; **EGTA** = ethyleneglycol-bis-β-aminoethyl ether)-N,N,N',N'-tetra-acetic acid; **FMLP** = formyl-methionyl-leucyl-phenylalanine; **LAD** = Leukocyte Adhesion Deficiency; **mAb** = Monoclonal Antibody; **MHC** = Major Histocompatibility Complex; **NK** = Natural Killer; **PKC** = Protein Kinase C; **PMA** = Phorbol Myristate Acetate

INTRODUCTION

Several observations indicate that the affinity/avidity of integrin receptors for their ligands can be modulated. Resting leukocytes or platelets do not adhere spontaneously, but a variety of stimuli can induce β1- (VLA-4, VLA-5, VLA-6), β2- (LFA-1, CR3) and β3 integrin (IIb/IIIa) mediated cell-cell interactions. Exposure of lymphocytes, myeloid cells or platelets to phorbol ester (PMA) strongly induces cell aggregation (Rothlein and Springer, 1986; Patarroyo et al., 1985). Similarly, FMLP can stimulate CR3-mediated adhesion of granulocytes to endothelial cells (Buyon et al., 1988; Vedder and Harlan, 1988) and activation of platelets by thrombin or ADP causes IIb/IIIa mediated aggregation (Marguerie et al., 1979). A prominent characteristic in all these observations is that adhesion is induced without an apparent increase in receptor expression. This suggests that changes in affinity of the receptor for its ligand, or changes in the avidity (for instance by alteration of the organization of the adhesion receptors at the cell surface), directly affect cell adhesion. Second messengers play a pivotal role in integrin activation, although at present the precise intracellular circuits that regulate integrin mediated cell adhesion are not completely understood.

Recently we (Van Kooyk et al., 1989), and others (Dustin and Springer, 1989) found that, except from PMA, also monoclonal antibodies directed against the cell surface molecules expressed by T cells, such as CD2 and CD3, can stimulate homotypic cell aggregation. In addition, also signals through CD43, CD44, and MHC class II can induce LFA-1 mediated adhesion of cloned T lymphocytes. These observations indicate that a number of surface molecules expressed by T cells can transduce signals and activate LFA-1 via intracellular signalling pathways. These findings imply the existence of at least two forms of LFA-1; an inactive and an active form of LFA-1. A strong argument in favor of the existence of these two states of LFA-1 is that if only one form of LFA-1 existed, spontaneous aggregation of peripheral blood leukocytes would cause injury of the microvascular network, induce micro-embolization, and might thereby compromise normal leukocyte circulation. Resting leukocytes do not tend to aggregate to each other, although they express significant levels of LFA-1 and ICAM-1, indicating that LFA-1 must be activated for high affinity ligand binding. Similarly, cloned cytotoxic T lymphocytes (CTL) and NK cells, which express extremely high levels of LFA-1 and ICAM-1, do not aggregate offhand (Van Kooyk et al., 1989). Only upon stimulation with antigen, PMA, or via CD2, CD3, CD43, CD44, or MHC class II by mAbs, rapid cell aggregation (< 20 min) of CTL can be observed. This cell clustering is LFA-1 dependent since it is abrogated by anti-LFA-1 antibodies. In addition, surface expression of LFA-1 and ICAM-1 does not change during activation of T cell or NK cell clones (Van Kooyk et al., 1989), demonstrating that cell aggregation is not caused just by augmented surface expression. These observations support the hypothesis that activation of LFA-1 can be induced through different surface receptors indicating that the LFA-1/ICAM-1 is a common adhesion pathway which may be used by leukocytes under quite different physiological conditions.

RESULTS AND DISCUSSION

Activation of LFA-1 by NKI-L16

We have previously described (Keizer et al., 1988; Van Kooyk et al., 1991) an anti-LFA-1 antibody, designated NKI-L16 (further called L16), that in contrast to other anti-LFA-1 antibodies, stimulates cell adhesion rather than inhibiting LFA-1 dependent cell interactions. L16 induced cell aggregation does not implicate "outside-in" signalling, but is merely thought to act by modulating the conformation of LFA-1 so that the affinity for ligand binding is greatly enhanced (Van Kooyk et al., 1991). Time course studies measuring cell aggregation induced by L16 or by F(ab)' fragments thereof, showed a kinetics strikingly similar to that observed when cells were stimulated with PMA (Keizer et al., 1988). This observation led us to suggest that stimulation of cells with PMA or through the CD3, CD2 or class II receptor, although by an entirely different mechanism, might ultimately also result in a conformational change of LFA-1 thus increasing the affinity for its ligands.

The L16 epitope is not within the ligand binding domain, since addition of the L16 antibody in itself induces LFA-1 dependent adhesion (Van Kooyk et al., 1989). This notion is supported by the observation that L16 induced adhesion is completely blocked by other anti- LFA-1α or -β antibodies (Van Kooyk et al., 1989; Van Kooyk et al., 1991). Since also F(ab)' fragments of L16 were capable of inducing adhesion, we can exclude that crosslinking of receptors or Fc receptor mediated phenomena are involved. The observation that PKC inhibitors (staurosporin, AMG) were unable to inhibit L16 induced adhesion, but unequivocally inhibited PMA stimulated adhesion, suggests that L16 does not induce signalling into the cell (Van Kooyk et al., 1991), although the L16-induced adhesion seems to depend on metabolic energy. In addition, L16-induced adhesion is not associated with a rise in $[Ca^{2+}]_i$ levels. Together, these results suggest that L16 induces a change in the tertiary structure of LFA-1 which may result in modulation of the ligand binding affinity.

We observed that the epitope on LFA-1 recognized by the L16 antibody depends on the presence of Ca^{2+} (Van Kooyk et al., 1991). Treatment of cloned T cells, which express L16 abundantly, with a metal chelating agent (EDTA or EGTA) results in a complete loss of this epitope (Table 1) and, more importantly, lose the capacity to bind to other cells in a LFA-1 dependent manner. Loss of the L16 epitope is not associated with a reduction of other epitopes of LFA-1, showing that the molecule is still present at the cell surface. In contrast to cloned T lymphocytes, resting PBL express LFA-1 on their cell surface but they generally lack L16 expression or exhibit only low levels of this epitope at the cell surface (donor-dependent).

CATIONS AND LFA-1 MEDIATED ADHESION

Integrin-mediated cell adhesion is a temperature- and energy-dependent process which requires an intact cytoskeleton and the presence of divalent cations, notably Mg^{2+} and/or Ca^{2+}. The leukocyte integrins express three cation binding domains (Larson, and Springer,

Table 1. Expression of the L16 Epitope Is Required for LFA-1 mediated Adhesion to L cells Expressing ICAM-1

	JS-136		
	L16 Expression	**Adhesion to L-ICAM-1 Cells**	
		Control	**Induced by PMA**
medium	+	-	+
medium + Mg^{2+}	+	-	+
medium + Ca^{2+}	+	-	+
medium + Sr^{2+}	+	-	+
medium + Mn^{2+}	+	+++	+++
EDTA	-	-	-
EDTA + Mg^{2+}	-	+	+
EDTA + Ca^{2+}	+	-	-
EDTA + Sr^{2+}	+	-	-
EDTA + Mn^{2+}	-	+++	+++
EGTA	-	-	-
EGTA + Mg^{2+}	-	+	+
EGTA + Ca^{2+}	+	-	+
EGTA + Sr^{2+}	+	-	+
EGTA + Mn^{2+}	-	+++	+++
dPBS	+	-	-
dPBS + Mg^{2+}	+	-	+
dPBS + Ca^{2+}	+	-	-
dPBS + Sr^{2+}	+	-	-
dPBS + Mn^{2+}	+	+++	+++

The Effect of Mg^{2+}, Ca^{2+}, Sr^{2+}, and Mn^{2+} Cations on L16 Epitope Expression and Binding to L cells Transfected with ICAM-1 after Stimulation with or without PMA (10 ng/ml). Medium was depleted from extracellular cations (dPBS) with 1% wt/vol Chelex 100 microspheres by rotary mixing for 2h at room temperature. Cation concentrations were restored by the addition of 1 mM $MgCl_2$, 1 mM $CaCl_2$, 1 mM $SrCl_2$, or 1 mM $MnCl_2$. Extracellular and receptor bound cations were removed by treatment of the cells with 5 mM EGTA or 5 mM EDTA for 10 min at 4° C. L16 expression on JS136 T cells were determined using FACScan analysis. Cell adhesion was performed as follows. Radiolabeled JS136 cells were allowed to adhere to a monolayer of L-ICAM-1 cells for 30 min at 37° C. Subsequently, the non adherent cells were removed by washing and the percentage of adherent cells was determined after measurement of amount of radioactivity in each well. Note that suspension of cells in cation depleted medium (dPBS) does not affect the L16 expression whereas the epitope is lost after treatment of the cells with EDTA or EGTA.

1990). This allows the possibility that occupancy by either Ca^{2+} or Mg^{2+} alone or a combination of both may have dramatic consequences for the functional status of the receptor. In some cases Ca^{2+} or Mg^{2+} can be substituted by other divalent cations. We observed that Mn^{2+} causes spontaneous activation of LFA-1 resulting in homotypic aggregation of T cells or binding of T cells to L-cells that express ICAM-1 (Table 1). The T cell clone JS136 which expresses high levels of LFA-1 and the L16 epitope was treated with EDTA or EGTA to remove Mg^{2+} and Ca^{2+} or Ca^{2+} respectively. In both cases the L16 expression and the capacity of the cells to bind to ICAM-1 is lost. Furthermore, we suspended the cells in medium (dPBS) completely depleted from cations by beads to which

EDTA is attached (Chelex). This treatment has the advantage that the cells are not exposed to EDTA. The results in Table 1 show that Mn^{2+} is always capable to induce LFA-1 mediated adhesion to ICAM-1, irrespective of the treatment of the cells (EDTA, EGTA, medium, dPBS).

The underlying mechanism and its physiological relevance is unknown, although recent observations by Altieri et al. (1991) indicate that the high affinity state of the CR3 receptor induced by Mn^{2+} on monocytes may be due to direct binding of Mn^{2+} to CR3 and cause a conformational change in the molecule, which results in an increased affinity for its ligand. Similarly, Dransfield et al. (Dransfield et al., 1992) also demonstrated that Mn^{2+} stimulates LFA-1-mediated adhesion.

Table 1 shows that expression of L16 is lost upon treatment with EDTA and EGTA but not after suspending the cells in dPBS. These findings suggest that Ca^{2+} binding is required for expression of the L16 epitope and that Ca^{2+} is not easily removed from LFA-1 by incubation of the cells in cation-free medium (dPBS). In addition the L16 epitope is expressed after addition of Sr^{2+} which is homologous to the Ca^{2+} ion. Stimulation of cell adhesion with PMA absolutely requires the presence of Mg^{2+} and exposure of cloned T lymphocytes to EDTA completely abrogates the capacity of these cells to bind to ICAM-1. We consistently observed that addition of Mg^{2+} to cells treated with EDTA or EGTA but not after treatment with dPBS, resulted in spontaneous binding of JS136 cells to L-ICAM-1. This may be caused by occupation of Ca^{2+} sites by Mg^{2+} ions, supporting the hypothesis that Mg^{2+} may induce an activated state of the receptor. However, if EDTA or EGTA-treated cells are exposed to Ca^{2+} ions prior to the addition of Mg^{2+} ions, to allow the Ca^{2+} ions to bind to LFA-1, spontaneous binding of JS136 cells to L-ICAM-1 was still observed, indicating that other mechanisms may play a role (Van Kooyk et al., submitted). In addition, we found that in the absence of Mg^{2+}, but in the presence of Ca^{2+}, L16 is still capable to induce LFA-1-mediated cell adhesion (not shown), demonstrating that LFA-1 mediated adhesion can also occur in the complete absence of Mg^{2+}. The observation that L16 induces LFA-1-mediated adhesion also in the presence of only Ca^{2+} suggests that this metal ion is not only required to express the epitope (Table 1), but is also sufficient to allow LFA-1-mediated function. In addition, it may indicate that both Ca^{2+} and Mg^{2+} ions bind to LFA-1, even when the molecule is active.

It must be stressed that the results obtained with chelators may affect other cell functions and not mimic the physiological situation. Therefore these results should be interpreted with caution. We favor to use cation depleted medium (dPBS) which prevents contact of the cells with chelating agents. The results shown underscore the importance of metal ions in integrin receptor activation. This notion is not only supported by our results, but also by those of Dransfield et al. (Dransfield et al., 1990) who recognized a Mg^{2+}-dependent epitope on the α chains of the $\beta 2$ integrins that seems to be expressed only when LFA-1 has bound its ligand.

EXPRESSION OF THE L16 EPITOPE IS REQUIRED FOR FUNCTION OF LFA-1

We have previously demonstrated that the L16 expression correlated with the capacity of cells to aggregate in an LFA-1-dependent manner (Van Kooyk et al., 1991). Here we demonstrate that expression of the L16 epitope is an absolute requirement for LFA-1-mediated adhesion. Figure 1 shows the expression of two epitopes of LFA-1 on JS136 T cells, on LFA-1 negative T cells (LAD6.6) from a leukocyte adhesion deficiency patient (Van de wiel van kemenade, E. et al., 1992) and on Jurkat cells. The data show that L7 (regular LFA-1 epitope) and L16 are equally well expressed on JS136 cells, but that Jurkat cells lack L16 expression although they express significant amounts of LFA-1 (L7). As expected, the LAD cells are completely LFA-1 negative. Table 2 shows the capacity of these cells to bind to L-ICAM-1 cells. Only JS136 cells expressing L16 are capable to bind after activation of LFA-1 by PMA. Jurkat cells behave like LFA-1 negative LAD cells and cannot bind. Similar results were observed when binding of these cells to TNFα stimulated endothelial cells (EC) was studied. Although all cells bound, only binding of JS136 cells was mediated by LFA-1 since it could be blocked by anti-CD18 antibodies. Binding of LAD6.6 and of Jurkat cells was mediated by VLA4, since anti-VLA4 antibodies inhibited binding to VCAM-1

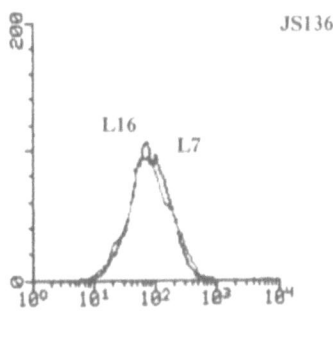

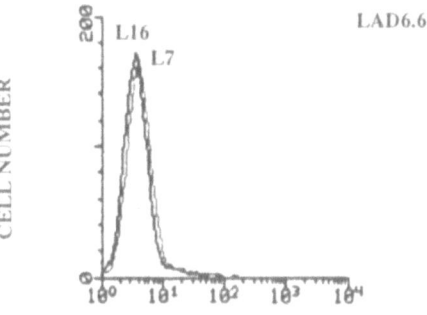

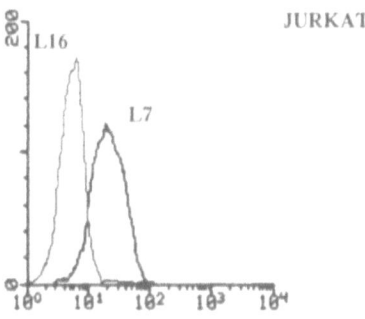

Figure 1. Expression of the L7 and L16 epitope of LFA-1 on JS136 T cells, LAD6.6 LFA-1 negative T cells and on Jurkat cells by immunofluorescence analysis.

which is expressed on TNFα stimulated EC. These data clearly support the hypothesis that expression of L16 is absolutely required for LFA-1 function.

Interestingly, we observed that VLA4 did not mediate adhesion of JS136 cells to EC. Anti-VLA4 antibodies alone or in combination with anti-LFA-1 were not inhibitory, despite the fact that VLA4 is abundantly expressed by JS136 (Table 3). These data indicate that there is a hierarchy in the usage of adhesion pathways (Vennegoor et al., 1992). The results suggest that if L16 is expressed, and LFA-1 is functional, VLA4-mediated adhesion is suppressed. The mechanism controlling this inhibition of VLA4-mediated adhesion is

Table 2. Binding of T cells to L-ICAM-1

	Binding of T cells (%)		
T cells	**Medium**	**PMA**	**PMA + αCD18**
JS136	14	54	15
Jurkat	11	12	10
LAD6.6	3	2	3

Binding of JS136 T cells, of LAD6.6 LFA-1 negative T cells and of Jurkat cells to L cells transfected with ICAM-1 after stimulation with or without PMA (10 ng/ml) was studied. Radiolabeled cells were allowed to adhere to a monolayer of L-ICAM-1 cells for 30 min at 37° C. Subsequently, the the percentage of adherent cells was determined (*see Table 1*). Anti-CD18 antibodies (CLB-LFA-1/1) were used to inhibit adhesion.

Table 3. Adhesion of T cells to 24h TNFα-stimulated EC

	Binding of T cells (%)			
T cells	**Control**	**+αCD18**	**+αVLA-4**	**+αCD18/VLA-4**
JS136	44	16	42	14
Jurkat	50	48	15	15
LAD6.6	39	40	8	10

Binding of JS136 T cells, of LAD6.6 LFA-1 Negative T cells and of Jurkat Cells to Confluent Monolayers of Human Umbilical Cord Endothelial Cells (EC) after Culture with TNFα (100U/ml) for 24h. Radiolabeled cells were allowed to adhere for 30 min at 37° C. Subsequently, the percentage of adherent cells was determined (*see Table 1*). Anti-CD18 antibodies (CLB-LFA-1/1) and anti-VLA4 (HP2/1 kindly provided by Dr. Sanchez-Madrid) were used to inhibit adhesion.

under investigation. In summary, these results demonstrate that cell adhesion is not only controlled by up- or down-regulation of adhesion molecules, or by activation/deactivation of adhesion structures, but that also a preference in the use of particular cell adhesion pathways may exist.

REFERENCES

Altieri, D.C., 1991, Occupancy of CD11b/CD18 (Mac-1) divalent ion binding site(s) induces leukocyte adhesion. *J. Immunol.* 147:1891.

Buyon, J.P., Abramson, S.B., Philips, M.R., Slade, S.G., Ross, G.D., Weissman, G., and Winchester, R.J., 1988, Dissociation between increased surface expression of Gp165/95 and homotypic neutrophil aggregation. *J. Immunol.* 140:3156.

Dransfield, I., Cabanas, C., Craig, A., and Hogg, N., 1992, Divalent cation regulation of the function of the leukocyte integrin LFA-1. *J. Cell Biol.* 116:219.

Dransfield, I., Buckle, A.-M., and Hogg, N., 1990, Early events of the immune response mediated by leukocyte integrins. *Immunol. Rev.* 114:29.

Dustin, M.L. and Springer, T. A., 1989, T cell receptor cross-linking transiently stimulates adhesiveness through LFA-1. *Nature* 341:619.

Keizer, G.D., Visser, W., Vliem, M., and Figdor, C.G., 1988, A monoclonal antibody (NKI-L16) directed against a unique epitope on the alpha chain of LFA-1 induces homotypic cell-cell interaction. *J. Immunol.* 140:1393.

Larson, R.S. and Springer, T. A., 1990, Structure and function of leukocyte integrins. *Immunol. Rev.* 114181.

Marguerie, G.A., Plow, E.F., and Edgington, T.S., 1979, Human platelets possess an inducible and saturable receptor specific for fibrinogen. *J. Biol. Chem.* 254:5357.

Patarroyo, M., Beatty, P.G., Fabro, J.W., and Gahmberg, C.G., 1985, Identification of a cell surface protein complex mediating phorbol ester induced adhesion (binding) among human mononuclear leukocytes. *Scand. J. Immunol.* 22:171.

Rothlein, R. and Springer, T.A., 1986, The requirement for lymphocyte function-associated antigen in homotypic leukocyte adhesion stimulated by phorbol ester. *J. Exp. Med.* 163:1132.

Van de wiel van kemenade, E., te Velde, A., Weening, R.S., Fischer, A., Borst, J., Melief, J.C.M., and Figdor, C.G., 1992, Both LFA-1-positive and -deficient T cell clones require the CD2/LFA3 interaction for specific cytolytic activation. *Eur. J. Immunol.* 22:1467

Van Kooyk, Y., van de Wiel-van Kemenade, P., Weder, P., Kuijpers, T.W., and Figdor, C.G., 1989, Enhancement of LFA-1 mediated cell adhesion by triggering through CD2 or CD3 on T lymphocytes. *Nature* 342:811.

Van Kooyk, Y., Weder, P., Hogervorst, F., Verhoeven, A.J., van Seventer, G., te Velde, A.A., Borst, J., Keizer, G.D., and Figdor, C.G., 1991, Activation of LFA-1 through a Ca^{2+} dependent epitope stimulates lymphocyte adhesion. *J. Cell Biol.* 112345.

Vedder, N.B. and Harlan, J.M., 1988, Increased surface expression of CD11b/CD18 (MAC-1) is not required for stimulated neutrophil adherence to cultured endothelium. *J. Clin. Invest.* 81:676.

Vennegoor, C.J.G.M., Van de Wiel-van Kemenade, E., Huijbens, R.J.F., Sanchez-Madrid, F., Melief, C.J.M., and Figdor, C.G., 1992, Role of LFA-1 and VLA-4 in the adhesion of cloned normal and LFA-1 (CD11/CD18)-deficient T cells to cultured endothelial cells indication for a new adhesion pathway. *J. Immunol.* 148:1101.

DISCUSSION

G. NICOLSON

It seems to me that your activation could also possibly be explained by the state of receptor distribution and cytoskeletal interactions, instead of just the conformation of the molecule itself. I think everything that you described could also be explained by the ability of the receptors to aggregate, and that aggregation then is stabilized by interactions of cytoskeletal elements. I would like you to comment on that point.

C. FIGDOR

That is a favorite theory I have written down in *Immunology Today*. We are looking into that. I think L16 has something to do with micro-clustering of receptors at the cell surface, but that still does not explain how this epitope is expressed. Because this epitope is not expressed at resting cells, only at a certain activation/differentiation stage you see expression of this epitope on these molecules. We are trying very hard to see whether there is an association, for instance, and that is also a nice theory that there is an association of LFA-1 with a secondary molecule, a calcium channel maybe or some other molecule, which induces this L16 epitope. Another explanation may be that this association of cytoskeletal elements with these molecules in itself induce this epitope, but we still do not know that. But I favor the idea that cytoskeletal reorganization is very important in this process.

N. HOGG

But what is the evidence that L16 causes micro-cluster formation?

C. FIGDOR

Well, we are doing now some fluorescence microscopy studies. It is still not clear what really goes on because it is different on different cell types and also the L16 expression varies very much depending on the donor you take. But I think that we are homing in on microclustering by that kind of immunofluorescence data.

G. TARONE

In one of your slides you showed that β_2-chain can be phosphorylated. Do you think that this is a possible mechanism of activation?

C. FIGDOR

Well I think that on the phosphorylation there are several data. I think the data of Tim Springer showed clearly that if you mutate the various potential sites of the β_2-chain, there is no effect with respect to the function of LFA-1, only if you mutate the threonines, you see a dramatic effect, but for instance with phorbol ester those sites are not phosphorylated. There is a serine which is phosphorylated, and if you mutate that serine, then you do not see any effects. On the α-chain it is less clear, again Tim showed that if you delete the α-chain, the cytoplasmic domain, then you seem to have no effect on function, and also L16 is still present in all cases. So, we tried very hard to see whether when we used one of the stimuli used in our studies, CD2, CD3, to see whether that induces phosphorylation on LFA-1. We see a signal, definitely, but it is so weak that I do not feel comfortable publishing that data. Maybe the reason why it is so weak is that it is very local, when you look at a specific T cell receptor interaction you may have only a minor part of the LFA-1 molecule become activated, and therefore you do not see this phosphorylation. The larger part of a number of LFA-1 molecules may not have to be involved in that process. We found that LFA-1α is actually constituitively phosphorylated and we do not see on the α-chain any effect with addition of these antibodies. We carried out many time course experiments and found essentially no effect on the α-chain.

M. HEMLER

Regarding that last point. I think it was fairly well shown by Francisco Sanchez-Madrid, and maybe less well by my laboratory, that integrins can have active and inactive conformations in solution, so that the cellular environment is not the whole story. But I want to get back to a point that you made that concerns me. You said that once the L16 epitope is expressed, VLA-4 adhesion is shut off. I think a lot of people have shown very clearly in fully active cells that VLA-4 is very functional and very readily able to mediate adhesion to VCAM or to fibronectin. So, I do not know if you want to say that it is shut off, but what you really have shown is that VLA-4 activity is not so easy to demonstrate in the context of your assay system using intact endothelium and added antibodies. Do you have any comments on that?

C. FIGDOR

Yes, I have not tested it on purified VCAM, for instance, or VCAM transfected cells. I have tested it in a cell-cell system where all molecules are present. And we have seen it not only for this T cell clone but for a number of T cell clones, and I can only say this is the data we have, and you can argue about it. But this is a consistent finding, and we always see this relationship with expression of this epitope.

N. HOGG

What about T lymphoblasts?

C. FIGDOR

Well, this is an interesting question because they express L16. But as it appears now, L16 is a reporter, as you know, of the cation binding state of the LFA-1 molecules. It is a calcium-dependent epitope. Now it seems that depending on the state of activation calcium can be bound to LFA-1, either very tightly or more loosely, and this also is related to this L16 expression. For instance if you have a T cell clone, you can wash this T cell clone in medium which is deprived from (free of) calcium so there is no calcium present, and you still see the L16 expression. However if you take resting lymphocytes or lymphocytes which are only briefly activated, then you can wash away the L16 expression, indicating that apparently calcium is not bound as tight to the LFA-1 molecule, as on a cell which is fully activated and in culture for a long period of time. This indicates that the expression of L16, correlates with different activation states; you have a state where calcium is very tightly bound and there VLA-4 is not involved in adhesion, whereas if calcium is bound very loosely where you can wash away the calcium and lose the L16 epitope, there is a role for VLA-4. For instance if we take resting lymphocytes we see that both LFA-1 and VLA-4 are involved in this process. So no, I do not want to argue that in T cell-mediated endothelial cell contact there is no role for VLA-4, no way. I only wanted to demonstrate that under certain conditions, when you have this epitope expressed, VLA-4 plays no role -- that is the point I want to make.

G. DOUGHERTY

You described in your talk a series of events in which the appearance of the L16 epitope preceded the ability of LFA-1 to bind to ICAM. Since it is likely that many of the T cell clones used in your studies express ICAM-1, can you rule out the possibility that interactions between LFA-1 and ICAM-1 occuring during the cultivation of these cells could contribute to the induction of L16 expression?

C. FIGDOR

Well the problem with this antibody is that it induces activation, so it is very difficult to perform that experiment.

J. CALVETE

This is regarding the possible regulation of LFA-1 by cellular calcium. You did not show any quantitative data. Do you know the range of the Kd and the intracellular calcium concentration which is needed for this? Do you have quantative data of the concentration of intracellular calcium in resting versus activated T cells.

C. FIGDOR

No, we used the indo-1 marker for the calcium measurements, so these are always relative measurements, and no absolute measurements.

J. CALVETE

I mean, if you measure the intracellular calcium in the resting T cells and then you measure the intracellular calcium in activated T cells what is the increase?

C. FIGDOR

In terms of maximums that you can reach, is that what you mean?

D. LIVINGSTON

But the question is, are there differences in resting calcium levels between activated and unactivated T cells?

C. FIGDOR

We did all these studies only on these T cell clones. We did not do those studies on resting T cells and I can also explain to you why we did not do them. When you use resting T cells and you want to have the same stimuli through CD2 and CD3, you need an extra cross-linker, you cannot only induce the signal by CD2 antibodies only, or CD3 antibodies only. Usually people use an extra cross-linker. Only then you see the signal. With the T cell clones it is different, and again I think it is because they are already in a rather active state, you only need anti-CD3 antibodies or anti-CD2 antibodies to see this phenomenon, and we only studied these phenomena therefore on the T cell clones not on the resting T cells. So I cannot answer that question.

N. HOGG

Can you really discriminate in those calcium flux studies between pre-LFA-1 activation events and post-activation events? As phorbol ester requires calcium for activation, so how were you able to get those effects, apparently, in the absence of calcium?

C. FIGDOR

I will answer the last question first. When you add phorbol ester in the absence of calcium you see still an effect of phorbol ester, but it is much lower. So, there is a calcium component, but there is also definitely a calcium-independent component. And, the first question, whether we can discriminate between post-binding and pre-binding, at this stage we cannot.

R. JULIANO

In terms of these calcium-transients seen during activation, I guess the presumption is that these would lead to activation of PKC, or other calcium-dependent kinases or phosphatases, but is there any direct data to support that, I mean have you actually measured either PKC activation or translocation, or activation of other kinases?

C. FIGDOR

No we have not done that, we are in the process of doing those studies at the moment. When you compare the CD3 and the CD2, you saw that there is a clear distinct pattern actually because the CD3-mediated calcium flux is fairly rapid, much more rapid than the CD2 flux you saw. And the CD3-mediated flux, and I have talked about this to many people who are in this field, completely parallels an IP3-mediated response. So probably this fast flux is an IP3-mediated response, whereas the second and much slower response is an effect of the incoming calcium through the calcium channels.

R. JULIANO

If your calcium levels are getting high enough, I think another process being brought into play is activation of calcium-dependent proteases, and I think that is something that has not been really thought about very much.

C. FIGDOR

Yes.

R. BANKERT

I would like to get back to a question that Garth raised earlier about the cytoskeleton, and also the question that Henry raised with regard to this. There is a paper recently out in *PNAS* by Gregorio and Repasky, (Gregorio et al., *Proc. Natl. Acad. Sci. USA* 89:4947-4951, 1992), where they show a rapid translocation of a specific cytoskeletal element, α-spectrin from the plasma membrane to a spot opposite the Golgi. I am struck by the kinetics of the change there with what you see. It is very rapid, it is calcium-dependent, it is coincident with the translocation of PKC. And, in fact, there is a coordinated movement of PKC and the α-spectrin. This is a major cytoskeletal element at the membrane and it moves away from the membrane. I think it is appropriate that people at this meeting pay attention to this work and relate this to studies with cell adhesion.

C. FIGDOR

Thank you very much for this comment. I was not aware of that.

E. BROWN

I would like to ask one more question about these calcium tracings. Maybe I got it wrong, but it looked to me like when you depleted extracellular calcium you actually delayed the increase from CD2 quite markedly. There was about a minute and a half delay. I just wondered if you had any insights as to what was going on there, since such delayed release from intracellular stores is, I think, relatively uncommon.

C. FIGDOR

Well, also when extracellular calcium is present there is a delay. But I think you are right, the delay, when you deplete for extracellular calcium, is much longer. And I have no explanation for that because you would expect, if it is really intracellular stores, why would not it be as rapid as the CD3 and MHC class II? I agree completely with you. But, at least in any case there is a different mechanism that is going on. That is all I can say at this moment. I have no idea why that delay is so long, but it is a consistent finding we see all the time in these experiments.

E. ROOS

In addition to the questions about PKC α-spectrin, I may say that LFA-1 when transfected into fibroblasts, is constitutively active. So it is a peculiar feature of a resting T lymphocyte that LFA function is blocked in some way. Could not it be that your signal, either via proteolytic action or via cytoskeletal redistribution, de-blocks LFA-1 rather than activates it.

C. FIGDOR

I can live with that theory. There is no evidence at all that either one is true, but the only point I want to make is that there is a clear distinction between LFA-1 resting cells and activated cells. Whether it is de-blocking or activation, I do not think I have any data which argue either for one or the other point.

N. HOGG

Our work would suggest that the calcium bound to LFA-1 in the resting state exerts inhibitory effects on its function (*J. Cell Biol. 11:219-226, 1992*).

F. SANCHEZ-MADRID

Carl, I would like to make a comment on the expression of L16, this intriguing L16 epitope. In your talk, you presented data that the L16 epitope is not expressed by resting T lymphocyte, but using your antibody with freshly isolated and purified T lymphocytes, we found quite various expression. Most of these freshly purified T lymphocyte expressed L16. My question is regarding development of this L16 epitope. Does L16 antibody react with conformational epitope, does it react with dissociated LFA-1 α-chains?

C. FIGDOR

I like that question very much, and I would like to answer it for this audience, because I send out a lot of these antibodies to many people, and there are several people, including you now, who say when I looked at resting T lymphocytes, I see L16 expressed, OK. Now, if you are going to wash those lymphocytes, and refer to my earlier remark, you will see that you lose L16 expression. When we do our experiments on the FACS we always use medium which has no cations present. So if you wash these lymphocytes you will see that you will lose L16 expression on these resting cells. However, as I said earlier, when you wash highly activated cells like these T cell clones you will not lose L16 expression. So it is a matter of how much calcium is present in your medium. Because in our earlier reports we said that there is a donor-dependency. Now I think it is dependent on the amount of cations that you have present in the medium. So I would advise you to wash your lymphocytes and you will see that you will lose expression.

F. SANCHEZ-MADRID

There is a second part to the question. It does react with the LFA-1 α-chain when it is dissociated from the β-chain? Do you have any experiments to show that?

C. FIGDOR

Yes, we have done experiments in which we enhance the pH to dissociate the chains. It is not clear at the moment. Earlier experiments indicated that it was only reacting with the α-chain, but experiments we did recently with another cell type demonstrate that it is not clear whether it is only the α-chain, or a combination of the α- and the β-chain. But I can certainly say that it is not the β-chain, because if you immunoprecipitate you see only the LFA-1 and you do not see CR3 or PI50. So I cannot exclude that this epitope is formed by either α- and β-chain. In addition, it is a non-blotting antibody, so you cannot do a Western blot with it. So, again indicating that it is at least probably conformation epitope which is formed. And I cannot exclude that it is an epitope which is formed by both α- and β-chains at this moment.

E. BROWN

I would just like to make a speculation which ties in your tyrosine kinase data with the fact that fibroblasts transfected with LFA-1 are constituitively active. Which is to say that in general to activate a tyrosine kinase requires first a tyrosine phosphotase to remove a phosphate at a regulatory site of the carboxy terminals. So it could be that LCA, which is present on lymphocytes but not present on the fibroblasts, is actually key in regulating the function of LFA-1. Do you know of any experiments where people have co-transfected the two?

N. HOGG

It is involved in activation with T cell receptor which presumably is an upstream event of what Carl is mentioning.

E. BROWN

Right, but you could actually test directly its role in this by just having two of them in a cell.

C. FIGDOR

Let me just say one word about this. I do not feel comfortable with all these blocking studies where you use agents to block the various routes of these signalling molecules. Because if you look very carefully at the data, many of these agents are not specific. For instance, if you use staurosporin it also has effects on the cytoskeletal elements.

N. HOGG

Yes, it dissociates actin.

C. FIGDOR

So I want to make this comment that you should never go only on these blocking studies, because these inhibitors are very, very tricky to use. And this is one of the reasons why the field of signalling is still such a dark field because everybody uses another cell line, and another blocker. I think it is a very tricky area.

S. GOODMAN

Yes, I will certainly go along with the comments about the staurosporin and inhibitors like that, you may well block protein synthesis with things like that, surprisingly, but that is not the point. I was interested in your comments about tightly bound and not tightly bound calcium, so what do you call washing? Does that mean resistant to chelators or not?

C. FIGDOR

I am very glad you asked this question because I am not in favor of using chelators. Many people who are using chelators, and apparently we do not know very much about these chelators except that they take away the cations we want to do away with. The way we do the washing in these experiments is as follows. First we deplete our medium by beads to which chelators are attached. So the cells never come into contact with chelating molecules, to be sure that the medium we use is completely depleted from cations. Then we wash the cells twice with this medium, then we do the FACS analysis. That is exactly the protocol we are using. There are no chelating molecules used to take away the cations. Because if you do that, if you use a chelating molecule, then you will lose also your L16 epitope even if you are using this cell line which normally stably binds LFA-1. So, you have a state of LFA-1 where this calcium ion is loosely bound to resting lymphocytes, if you wash your cells you will lose the L16 expression. If you take these activated T cell clones you have stable expression of L16, if you wash your cells you do not lose L16 expression. If you take those cells and treat them with chelating agents you lose L16 expression and you will lose LFA-1 function.

S. GOODMAN

Yes, and I am sure Garth will go along with this and might have a comment on it, but this is a very, very tacky point. Because if the cells do not see a chelator, so we are talking about EGTA, and EDTA usually, where we are talking about 10-12 M affinity for calcium. The surface of the cell is classically coated with negative charge. I mean there is a calcium pool not only internally, but also externally. So we are talking about a very, very dodgy area here, about whether the LFA molecules actually have access to an external calcium pool which is presented in the same sort of way that growth factors do. We are not talking about free calcium. I mean calcium ions bound to the washed surface, to the membrane of the cell. There was a theory talking about calcium being necessary to stabilize cell membranes in the classical membrane literature, and so if you do not wash the cell with chelators then you have potentially accessible calcium. As you say, we do not know anything about the affinity of the integrins for cations.

C. FIGDOR

Sure, I completely agree with that because this epitope was only expressed as calcium bound, so there should be calcium still present otherwise you do not see this epitope expressed.

V. QUARANTA

Did anyone ask if you can immunoprecipitate LFA-1 from solubilized extract of Jurkat cells with the L16 antibody. In other words, L16 epitope is not on LFA-1 on the surface of intact cells, but once you solubilized the membrane can you make LFA-1 reactive with L16?

C. FIGDOR

We have not done that experiment, that is a good suggestion.

V. QUARANTA

There may be something on Jurkat cells that prevents the LFA-1 chain from expressing the L16 epitope.

C. FIGDOR

Yes, it makes sense to do that experiment but we just did not do that so I cannot give you that information. But it is true that Jurkat cells do not express very high levels of LFA-1.

ACTIVATION-DEPENDENT REGULATION OF β1 INTEGRIN EXPRESSION AND FUNCTION IN HUMAN NATURAL KILLER CELLS

Angela Gismondi, Fabrizio Mainiero, Gabriella Palmieri**, Stefania Morrone, Michele Milella, Mario Piccoli, Luigi Frati, and Angela Santoni*

Department of Experimental Medicine
University of Rome "La Sapienza," ITALY
*Laboratory of Pathophysiology
Regina Elena Cancer Institute, ITALY

**Institute of Biomedical Technologies, CNR, Rome, ITALY

ABBREVIATIONS USED

RGD = Arg-Gly-Asp; **Coll I** = Collagen I; **ECM** = Extracellular Matrix; **FN** = Fibronectin; **LGL** = Large Granular Lymphocytes; **LM** = Laminin; **NK** = Natural Killer Cells; **PBL** = Peripheral Blood Leukocytes; **PKC** = Protein Kinase C; **TNF-α** = Tumor Necrosis Factor-α

INTRODUCTION

Lymphocytes express a wide variety of cell surface receptors which mediate their adhesion to other cells and to extracellular matrix (ECM) and are critical for recirculation and homing and immunological recognition (Springer, 1990; Shimizu and Shaw, 1991). The majority of these adhesion molecules belongs to the three different supergene families of immunoglobulins, integrins and selectins, and interacts with an ECM ligand. Recently, an increasing number of lymphocyte ECM receptors has been identified in the integrin superfamily, mostly belonging to the β1 integrin family (Hemler, 1990) (Figure 1).

In the β1 family, at least four heterodimers have been implicated as LM receptors on lymphocytes: α1β1 (VLA-1), α2β1 (VLA-2), α3β1 (VLA-3), α6β1 (VLA-6). VLA-1, originally described on long-term activated T lymphocytes, is an ubiquitous 200/110 KD heterodimer capable of binding to both LM and Coll I (Hemler et al., 1985; Ignatius, 1990). Among lymphoid cells, VLA-1 is not detectable on PBL, while it is expressed on T lymphocytes isolated from inflamed synovium (Hemler et al., 1986) and from the epithelium of the lower respiratory tract (Saltini et al., 1988). VLA-2 is a 150/110 KD cell surface receptor for both LM and Coll I (Elices and Hemler, 1989). Expression on lymphocytes is restricted only to activated cells. VLA-3 is a 150/110 KD integrin implicated as receptor for FN, LM, and Coll I (Elices et al., 1991). Resting peripheral blood human T lymphocytes express low, but detectable levels of VLA-3, which increase after *in vitro* activation. VLA-6 is the major LM receptor; it was first identified on platelets (Sonnenberg et al., 1988) but it is also present on monocytes, lymphocytes and thymocytes. Differently from the other β1 integrins, VLA-6 expression on lymphocytes decreases upon *in vitro* activation.

In addition to VLA-3 two other lymphocyte receptors for FN belong to the β1 integrin family: α4β1 (VLA-4) and α5β1 (VLA-5) (Takada et al., 1987; Wayner et al., 1989). α5β1 is a 150/110 KD heterodimer recognizing the RGD sequence in the central region of FN, mainly expressed among lymphoid cells by peripheral blood T lymphocytes (Hemler, 1990) and NK cells (Gismondi et al., 1991a). Peripheral blood B cells and thymocytes bear none or only negligible levels of α5β1. α4β1 (150/110 Kd) is the receptor for the carboxy-terminal cell adhesion region containing the heparin II domain and the III connecting segment (IIICS) of FN. α4β1 ligands include CS1 and CS5 regions in the IIICS of FN (with LDV being the sequence recognized in CS1), and VCAM-1, a vascular adhesion molecule induced by inflammatory cytokines such as TNF-α and IL-1 on endothelial cells (Elices et al., 1990). α4β1 is expressed by peripheral blood T, B, and NK cells, and by thymocytes (Hemler, 1990; Gismondi et al., 1991a).

Lymphocyte adhesion is a dynamically regulated process which, at the receptor level, could be the result of quantitative changes in expression or qualitative changes in function.

In vitro and *in vivo* evidence suggests that β1 integrin expression is modulated during antigen-dependent lymphocyte differentiation. Thus, greater levels of VLA-4, VLA-5 and VLA-6 are expressed on memory as compared with naive CD4+ T cells (Shimizu et al., 1990a). *In vitro* long-term stimulation of lymphocytes results in ECM binding by increasing the overall levels of receptor expression (Hemler, 1990).

The adhesive function of integrin receptors is also highly regulated, thus allowing lymphocytes to rapidly pass from a nonadherent state to an adherent one. It has been shown that activation of T lymphocytes through the T cell receptor complex (TcR/CD3) or via CD2

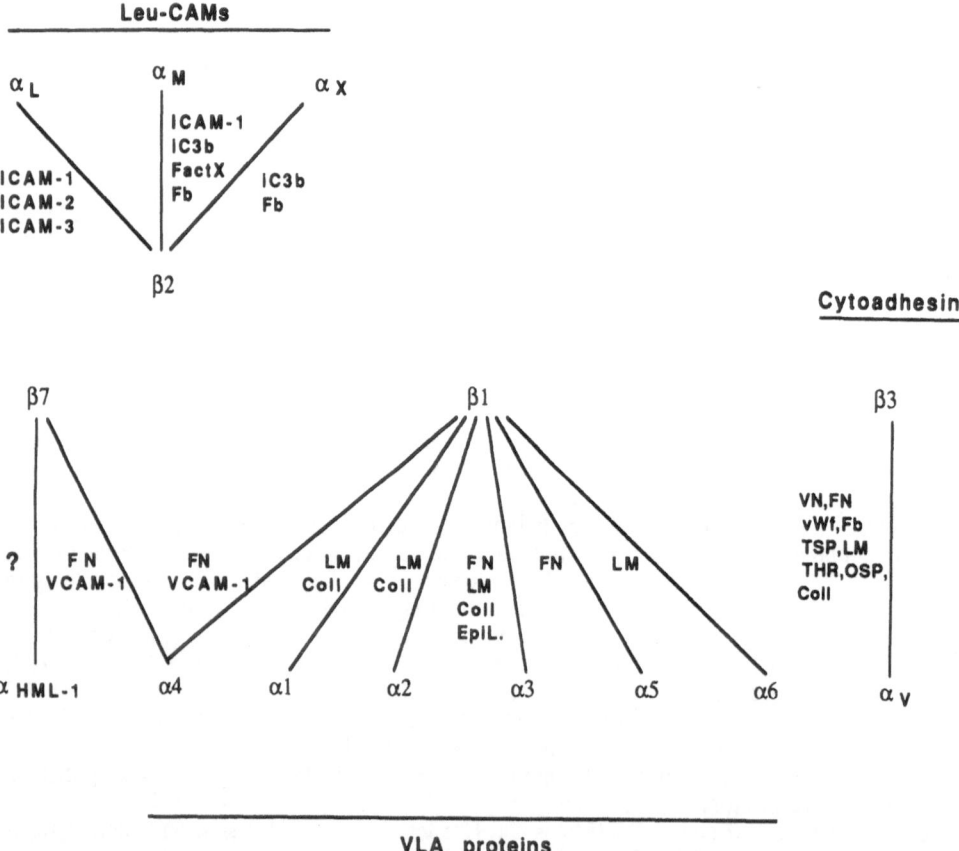

Figure 1. Leukocyte Integrins. Schematic representation of integrin receptors expressed by lymphoid cells and their ligands. C3b component of complement (inactivated) (iC3b); Epiligrin (Epil); factor X (FactX); Fibrinogen (FB); Human Mucosal Lymphocyte 1 (HML-1); Intercellular Adhesion Molecule (ICAM)-1, 2, 3; Osteopontin (OSP); Thrombin (THR); thrombospondin (TSP); Vascular Cell Adhesion Molecule 1 (VCAM-1); Vitronectin (VN); von Willebrand factor (vWf).

(Shimizu et al., 1990b; Matsuyama et al., 1989), CD7 and CD28 surface antigens (Shimizu et al., 1992) rapidly enhances the ability of VLA-4, VLA-5 and VLA-6 to bind to their specific ligands. The enhanced adhesiveness can be transient depending on the activation signal and on the integrin involved, and it is not associated with changes in the surface receptor expression.

Natural killer (NK) cells are a heterogenous population of large granular lymphocytes (LGL) with the ability to mediate MHC-unrestricted and antibody-dependent cytotoxicity and to produce a large variety of cytokines (Trinchieri, 1989). They are easily identified by the expression of surface antigens such as CD16 (FcγRIII) and CD56 (NCAM), with CD16 representing one of the most relevant structures on NK cells capable of triggering their functional program (Perussia et al., 1984; Cassatella et al., 1989; O'Shea et al., 1991). NK cells mainly recirculate in the peripheral blood and localize in several lymphoid and nonlymphoid tissues under physiological and inflammatory conditions (Trinchieri, 1989).

We have previously shown that fresh human peripheral blood NK cells express a4β1 (VLA-4), a5β1 (VLA-5) and a6β1 (VLA-6) integrin receptors which mediate their adhesion to FN and LM respectively (Gismondi et al., 1991a; Morrone et al., 1991). In this study we investigated the modulation of VLA expression and function following short- and long-term NK cell activation.

RESULTS AND DISCUSSION

Short-term NK Cell Activation Results in Rapid Enhancement of NK Cell Adhesion to FN and LM without Affecting β1 Integrin Expression

It has been previously shown in several cell systems that integrin functions can be rapidly enhanced by cell activation (Hynes, 1992). It was therefore of interest to investigate whether stimuli able to trigger NK cell functional program could affect their adhesion to FN and LM. Treatment of ^{51}Cr-labeled NK cells (10 min at 37° C) with TPA (20 ng/ml) or with mAb directed against the CD16 antigen (B73.1, 3G8) resulted in a marked enhancement of adhesion to both FN and LM. Increased adhesion, as the constitutive one (Gismondi et al., 1991a; Morrone et al., 1991), was mediated by VLA-4, VLA-5 and VLA-6 (Table 1). In an attempt to understand the mechanisms responsible for the activation-dependent adhesion to FN and LM, we evaluated whether expression of β1 integrin receptors was affected upon

Table 1. Adhesion Properties of Fresh and Activated NK Cells

| Substrate | Fresh [a] | | Short-term activated [b] | | Long-term activated [c] | |
	adhesion	receptor	adhesion	receptor	adhesion	receptor
FN	++[d]	α4β1,α5β1	+++	α4β1,α5β1	++	α4β1,α5β1,?
LM	+	α6β1	++	α6β1	-	nd [e]
Coll I	-	nd	-	nd	++	α2β1,
Coll IV	-	nd	-	nd	-	nd
VN	-	nd	-	nd	+	αvβ3

a) CD3⁻ CD56+ CD16+ peripheral blood NK cells isolated by Percoll fractionation and immunomagnetic negative selection.

b) CD3⁻ CD56+ CD16+ NK cells stimulated for 10 minutes at 37°C with TPA or anti-CD16 mAb.

c) CD3⁻ CD56+ CD16+ NK cells generated from co-culture of PBL with an EBV+ lymphoblastoid cell line (RPMI 8866).

d) Data are presented as % of cell adhesion: +++ = 30-40%; ++ = 20-30%; + = 10-20%; - = <3%.

e) nd = not determined

Table 2. β1 Integrin expression on fresh and activated NK cells [a]

Integrin subunit	Fresh	Short-term activated [b]	Long-term activated [c]
α1	1.2 [d]	1.3	5.0
α2	1.3	1.2	2.8
α3	1.8	1.5	1.5
α4	18.6	18.1	49.4
α5	5.1	5.2	6.4
α6	2.9	2.6	1.6
β1	24.8	21.3	27.7

a) Expression was evaluated by double immunofluorescence and FACS analysis.

b) CD3- CD56+ CD16+ NK cells stimulated for 10 minutes at 37°C with TPA or anti-CD16 mAb.

c) CD3- CD56+ CD16+ NK cells generated from coculture of PBL with an EBV+ lymphoblastoid cell line (RPMI 8866).

d) Data were evaluated on CD56+ cells as the ratio between MFI of experimental group and MFI of control (second step reagent).

short-term NK cell activation. As shown by double immunofluorescence and cytofluorimetric analysis, treatment with TPA (20 ng/ml for 10 min at 37°C) did not induce any detectable changes in the levels of β1 integrin expression. Similarly, β1 expression did not change after stimulation of NK cells with anti-CD16 mAb (Table 2).

To investigate whether PKC could play a role in the enhancement of adhesion to FN and LM, NK cells were pretreated with different doses of staurosporin and then activated with TPA or anti-CD16 mAb. As shown in Figure 2, pretreatment with staurosporin (at the concentrations known to inhibit PKC) resulted in complete inhibition of the increased adhesion to both FN and LM.

The rapid enhancement of NK cell adhesion to FN and LM, involving PKC activation and associated with no changes in the levels of FNr or LMr expression, suggests that

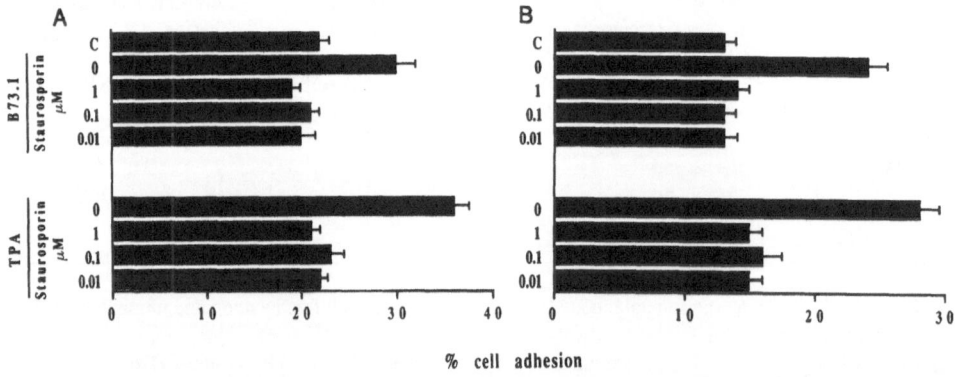

Figure 2. Staurosporin Inhibits TPA and CD16-induced Enhancement of NK Cell Adhesion to FN or LM. [51]Cr labelled highly purified NK cells, pretreated for 10 min at 37° C with different doses of staurosporin (from 1 to 0.01 μM) were stimulated for 10 min at 37° C with TPA or B73.1 (anti-CD16) mAb and then tested for the ability to bind to FN (panels A) or LM (panels B) in a 2h adhesion assay.

qualitative alterations of integrins themselves and/or of integrin-associated molecules are likely responsible for activation-induced enhanced adhesiveness. Increased avidity for counter ligands as a result of cell activation is a behaviour common to many integrin receptors (Hynes, 1992). The mechanisms involved in the regulation of activation-dependent integrin receptor functions are still unclear, but posttranslational modifications (i.e. phosphorylation) of either α and/or β integrin subunits, as well as of other integrin-associated molecules such as cytoskeleton components, are likely to occur. On the basis of the molecular alterations, several mechanisms including cytoskeleton association, interaction with other intracellular moieties, receptor clustering, exposure of activation epitopes, may underlie the increased integrin avidity.

Long-term NK Cell Activation Results in Modulation of Integrin Expression and Function

Much evidence indicates that integrin expression and function is regulated during cell growth and differentiation (Hynes, 1992). We investigated whether long-term activation of NK cells results in modulation of integrin expression and function. It has been recently described that coculture of nylon-wool-nonadherent peripheral blood mononuclear cells with irradiated EBV[+] lymphoblastoid B cell line (RPMI-8866) results in selective proliferation and expansion of NK cells (Perussia et al., 1987). These cells were stained at day 10 of culture with a panel of mAb directed against α1 (TS2/7), α2 (P1E6), α3 (MKID-2), α4 (P4G9), α5 (P1D6), α6 (GoH3), and β1 (4B4). Two-color immunofluorescence analysis on gated CD56+ cells showed induction of α1 and α2, increased levels of α4, α5 and β1 and a decline of α6 on cultured cells with respect to fresh NK cells; expression of α3 remained undetectable (Table 2). The presence of the different α's and their association with β1 subunit was confirmed by biochemical analysis (data not shown). To investigate whether long-term activation could also affect integrin function, we assayed the ability of 10 day-cultured NK cells to bind to LM, FN, Coll I and Coll IV. We found that long-term activated NK cells, differently from fresh cells, do not adhere to LM. Loss of adhesion to LM parallels α6 decline and indicates that VLA-1 and VLA-2 are not functional receptors for this glycoprotein. Long-term activated NK cells adhered to FN at the same degree (20-30%) of fresh NK cells but their adhesion was only partially mediated by β1 integrins. Moreover, long-term activated NK cells, but not fresh NK cells, adhered to Coll I and this adhesion was mediated by VLA-2. Neither fresh nor long-term activated NK cells bound to Coll IV.

Overall these results indicate that adhesion molecules other than β1 integrins could play a role in the adhesion of long-term activated NK cells to both FN and Coll I.

The mechanisms regulating integrin expression are still largely undefined. Several cytokines affect cell adhesion receptors: TGF-β1 increases the expression of several β1 integrins in fibroblasts (Heino et al., 1989) and IL-1β in osteosarcoma cells (Dedhar, 1989); moreover TNFα by itself or in combination with IFNγ modulates the expression of some β1 or β3 integrins on endothelial cells (Defilippi et al., 1991a; Defilippi et al., 1991b). Therefore it is conceivable that cytokines released during immune response or inflammation may be responsible for the regulation of integrin expression and function on lymphocytes.

CONCLUSION

Taken together our results indicate that while short-term NK cell activation results in modulation of β1 integrin function without affecting integrin expression, long-term activation causes modulation of both β1 integrin expression and function. Thus the ability of NK cells to interact with ECM components can be regulated by their activation. This could result in a rapid change in their migration and tissue localization capacity during inflammatory or immune responses.

ACKNOWLEDGEMENTS

Fabrizio Mainiero is the recipient of an AIRC fellowship.

REFERENCES

Cassatella, M., Anegon, I., Cuturi, M.C., Griskey, P., Trinchieri, G., and Perussia, B., 1989 FcγR (CD16) interaction with ligand induces Ca^{++} mobilization and phosphoinositide turnover in human natural killer cells. Role of Ca^{++} in FcγR (CD16)-induced transcription and expression of lymphokine genes. *J. Exp. Med.* 169:549.

Dedhar, S., 1989, Regulation of expression of the cell adhesion receptors, integrins, by recombinant human interleukin-1β in human osteosarcoma cells: Inhibition of cell proliferation and stimulation of alkaline phosphatase activity. *J. Cell. Physiol.* 138:291.

Defilippi, P., van Hinsbergh, V., Bertolotto, A., Rossino, P., Silengo, L., and Tarone, G., 1991a, Differential distribution and modulation of expression of alpha1/beta1 integrin on human endothelial cells. *J. Cell Biol.* 114:855.

Defilippi, P., Truffa, G., Stefanuto, G., Altruda, F., Silengo, L, and Tarone, G., 1991b, Tumor necrosis factor α and interferon γ modulate the expression of the vitronectin receptor (integrin β3) in human endothelial cells. *J. Biol. Chem.* 266:7638.

Elices, M.J. and Hemler, M.E., 1989, The human integrin VLA-2 is a collagen receptor on some cells and a collagen/laminin receptor on others. *Proc. Natl. Acad. Sci. USA* 86:9906.

Elices, M.J., Osborn, L., Takada,Y., Crouse, C., Luhowskyj, S., Hemler, M.J., and Lobb, R., 1990, VCAM-1 on activated endothelium interacts with the leukocyte integrin VLA-4 at a site distinct from the VLA-4/fibronectin binding site. *Cell* 60:577.

Elices, M.J., Urry, L.A., and Hemler, M.E., 1991, Receptor functions for the integrin VLA-3: Fibronectin, collagen, and laminin binding are differentially influenced by arg-gly-asp peptide and by divalent cations. *J. Cell Biol.* 112:169.

Gismondi, A., Morrone, S., Humphries, M.J., Piccoli, M., Frati, L., and Santoni, A., 1991a, Human natural killer cells express VLA-4 and VLA-5, which mediate their adhesion to fibronectin. *J. Immunol.* 146:384.

Heino, J., Ignotz, R.A., Hemler, M.E., Crouse, C., and Massagué, J., 1989, Regulation of cell adhesion receptors by transforming growth factor-β. J. Biol. Chem. 264:380.

Hemler, M.E., Jacobson, J.G., Brenner, M.B., Mann, D., and Strominger, J.L., 1985, VLA-1: A T cell surface antigen which defines a novel late stage of human T cell activation. *Eur. J. Immunol.* 15:502.

Hemler, M.E., Glass, D., Goblyn, J.S., and Jacobson, J.G., 1986, Very late activation antigens on rheumatoid synovial fluid T lymphocytes: Association with stages of T cell activation. *J. Clin. Invest.* 78:696.

Hemler, M.E., 1990, VLA proteins in the integrin family: Structures, functions, and their role on leukocytes. *Ann. Rev. Immunol.* 8:365.

Hynes, R., 1992, Integrins: versatility, modulation, and signaling in cell adhesion. *Cell* 69:11.

Ignatius, M.J., Large, T.H., Houde, M., Tawil, J.W., Barton, A., Esch, F., Carbonetto, S., and Reichardt, L.F., 1990, Molecular cloning of the rat integrin α1-subunit: A receptor for laminin and collagen. *J. Cell Biol.* 111:709.

Matsuyama, T., Yamada, A., Kay, J., Yamada, K.M., Akiyama, S.K., Schlossman, S.F., and Morimoto, C., 1989, Activation of CD4 cells by fibronectin and anti-CD3 antibody. A synergistic effect mediated by the VLA-5 fibronectin receptor complex. *J. Exp. Med.* 170:1133.

Morrone, S., Gismondi, A., Mainiero, F., Milella, M., Santoni, G., Falcioni, R., Piccoli, M., Frati, L., and Santoni, A., 1991, Expression of β1 integrin laminin receptors by fresh and cultured NK cells. *Nat. Immun. Cell Growth Regul.* 10:135.

O'Shea, J.J., Weissman, A.M., Kennedy, I.C.S., and Ortaldo, J.R., 1991, Engagement of the natural killer cell IgG Fc receptor results in tyrosine phosphorylation of the ζ chain. *Proc. Natl. Acad. Sci. USA* 88:350.

Perussia, B., Trinchieri, G., Jackson, A., Warner, N.L., Faust, J., Rumpold, H., Kraft, D., and Lanier, L.L., 1984, The Fc receptor for IgG on human natural killer cells: Phenotypic, functional and comparative studies using monoclonal antibodies. *J. Immunol.* 133:180.

Perussia, B., Ramoni, C., Anegon, I., Cuturi, M.C., Faust, J., and Trinchieri, G., 1987, Preferential proliferation of natural killer cells among peripheral blood mononuclear cells cocultured with B lymphoblastoid cell lines. *Nat. Immun. Cell Growth Regul.* 6:171.

Saltini, C., Hemler, M.E., and Crystal, R.G., 1988, T-lymphocytes compartmentalized on the epithelial surface of the lower respiratory tract express the very late activation antigen complex VLA-1. *Clin. Immunol. Immunopathol.* 46:221.

Shimizu, Y., van Seventer, G.A., Horgan, K.J., and Shaw, S., 1990a, Regulated expression and binding of three VLA (β1) integrin receptors on T cells. *Nature* 345:250.

Shimizu, Y., van Seventer, G.A., Horgan, K.J., and Shaw, S., 1990b, Roles of adhesion molecules in T-cell recognition: Fundamental similarities between four integrins on resting human T cells (LFA-1, VLA-4, VLA-5, VLA-6) in expression, binding and costimulation. *Immunol. Rev.* 114:109.

Shimizu, Y.and Shaw, S., 1991, Lymphocyte interactions with extracellular matrix. *FASEB J.* 5:2292.

Shimizu, Y., van Seventer, G.A., Ennis, E., Newman, W., Horgan, K.J., and Shaw, S., 1992, Crosslinking of the T cell-specific accessory molecules CD7 and CD28 modulates T cell adhesion. *J. Exp. Med.* 175:577.

Sonnenberg, A., Modderman, P.W., and Hogervorst, F., 1988, Laminin receptor on platelets is the integrin VLA-6. *Nature* 336:487.

Springer, T.A., 1990, Adhesion receptors of the immune system. *Nature* 346:425.

Takada, Y., Huang, C., and Hemler, M.E., 1987, Fibronectin receptor structures in the VLA family of heterodimers. *Nature* 326:607.

Trinchieri, G., 1989, Biology of natural killer cells. *Adv. Immunol.* 47:187.

Wayner, E.A., Garcia-Pardo, A., Humphries, M.J., McDonald, J.A., and Carter, W.J., 1989, Identification and characterization of the lymphocyte adhesion receptor for an alternative cell attachment domain (CS-1) in plasma fibronectin. *J. Cell Biol.* 109:1321.

DISCUSSION

S. HASKILL

Have you looked to see if you are inducing any genes with this integrin engagement, since you get degranulation as well?

A. SANTONI

We have very preliminary results which I do not know whether I should talk about. In certain experimental conditions, fibronectin seems to induce γ-interferon gene and BLT esterase release.

S. HASKILL

It could be selective. So, if you adhere to collagen type-1, did you see degranulation?

A. SANTONI

The results on the expression of collagen receptor on human NK cells are quite recent. We do not know yet whether collagen too, activates NK cell functions.

R. JULIANO

How do you visualize the role of the integrins in terms of the killer function of the cells? Does the cell simply get a better perch on the substratum, and is then able to attack the target, or do you really see some sort of sandwich formation that occurs between integrins on the target, integrins on the killer cell and intervening ECM proteins?

A. SANTONI

The integrin receptors are endowed with an extreme versatility and signalling capacity. They have been recently shown to be capable of signalling. An interesting possibility is that integrins are involved in the NK cell recognition of certain targets (solid tumor cells, microorganisms, etc.). So far, however, most of the studies on the role of integrins in the NK-target cell interactions have been performed using target cells of hematopoietic origin (K562, MOLT-4). I would like to postulate that at least for certain targets, i.e, ECM-producing tumors, endothelial cells, microorganisms, a combination of different adhesion/activating molecules could be sufficient to trigger NK activity.

V. QUARANTA

Have you done pre-clearing experiments between the anti-α_V and the anti-β_3 antibodies? Do you know whether there is any β_5 in these cells?

A. SANTONI

I do not know. I do not know whether β_5 is expressed by NK cells.

V. QUARANTA

But is the vitronectin adhesion entirely knocked out by the anti-β_3 antibody?

A. SANTONI

I do not know.

M. HEMLER

Over on this side of the room, I detected a wave of perhaps excitement or perhaps skepticism at the apparent violation of the current dogma regarding integrin specificity. I am referring to the $\alpha_V\beta_3$ collagen. If you could expand on that a little bit. Which particular antibodies were utilized?

A. SANTONI

It was used a rabbit polyclonal antibody anti vitronectin receptor purchased from the Calbiochem Co.

M. HEMLER

Polyclonal against what?

A. SANTONI

It is a commercial reagent. In our experimental conditions, this antibody does not inhibit the β_1-dependent adhesion of fresh NK cells to fibronectin. Moreover, several reports indicate that $\alpha_V\beta_3$ may bind to collagen type I in a RGD-dependent manner.

M. HEMLER

So do the NK cells express α_2 at any stage?

A. SANTONI

Long-term activated, but not fresh NK cells express $\alpha_2\beta_1$. This receptor, however, only partially mediates NK cell adhesion to collagen type I since only 50% inhibition of binding is observed in the presence of anti-β_1 or anti-α_2 monoclonal antibodies. I would like to point out that these reagents, in our hands, completely block adhesion to collagen type I in other cell systems.

M. HEMLER

As I recall collagen type-1 does have a few RGD sequences, is that right?

A. SANTONI

Yes, in the α_1- and α_2-chains.

S. GOODMAN

How did you prepare your collagen substrate? Under what conditions did you coat the type-1 collagen and at what temperature?

A. SANTONI

At 4°C, as for fibronectin and vitronectin.

E. ROOS

I have a question concerning the dissemination pattern of your NK cells. You showed that the activated NK cells moved to lungs and liver, you specifically referred to pit cells in liver, which I think are present in the blood vessels and not outside the blood vessels. Do these matrix receptors have a function in getting the NK cells to an extravascular localization. This is interesting also because it is well established in the cancer metastasis literature that NK-killing of tumor cells takes place in the circulation and not outside of the circulation.

A. SANTONI

There is evidence in the literature showing that NK cells localize both in the circulation and outside of circulation. Several years ago Carlo Riccardi and myself demonstrated that NK cells are quite efficient in the clearance of tumor cells from the blood. Following *in vivo* treatment with several BRMs or during viral infections, accumulation of NK cells in liver parenchyma has been shown. NK cells have been also found in the site of allograft rejection.

E. ROOS

So more specifically, do you think that activated NK cells do move outside of blood vessels in the tissues that they disseminate to?

A. SANTONI

Yes. The difficulty to find NK cells in the tissues may be also related to the time when they are analyzed during the inflammatory reactions. I think that NK cells could be one of the very early inflammatory lymphoid cells crossing blood vessels.

F. SANCHEZ-MADRID

My question is related to the one Dr. Juliano asked you about the ability of the anti-α_4 antibody to inhibit cytolytic mechanism. And this is probably a comment more general for not only NK cells but also for cytolytic T lymphocyte, because there are two reports about it. I think there is some controversy on the VLA-4 involvement on the cytolytic T cell-killing mechanism. The question for you is whether you have found any inhibitory effect of anti-α_4 in your system, and then probably you will relate the results other people have found with CTL.

A. SANTONI

We do not yet know whether $\alpha_4\beta_1$ may play a role in the NK cytotoxicity. The major problems are related to lack of suitable reagents. NK cells are FcRγIII positive and to correctly evaluate the role of integrin receptors in the effector-target cell interaction, Fab fragments of anti-α or anti-β subunits antibodies should be used. We just got enough amount of an anti-β_1 antibody to perform these studies. In regard to the paper by Takada et al. in the *EMBO Journal,* in which it has been reported that VLA-4 plays a role in T cell-mediated cytotoxicity, I remember that a CTL system was used which was particularly sensitive to inhibition with antibodies (i.e., anti-CD8).

M. HEMLER

If I could comment on that, we observed the initial report by Krensky, Clayberger and McIntyre in which antibody L25 to VLA-4 was selected based on the ability to inhibit cytotoxic T cell function. We were quite surprised and we tried 10 or 20 times and we were unable to reproduce those results using all available T cell clones that we had at our disposal. So, upon contacting Dr. McIntyre, we were advised that you have to use relatively weak killer cells. So we used bulk cultures that had only been cultured 1-2 weeks, that had sufficiently low cytotoxic activity and only under those conditions we were able to see even partial inhibition by anti-α_4 antibody. But, we did see that consistently.

A. SANTONI

There is also an abstract from McIntyre's group presented three years ago at the Integrin UCLA meeting reporting that the anti-α_4 L25 monoclonal antibody inhibits the cytolytic activity of a NK clone against MOLT-4, but not K562 target cells. So, I think that the effector-target cell system employed to study this issue is quite crucial. It is clear that the adhesion pathways involved in this interaction largely depend both on the type of the effector (endogenous, activated, or cytolytic cell clone) and target cell.

F. SANCHEZ-MADRID

I would like to make a further comment to Martin's. At the time you published the paper, and the one by McIntyre, there was no evidence that this L25 induces homotypic cell aggregation. How could you rule out a simple alternate explanation that the inhibition was due to cell aggregation?

M. HEMLER

I think that could be a good explanation for anti-α_4 inhibition of cytotoxicity.

F. SANCHEZ-MADRID

I think that that is important because you know that that implies that there are other ligands, because VCAM-1 is not there, or anti-VCAM-1 are not affecting this CTL-target interaction. And then that is not only homotypic cell aggregation, there is something else there, maybe another ligand. I think that an important clue in this issue is to clarify that. Anyway, we have not observed any inhibitory effect with a huge panel of our anti-α_4 antibodies.

A. SANTONI

What type of CTL system has been used?

F. SANCHEZ- MADRID

CTL clones.

G. NICOLSON

This goes back to the question from Ed Roos and I think you have partially answered it. It concerns whether these integrins are more important for extravasation of activated cells, or are they important in recognition and killing? I am skeptical, or kind of doubtful of the latter, just based upon talking to people in the NK area. But, in fact, the former possibility could be quite important and quite dependent upon the targets that you analyzed. If you analyze a target that grows on a substrate, or makes extracellular matrix components, then killing could be enhanced by binding to both the matrix and the cell at the same time. That would be a proximity effect. Thus the extracellular matrix binding may actually aid in association with the target. On the other hand, during extravasation it is very important that these cells recognize matrix-type components, and the integrins are certainly important in that type of interaction. Could you comment on those possibilities?

A. SANTONI

I feel that both the roles are possible. I certainly believe that integrins are important for NK cell extravasation. A paper by Paola Allavena et al. in *J.E.M.* reported that NK cells may adhere to activated endothelial cells through the VLA-4/VCAM-1 pathway. Endothelial cells, at least those derived from umbelical vein, may be also lysed by IL2-activated NK cells.

G. NICOLSON

Can I interrupt you at that point, because it is my understanding in talking to people who work in this area that it depends how you handle these so-called normal cells when you put them into culture. If you put them into short-term culture they are completely refractory to NK, but if they are left in culture for a period of time, then they acquire NK target sensitivity.

A. SANTONI

In regard to the role of integrins in the NK cytotoxicity, I like to hypothesize that a combination of several adhesion molecules may be sufficient to trigger cytolytic activity and eventually explain NK selectivity toward certain target cells.

G. NICOLSON

OK, that clarifies it. I agree with you on the specificity.

A. SANTONI

I discussed this possibility at the last NK meeting.

G. NICOLSON

None of the NK-ologists have been able to produce a target molecule.

A. SANTONI

How triggering or recognition molecules are defined is sometimes confusing. It should be established whether recognition implies also the concept of specificity as for the antigen TcR-complex. Many molecules expressed on the NK cells are able to trigger their functions (CD16, CD2, etc.).

G. NICOLSON

Unless the target is a carbohydrate, or an incomplete carbohydrate chain, which of course can be altered during the various physiologic states of the cell without altering the polypeptide components.

C. FIGDOR

I want to go back to your phosphorylation experiments. You showed that upon PMA-exposure phosphorylation of the α-chain of the α_6 occurs. Was that done when the cells were bound to laminin or was it just by a PMA stimulation, and what about VLA-4 and VLA-5? You did not find phosphorylation with them. So how do you interpret those data?

A. SANTONI

We stimulated the NK cells for different times (5-20 min) with PMA or anti-CD16 moAb in the absence of the ligand. In regard to the lack of α_4- or α_5-phosphorylation, it could be due to the experimental conditions used, which are suitable to analyze α_6-, but not α_4- or α_5-phosphorylation. It is also conceivable that different kinases or phosphatases could regulate the phosphorylation status of different integrin subunits. Experiments are undergoing to analyze phosphorylation in the presence of phosphatase inhibitors such as pervanadate. Moreover, I like to stress that these experiments performed with fresh cells are quite difficult.

N. HOGG

Could I just ask a subsidiary question? Did the phosphorylation, the kinetics of phosphorylation, correlate with the transiency of the adhesion?

A. SANTONI

Not completely. Indeed CD16-induced phosphorylation declined 20 minutes after stimulation, when enhanced adhesion is still observed. These results could be explained by postulating that the conformational changes in the integrin receptor possibly induced by phosphorylation persist longer than receptor phosphorylation.

C. FIDGOR

But maybe it is a good idea to do these experiments in the presence of the ligands?

A. SANTONI

Yes, it is a very good suggestion.

G. TARONE

You showed that integrin expression changed quite deeply in long-term activated NK. Do you have any idea what can be the signal that triggered this altered integrin expression during this long-term activation?

A. SANTONI

We would like to reproduce some of the results observed activating NK cells in the presence of the EBV + lymphoblastoid cell line RPMI-6688, using different cytokines. We chose this system of activation for its convenience. It is indeed possible to obtain large quantities of highly purified NK cells. It is evident that several cytokines are produced during NK cell culture and among these, certainly IFNγ and TNFα. I am aware of your results on the modulation of integrin expression by IFNγ and TNFα on endothelial cells. It is really possible that these cytokines are also responsible for the changes in integrin expression and function we observed in the NK cells following long-term activation.

R. JULIANO

Getting back to the question of the activation process and possible role of integrins, have you ever tried just simply putting in fragments of extracellular matrix proteins, or peptides, to see if they might trigger a release of perforin or other elements of the activation process?

A. SANTONI

As I already said, these experiments just started. We have no results using fibronectin fragments yet.

R. JULIANO

But not the fragment in solution?

A. SANTONI

No, experiments were performed only with immobilized fibronectin.

S. HASKILL

It has been a long time since I worked in tumor systems but I remember that there were a lot of studies looking for NK cells in solid tumors. And you could find NK cells that were not active any more at least in terms of having markers of NK cells. Looking at lung lesions before and after BCG, stimulation of tissue localization of active NK cells was noted in the past. Do you know anything about whether you activate different integrins, selectively or not, *in vivo*? And second, really an off-the-wall question, I remember also that it seemed like NK cells could be reciprocally killed by interaction with target cells and that there were a lot of effector cells dying at the same time that the target cells were dying.

A. SANTONI

It is not clear whether NK cells are autocytotoxic. It seems likely that they can protect themselves from autolysis; however, the mechanisms of protection are undefined. In regard to the *in vivo* studies, we have not performed these kind of experiments. I would like to recall that only a few reagents are available against integrin subunits in the mouse. It is likely that integrin expression is modulated following *in vivo* activation.

J. CALVETE

My question is regarding the possible involvement of phosphorylation in VLA-6 activity. If there is any relationship then it should be a stoichiometric ratio of phosphorylation. Have you tried to measure the increase in phosphorylation? Was it a 2%, was it a 50% increase?

A. SANTONI

We did not measure the increase in phosphorylation.

N. HOGG

Did you mention whether the cultured T cells had to be activated in order to bind to the various matrix molecules? Activated by CD16? Or are they constitutively active?

A. SANTONI

Fresh NK cells constitutively bind to fibronectin and laminin. The percent of binding depends on the donor and ranges between 20-40% for fibronectin and 10-20% for laminin.

N. HOGG

And can that be activated further by CD16 and by how much?

A. SANTONI

Yes, a 2-fold increase can usually be observed in cell adhesion following stimulation with anti-CD16 or PMA. It will be of interest to test whether enhancement induced by anti-β_1 monoclonal antibody (TS2/16) could be additive to that induced by triggering via CD16.

E. MIHICH

When you talk about activated NK cells, do you refer to cells which are different from LAK cells? You know the whole concept of LAK cells being a form of activated NK cells and the corollary questions? Have you ever compared the characteristics that you are studying in NK cells with those of corresponding LAK cells, even as a way of clarifying whether that hypothesis is correct or not?

A. SANTONI

The culture system I described is different from LAK cells. Our long-term activated cells have a different pattern of target specificity in that they are not capable to lyse certain LAK-susceptible targets such as RAJI. Moreover, while LAK cells express high levels of activation markers such as IL2 receptor α-chain and MHC II antigens, the long-term activated cells generated in our culture system display only low levels of these markers. Thus, LAK cells and our long-term activated NK cells may differ in their activation status. We have performed a few experiments analyzing the modulation of integrin expression and function on LAK cells and similar results have been obtained.

INTERACTION OF CYTOTOXIC T LYMPHOCYTES WITH AUTOLOGOUS MELANOMA: ROLE OF ADHESION MOLECULES AND β1 INTEGRINS

Andrea Anichini, Roberta Mortarini, and Giorgio Parmiani

Division of Experimental Oncology D
Istituto Nazionale per lo Studio e la Cura dei Tumori
Via Venezian 1, 20133
Milan, ITALY

ABBREVIATIONS USED

CTL = Cytotoxic T Cell; **ICAM-1** = Intercellular Adhesion Molecule-1; **IFN-γ** = Interferon gamma; **LAK** = Lymphokine-activated Killer Cell; **LFA-1** = Lymphocyte-function Associated Antigen-3; **MHC** = Major Histocompatibility Complex; **PBL** = Peripheral Blood Lymphocyte; **TAA** = Tumor-associated Antigen; **TcR** = T Cell Receptor; **TIL** = Tumor-infiltrating Lymphocyte; **TNF-α** = Tumor Necrosis Factor Alpha; **VLA** = Very Late Antigen

INTRODUCTION

The clonal analysis of immune response to autologous tumors has indicated that CTLs expressing the αβ-type of TcR and able to lyse autologous neoplastic cells can be grouped in two main functional subsets on the basis of the specificity and mechanism of lymphocyte-tumor interaction (Anichini et al., 1987).

The first type of CTL is characterized by specificity restricted to the autologous tumor and, in some instances, by ability to lyse allogeneic tumors of the same histology. In addition, these effectors use the TcR in the interaction with neoplastic cells to recognize TAA in association with products of the MHC complex. The MHC restriction also is valid for the lysis of allogeneic tumors. In fact, the allogeneic cells have to share the appropriate HLA allele with the autologous tumor in order to be recognized (Crowley et al., 1991). The second group of effectors is found at higher frequency than the first one when screening large panels of CTL clones and is characterized by lysis of allogeneic tumors of different histological origin and by absence of MHC-restriction and of TcR-engagement in the interaction with neoplastic cells (Anichini et al., 1987). In addition, the receptor/ligand interactions that are responsible for triggering the cytotoxic activity of the second subset of effectors are not known.

Both types of effectors could be useful as anti-tumor reagents. In fact tumor-specific CTL clones might represent the ideal tool to be used for the adoptive immunotherapy of cancer. On the other hand, the cytotoxic potential of non-specific CTL clones might be therapeutically exploited by association with bi-specific antibodies able at the same time to activate the T cell and to recognize tumor-associated antigens (Nistico' et al., 1992).

However, the outcome of the interaction of CTL with autologous tumor cells is not dependent only on the TcR/TAA pathway (effective for specific CTL) and on the unknown receptor/ligand pathway hypothesized for non-specific CTL. In fact additional families of known cell surface proteins, expressed on neoplastic cells can contribute to the efficacy of the CTL-tumor interaction. These are the adhesion molecules, such as ICAM-1 and LFA-3, and the VLA antigens, also known as β1 integrins. The aim of our study was to evaluate the role of these molecules in the interaction of specific and non-specific CTL with autologous human melanoma cells. The results indicate that adhesion molecules and integrins participate in the CTL-tumor interaction and suggest that ICAM-1 and at least three different β1 integrins (VLA-2, VLA-5, and VLA-6), expressed on melanoma cells, may act not only in regulating the CTL-tumor adhesion but also as ligands for T cell co-receptors that mediate activation signals.

RESULTS AND DISCUSSION

Susceptibility of Melanoma Clones to Lysis by Specific and Non-specific T cell Clones Is Associated with High Expression of ICAM-1 and Presence of VLA-2, VLA-5, and VLA-6

Table 1 shows the VLA, ICAM-1 and LFA-3 profile of representative tumor clones isolated from the same subcutaneous metastasis of melanoma. The tumor clones can be clearly split in two groups with high and low expression of ICAM-1 and of at least three different VLA antigens (VLA-2, -5, and -6).

Table 1. VLA, ICAM1, and LFA3 Phenotype of Melanoma Clones Isolated by the Soft-agar Technique (Anichini et al., 1986) from Metastasis Me665/2

Tumor Clone	VLA1	VLA2	VLA3	VLA4	VLA5	VLA6	ICAM1	LFA3
2/4	27*	47	86	85	73	46	173**	129
2/14	85	68	91	94	35	60	175	143
2/17	86	89	48	27	62	42	166	124
2/51	51	37	94	78	48	34	173	128
2/60	74	30	96	3	51	90	176	142
2/21	26	0	46	1	0	6	79	90
2/30	42	7	56	2	1	6	91	140
2/33	3	8	46	2	0	1	93	77
2/39	33	9	61	14	0	17	100	136
2/56	36	2	32	0	0	0	83	90

The phenotype was performed by indirect immunofluorescence and cytofluorimetric analysis with the following antibodies: TS2/7.1.1 (VLA1), CLB/thr4 (VLA2), J143 (VLA3), B5G10 (VLA4), B1E5 (VLA5), GOH3 (VLA-6), 84H10 (ICAM1), and TS2/9.1.1 (LFA3) (Anichini et al., 1990; Mortarini et al., 1991).(*) Results are expressed as % of positive cells for the VLA antigens.(**) All tumor clones are positive for ICAM-1 and LFA-3 and the results of expression of these two molecules are expresssed as mean channel intensity of fluorescence in arbitrary units on a 256 channel log scale.

Two melanoma clones (2/14 and 2/30) representative of the two groups with high and low expression of adhesion molecules and integrins were tested for lysability by specific and non-specific CTL clones isolated from either TIL or PBL. As shown in Table 2 all CTL clones expressed higher lytic activity on the tumor clone 2/14 than on tumor clone 2/30. These results are in agreement with the differential lysability of the two groups of melanoma clones observed using as effectors IL-2 activated CD3+ or CD3- lymphocytes (Anichini et al., 1990) and indicate that a correlation exists between susceptibility to lysis and the expression of ICAM-1 and of some β1 integrin.

Table 2. Comparison of Lysability of Melanoma Clones 2/14 and 2/30 was Performed by Measuring the Release of ^{51}Cr from Radiolabelled Tumor Cells in a 4h Assay

CTL clone	TcR engagement	Lysis of:	
		Tumor clone 2/14	Tumor clone 2/30
TIL 10C7	yes	82	35
TIL 8D9	yes	22	3
TIL 6E1	yes	54	14
PBL 13H7	no	29	1
PBL 14A8	no	68	0
PBL 15B2	no	29	0
PBL 11A2	no	40	0
PBL 14A4	no	36	1
PBL 15B1	no	79	29
PBL 15E1	no	57	0
PBL 15G3	no	22	0
PBL 15E12	no	21	0
PBL 14F12	no	24	0
PBL 15G1	no	6	0
PBL 15D4	no	12	2
PBL 15F1	no	16	5
PBL 15D8	no	40	0
PBL 15A3	no	75	5
PBL 15D9	no	36	0
PBL 13H2	no	24	2
PBL 14E11	no	48	0
PBL 14A7	no	26	5
PBL 15C5	no	24	0
PBL 15E8	no	66	11
PBL 15F12	no	6	0
PBL 14E3	no	55	4
PBL 15D3	no	56	0
LAK CELLS	no	79	6

All the CD3+, CD8+. WT31+ CTL clones were used at the Effector:Target ratio of 10:1.(*) TIL clones 10C7, 8D9, and 6E1, but not all PBL-derived CTL clones, use their TcR in the interaction with autologous neoplastic cells and are MHC-restricted, as indicated by blocking experiments with monoclonal antibodies to the TcR-associated CD3 complex and to HLA class I molecules (Anichini et al., 1989). LAK cells were produced by culturing allogeneic peripheral blood lymphocytes with 500 U/ml of rIL-2 for two weeks.

Analysis of conjugate formation by two color cytometry (data not shown) between CTL clones and the two groups of tumor clones indicated that all melanoma clones formed conjugates with the same efficiency with the T cells. These data suggested that the differential lysability of the two groups of clones was not due to differences in the adhesion step of the CTL-tumor interaction. In addition, monoclonal antibody (MAb) triggering of the TcR of any CTL (by pre-treatment with anti-CD3 mAb) induced similar lysis of all tumor clones, thus indicating that the differences of lysability among melanoma clones were not due to intrinsic resistance of some neoplastic cells to the lytic mechanism of the effectors (data not shown).

Cytokine Treatment Enhances ICAM-1 Expression and Lysability of Melanoma Clones

The role of ICAM-1 in the differential lysability of tumor clones was verified by up-regulating its expression by cytokine-treatment and then testing the effect on lysability. As shown in Table 3, IFN-γ or TNF-α-mediated tumor ICAM-1 up-regulation on melanoma clone 2/39 could enhance the lysability of these cells but only by a tumor-specific effector, TIL clone 8G4, and not by non-specific PBL clones 13B4 and 13C2 or by LAK cells.

Table 3. Cytokine Treatment of Melanoma Cells Enhances ICAM1 Expression and Lysability but Only by Tumor Specific CTL Clones.

Tumor clone	Cytokine	ICAM-1 MCI*	TIL 8G4	PBL 13B4	PBL 13C2	LAK Cells
2/39	NO	91	22 (**)	9	0	75
	IFN-γ	160	40	2	9	67
	TNF-α	127	42	4	4	50

Tumor clone 2/39 was treated for 48h with or without 500 U/ml of rIFN-γ or rTNF-α (Mortarini et al., 1990) and then tested for ICAM-1 expression as well as for lysability.(*) Expression of ICAM1 is as mean Channel Intensity of Fluorescence (MCI).(**) % lysis at E:T ratio of 50:1. TIL clone 8G4 is TcR-dependent and MHC-restricted in the interaction with the autologous tumor (data not shown) while PBL clones 13B4 and 13C2 are non-specific effectors. LAK cells were produced by culturing allogeneic peripheral blood lymphocytes with 500 U/ml of rIL-2 for two weeks.

The role of ICAM-1 in the CTL-melanoma interaction also was verified in blocking experiments by pre-incubating the tumor clones with an anti-ICAM-1 antibody (84H10). These experiments indicated that the lysis of neoplastic cells by either specific and non-specific effectors could be significantly reduced by preventing the LFA-1/ICAM-1 interaction (data not shown). Taken together these data indicate that the differential expression of ICAM-1 on neoplastic targets contribute to the outcome of the CTL-target interaction not by changing the efficiency of conjugate formation but by contributing a co-activation signal, delivered through the LFA-1 pathway, that cooperates with the TcR-dependent signal in the induction of the cytolytic program of the effector (Van Seventer et al., 1990). This interpretation explains why non-specific effectors do not change their lytic efficiency after tumor ICAM-1 up-regulation. In fact, in the interaction with the autologous tumor, these effectors do not engage their TcR. In the absence of TcR triggering there is no possibility for activating the LFA-1/TcR cooperation that leads to enhanced lytic activity.

Antibodies To VLA-2, -5, and -6 Inhibit Lysis of VLA-2-, VLA-5-, and VLA-6-Positive Tumor Clones by Specific and Non-specific Effectors.

The role of ICAM-1 in the lysability of melanoma clones does not explain why a subset of tumor cells can be recognized with higher lytic efficiency by either specific or non-specific effectors. One possibility is that tumor clones with enhanced lysability express surface structures which can act in delivering co-activation signal to the T cell. The possible role of integrins in this system was verified directly by blocking experiments. As shown in Table 4 pre-incubation of the VLA-2+, VLA-5+, VLA-6+ melanoma clone 2/60 with antibodies to VLA-2, -5, and-6 could significantly reduce the lysis of these cells, but not the lysis of a VLA-2-, VLA-5-, and VLA-6- melanoma clone (2/56), by either specific (TIL clone 10C7) and non-specific (PBL clones 15A3 and 15B1).

Table 4. Lysis of Tumor Clone 2/60, but Not of Tumor Clone 2/56, by Specific and Non-specific CTL Clones Can Be Inhibited by Anti-VLA Antibodies

Effector	MAb	Clone 2/60	Clone 2/56
TIL 10C7	NO	37(*)	26
	anti-VLA-2	25**	28
	anti-VLA-5	20**	20
	anti-VLA-6	17**	25
PBL 15A3	NO	56	8
	anti-VLA-2	34**	10
	anti-VLA-5	27**	10
	anti-VLA-6	26**	5
PBL 15B1	NO	67	4
	anti-VLA-2	40**	9
	anti-VLA-5	35**	10
	anti-VLA-6	33**	7
LAK cells	NO	62	11
	anti-VLA-2	58	15
	anti-VLA-5	53	26**
	anti-VLA-6	36**	11

Tumor clones were pre-incubated with antibodies to VLA-2, -5, and -6. Effectors were then added without washing the antibodies and the cytotoxic assay was performed as described in the legend to Table II.(**) Lysis of tumor cells in the presence of antibodies is significantly different from lysis in the absence of antibodies (SNK test, p=0.01).(*) Results expressed as % lysis. TIL clone 10C7 is TcR-dependent and MHC-restricted in the interaction with autologous tumor cells (data not shown). PBL clones 15A3 and 15B1 are non-specific CTL clones. LAK cells were produced by culturing allogeneic peripheral blood lympocytes with 500 U/ml of rIL-2 for two weeks. Effector to target ratio was 50:1.

Taken together these data indicate that integrins such as VLA-2, -5, and-6, expressed on the tumor targets can regulate the interaction with either specific and non-specific CTL clones. Further experiments are clearly needed to assess which receptors are engaged on the T cells in response to integrin binding and whether integrins on melanoma cells are contributing an adhesion or an activation signal for the T cell. Preliminary evidence, obtained by evaluation of conjugate formation, indicates that presence or absence of VLA-2, -5, and -6 does not affect the efficiency of conjugate formation between CTL clones and tumor clones. This suggests that integrins on melanoma cells might act as part of ligand/receptor pathways involved in co-activation of cytotoxic T cells. Finally, it cannot be ruled out that VLA integrins expressed on CTL clones may play a role in the interaction with tumor cells.

REFERENCES

Anichini, A., Mortarini, R., Fossati, G., and Parmiani, G., 1986, Phenotypic profile of clones from early cultures of human metastatic melanomas and its modulation by recombinant interferon-γ. *Int. J. Cancer* 38:505-511.

Anichini, A., Fossati, G., and Parmiani, G., 1987, Clonal analysis of cytolytic T cell response to human tumors. *Immunol. Today* 8: 385-389.

Anichini, A., Mazzocchi, A., Fossati, G., and Parmiani, G., 1989, Cytotoxic T lymphocyte clones from peripheral blood and from tumor site detect intra-tumor heterogeneity of melanoma cells. Analysis of specificity and mechanism of interaction. *J. Immunol.*, 142:3692-3701.

Anichini, A., Mortarini, R., Supino, R., and Parmiani, G., 1990, Human melanoma cells with high susceptibility to cell-mediated lysis can be identified on the basis of ICAM-1 phenotype, VLA profile and invasive ability. *Int. J. Cancer* 46: 508-515.

Crowley, N.J., Narrow, T.L., Quinn-Allen, M.A., and Seigler, H.F., 1991, MHC-restricted recognition of autologous melanoma by tumour-specific cytotoxic T cells. Evidence for restriction by a dominant HLA-A allele. *J. Immunol.* 146:1692-1699.

Nistico', P., Mortarini, R., De Monte, L.B., Mazzocchi, A., Malavasi, F., Parmiani, G., Natali, P.G., and Anichini, A., 1992, Cell retargeting by bispecific monoclonal antibodies: evidence of bypass of intra-tumor susceptibility to cell lysis in human melanoma. *J. Clin. Invest.* 90:1093-1099.

Mortarini, R., Belli, F., Parmiani, G., and Anichini, A., 1990, Cytokine-mediated modulation of HLA-class II, ICAM-1, LFA-3 and tumor-associated antigen profile of melanoma cells. Comparison with the anti-proliferative activity by rIL-1β, rTNF-α, rIFN-γ, rIL-4 and their combinations. *Int. J. Cancer* 45:334-341.

Mortarini, R., Anichini, A., and Parmiani, G., 1991, Heterogeneity for integrin expression and cytokine-mediated VLA modulation can influence the adhesion of human melanoma cells to extracellular matrix proteins. *Int. J. Cancer* 47:551-559.

Van Seventer, G.A., Shimizu, Y., Horgan, K.J, and Shaw, S., 1990, The LFA-1 ligand ICAM-1 provides an important costimulatory signal for T-cell receptor-mediated activation of resting T cells. *J. Immunol.* 144:4579-4586.

DISCUSSION

C. FIGDOR

In your last experiment you indeed demonstrated that when you increase ICAM-1 levels it induces a higher killing of the tumor by the specific CTL, but would you also want to argue from that experiment that other possibilities, for instance higher MHC class-1 or class-2 expression, as well as processing are not important in this respect?

A. ANICHINI

If I understood your question you're saying that I cannot exclude that the differences in the expression of antigens are relevant in this system? The answer is yes, I am sure that heterogeneity in the expression of antigens might be as relevant, and may be even more relevant than heterogeneity in the expression of adhesion molecules. As a matter of fact, the reactivity of the TIL clone 8B3 precisely suggests that the antigen or the restricting elements, are defective in those tumor clones that are not recognized. So just by changing the expression of an adhesion molecule nothing happens. The idea here, if we want to move towards a therapeutic use of these reagents, is that first we have to recognize that there are a number of possible tumor-associated antigens expressed on the same melanoma, and this evidence comes from a number of laboratories. But heterogeneity for the expression of all molecules that are relevant to the interaction. That is, accessory molecules, adhesion molecules and the antigen, is strong, so we would have to use a number of different specific effectors with different specificity, and plus, maybe, cytokines, to increase the expression of the antigens and of the adhesion molecules if we want to boost the efficacies of these reagents. I do not know whether this answers your question?

C. FIGDOR

You have been able to generate quite a number of specific autologous CTL against melanoma, which I think is beautiful, because a lot of people have tried that and were not as successful as you. Another question is whether you have any idea about the antigens recognized, for instance, by these HLA-A21 clones?

A. ANICHINI

We are moving in that direction, but we have no molecular evidence. The only thing that we know, because that has been published in different systems, is that peptides which bind to class-1 molecules have a motif. A motif that is acting in the binding of peptides to HLA A2 is considered to be one that should have leucine in position 2 and valine in position 9, something like that. So it is possible that the antigen that is being recognized has that kind of structure in those two positions. How about the rest? We do not know. There are a number of possibilities. One of the possibilities is to follow the technique that Thierry Boon used to isolate the antigen, and we are collaborating with his laboratory. The other possibility is to elute the peptides bound to the same HLA class-1 molecule and then fractionate them and add the HPLC peaks to an autologous cell that is normally not recognized, such as a lymphoblastoid cell line, and see whether you can get killing. At that point you should sequence those peptides that you found in the HPLC peak. But there is a complication because there is a recent paper in *Science* where people have been able to identify at least 200 different peptide species bound to the same HLA A2 molecule, and they were able to sequence only 8 of them. Finding the right peptide will not be easy. Even if you think in terms of how many molecules of a peptide are necessary to sensitize a T cell, maybe they are in the order of a few hundred.

C. FIGDOR

Fishing in peptide banks may also be an opportunity to detect it.

A. ANICHINI

Yes.

G. NICOLSON

This was really a very intriguing lecture for me, because I think you brought out very nicely that multiple molecules are required in the recognition process of CTLs, and part of that was known before. During progression with an increase in ICAM-1, obviously in metastatic tumors those cells were escaping CTL surveillance.

A. ANICHINI

There is a paradox. You have ICAM-1, you have everything you need, but it does not work.

G. NICOLSON

Yes, but let me finish my question. My question is that in the metastases from the patients that you examined, are the CTL responses being suppressed by some mechanism? Or does the tumor microenvironment alter the expression of the accessory molecules, such as the TCR and HLA-A2, such that they are not being expressed at the appropriate quantitive level for recognition *in vivo*?

A. ANICHINI

The literature, and also our experience, suggests that all those mechanisms can be working at the same time. At the level of suppression, for example, a number of years ago we showed that metastatic melanoma cells, but not primary melanoma cells, and mainly those metastatic melanoma cells that expressed HLA DR antigens, by a mechanism that we still do not understand, have the ability to suppress the proliferation of autologous lymphocytes to polyclonal stimuli such as IL2 or PHA. More recently, in the laboratory of Dupont Guerry, they have been able to show that DR positive metastatic melanoma cells cannot present a conventional antigen such as a tetanus toxoid to a specific CTL, while primary melanoma cells expressing the same HLA-DR restricting molecule from the same patient can. So something happens in the progression of the disease, such that the melanoma cells not only acquire the ability to suppress the response, a number of responses, of autologous lymphocytes. Perhaps something happens also to the structure of the HLA molecules, for example, in terms of the structure of HLA class-1 molecules. It has been described by Natali and others in Rome, that *in vivo* metastatic melanoma cells are heterogeneous for the expression of HLA-2, from HLA-2 patients, that there is down-regulation.

G. NICOLSON

In vivo?

A. ANICHINI

Yes, *in vivo*.

G. NICOLSON

Because that is the second part of my question. Is there a difference in the expression of these accessory molecules *in vivo?*

A. ANICHINI

It can be lost *in vivo*.

N. HOGG

Could I suggest there is another player in this matter, and that is secreted ICAM. I think there have now been three different reports showing that it, in fact, correlated with advanced melanoma.

A. ANICHINI

Exactly, that is true.

N. HOGG

There are different levels of secreted ICAM.

C. FIGDOR

That could be of importance to block the LFA-l molecule.

A. ANICHINI

Yes, it could just block all LFA-1 positive T cells around inhibiting them from the right interaction.

N. HOGG

Soluble ICAM is apparently not supposed to be able to interact with LFA-1, however, we do not really know the state or size of the secreted material.

E. ROOS

I was intrigued by your finding that the CTL clone is very effective in killing the whole tumor and then is not very effective in killing many of the clones. That seems to imply that some of the clones that are not killed in isolation are being killed in the tumor. So, could it be that the engagement of CTL clone with ICAM-1+ cells increases the efficiency of the killing by CTL clones so that it can now kill the other clones in the tumor.

A. ANICHINI

Well, there are a number of possible explanations. First of all, when we isolate tumor clones we are inducing a selection which means that if I find that five tumor clones can be recognized, and fifteen cannot, that does not mean that the same percentage of tumor clones will be represented in the uncloned population. So this is something we do not know. There are possibilities such as that there might be innocent bystander killing by T cells bound at the same time to an antigen-positive melanoma cell and also to an antigen-negative melanoma cell.

G. NICOLSON

There is a more simple, straight-forward possibility and that is that the cell-cell interactions that take place in the polyclonal population are different from single cells derived from that population and grown *in vitro*. In the case of transferrin receptor and other growth factor receptors, we have found that mouse melanoma cells growing as a coculture of different cell clones maintain their expression of these receptors, but when cultured as single cells they lose expression of them quite quickly. Thus these interclonal interactions are very important in maintaining the expression of certain molecules.

A. ANICHINI

That is absolutely right. There I absolutely agree with you. The fact is that if we check, for example, cytokine gene expression in melanoma clones from the same metastasis we found differential expression of TNFα, IL1, IL6. It is possible that local production of these cytokines in the tumor mass may contribute to maintaining the phenotype even of tumor clones which, once they are separated, without having the ability to produce the same cytokine, might down modulate, for example, an adhesion molecule.

N. HOGG

Did you not say that you have evidence for extremely limited clonal development in the tumors?

A. ANICHINI

I said we found all T cell receptor repertoire at the tumor site. All T cell receptors, the α and the β genes, are expressed.

G. DOUGHERTY

We have used retroviral gene transfer to introduce and express ICAM-1 in a variety of murine tumor cell lines differing in immunogenicity. And in no case have we found that when you inoculate those into animals differences occur either in tumor take or growth rate. So what I want to emphasize is the difference between what you see working with highly selected T cell clones and what happens in the very heterogeneous situation that exists in animals. When you inject these modified tumor cells i.v., instead of being rejected, they show a dramatically increased incidence of metastasis, presumably because they are binding LFA-1 positive cells which do not kill and are therefore protected, or trapped. So there are clear differences between *in vitro* and *in situ* phenomena.

A. ANICHINI

As a matter of fact one of the possible reasons why ICAM-1 melanoma, metastatic melanoma, is so frequent is that these circulating melanoma cells might aggregate with LFA positive cells in the circulation, and that might facilitate arrest and then extravasation at distant sites.

V. QUARANTA

Were those all mouse melanoma cell lines, or different types of tumors?

G. DOUGHERTY

They were all fibrosarcomas, some of which were spontaneous, but mainly the usual kind of carcinogen-induced, well characterized ones.

V. QUARANTA

So, is the ICAM expression a unique feature of melanoma tumors or also of other tumors?

A. ANICHINI

We find ICAM expression on a number of different neoplastic lesions. The nice thing about ICAM-1 in melanoma is that it is one of the best markers of progression in that it appears in a significant proportion of cases when the primary melanoma has a degree of invasion. I would say a thickness higher than 1.5 millimeters, and the expression of ICAM-1 in addition to the thickness of the primary tumors is one of the best prognostic indicators in melanoma. That is, if your primary tumor has ICAM-1 and has a thickness higher than 1.5m the prognosis will be very bad.

V. QUARANTA

Is not there something strange? One would expect that in a proliferating population of cells, a molecule that would favor killing should be selected against. In reality, *in vivo*, the opposite is happening. That is, ICAM-1 is selected for in proliferating melanoma. Should we think of other explanations for the presence of ICAM-1?

A. ANICHINI

I know this is a paradox and has been discussed for quite some time, but one of the possible explanations is that these sets of cytotoxic T cell clones that we can find, even at the tumor site, *in vivo*, are not activated, maybe they are not differentiated to active T cells, they are just precursor cells. They might not get those signals. We provide them *in vivo* by restimulating them with the tumor in the presence of IL2. So these T cells, even though they see ICAM-1, maybe they see the antigen, they might be suppressed *in vivo*, they might not get enough signals in order to differentiate to active T cells.

V. QUARANTA

ICAM-1 is expressed constitutively in most metastatic melanoma cells, right? So is it as if they are perhaps using it for something?

A. ANICHINI

It is possible, as I said before, that ICAM-1 positive melanoma can inhibit by shedding ICAM-1 or by making aggregates in the circulation. We do not know. We do not have to think to tumor biology in terms of immunology alone.

M. HEMLER

Maybe I missed it, but did you speculate on why ICAM gives you greater susceptibility but not greater conjugate formation? I mean how is it working?

A. ANICHINI

There are two possibilities. First of all differential expression of ICAM-1 has been shown to result in differential conjugate formation, if you compare ICAM-1 negative with ICAM-1 positive melanoma cells. The difference in the expression we found in ICAM-1, maybe it is subtle, it is not big enough to induce differences of conjugate formation, maybe because there are also a number of other accessory molecules helping in the conjugate form, and they appear to be equally expressed. So it is possible that at the level of conjugate formation the difference is not enough. But the specific T cell will see the difference, because it will receive two activation signals at the same time; the T cell receptor and the LFA-1. And so if the T cell receptor is activating the LFA-1 molecule, and if the ICAM-1 has higher expression, the LFA-1 signaling will be working at higher efficiency, I think.

TUMOR CELL-ENDOTHELIAL CELL INTERACTIONS DURING BLOOD BORNE METASTASIS: ROLE OF SPECIFIC ADHESION, MOTILITY, AND GROWTH MOLECULES

Garth L. Nicolson, Timothy J. Yeatman, Robert J. Tressler,
Timothy V. Updyke, Jun-ichi Hamada, and Phillip G. Cavanaugh

Department of Tumor Biology
The University of Texas M.D. Anderson Cancer Center
Houston, Texas, U.S.A.

ABBREVIATIONS USED

HSE = Hepatic Sinusoidal Endothelial Cell; **HX** = High; **LDGF-1** = Lung-derived Growth Factor-1; **LX** = Low; **P** = Parental Cell Line; **Tf** = Transferrin Family of Growth Factors

INTRODUCTION

The metastatic spread of blood borne tumor cells to near and distant sites does not occur randomly. Certain tumors tend to metastasize to particular organ sites, and this process appears to be due to differences in tumor cell and host organ molecular properties (Nicolson, 1988a; Nicolson, 1989; Nicolson, 1991; Zetter, 1990). Once in the blood, tumor cells can circulate to virtually every organ, yet metastases of many common tumor types form only at certain sites (Sugerbaker, 1983). This brief review will consider some of the properties of tumor cells and some of the properties of host organ cells and stroma in the organ preference of metastasis. For additional information, the reader is refered to more extensive reviews (Nicolson, 1988a; Nicolson, 1989; Nicolson, 1991; Liotta et al., 1991b).

TUMOR CELL-ENDOTHELIAL CELL ADHESION AND THE IDENTIFICATION OF CATION-DEPENDENT ADHESION COMPONENTS

Since the first barrier the circulating tumor cell encounters in homing to a specific organ site is the endothelial lining of the microvasculature, tumor cell-endothelial cell adhesion and its stablilization are the initial steps in the process of nonrandom metastasis formation (Nicolson, 1988a; Nicolson, 1989; Menter et al., 1992; Pauli et al., 1990). Multiple molecular interactions appear to be involved in tumor cell-endothelial cell adhesion (Nicolson, 1988a; Nicolson, 1989; Pauli et al., 1990). Some of the molecules appear to function in the preferential adhesion of tumor cells to specific types of endothelial cells found in specific organs or tissues, whereas other molecules appear to be associated with the adhesion of tumor cells to endothelial cells isolated from a variety of tissues (Tressler and Nicolson, 1992).

Different adhesion molecules have been described that participate not only in normal developmental processes, but also in tumor cell metastatic processes (Zetter, 1990; Nicolson, 1992; Pauli et al., 1990). The molecules important in this process have been classified into various groups, such as the integrins, cadherins, selectins, and immunoglobulin-like and other molecules, based on their unique properties (Nicolson, 1991). We have recently identified a new class of tumor cell surface components that is associated with divalent cation-dependent adhesion of murine large-cell lymphoma cells to murine microvascular endothelial cells (Tressler et al., 1992). This class was identified by its cation-dependent binding to syngeneic organ-derived microvessel endothelial cells.

For our studies, we used the highly liver-metastatic large-cell lymphoma RAW117 and microvessel endothelial cells isolated from syngeneic lung and liver (Belloni et al., 1992). Variant sublines of this lymphoma are available that are poorly (RAW117-P) or highly metastatic to liver (RAW117-H10) or lung and liver (RAW117-L17) (Nicolson et al., 1989). In this system, the highly liver metastatic H10 subline is significantly more adherent to unfixed or fixed hepatic sinusoidal endothelial cell (HSE) monolayers than the poorly metastatic RAW117-P cell line, whereas the highly lung-metastatic subline L17 is significantly more adherent to lung microvessel endothelial (LE) cell monolayers than to other microvessel endothelial cells (Nicolson et al., 1989). All of the RAW117 cell lines were significantly more adherent to unfixed or fixed HSE and LE monolayers than to bovine aortic endothelial cell monolayers (Nicolson et al., 1989; Tressler and Nicolson, 1992). Although the rates of adhesion were lower with the fixed endothelial cell monolayers, the quantitative differences in tumor cell adhesion of the RAW117 cell sublines paralleled their degree of liver colonization (Nicolson et al., 1989). When Ca^{2+} and Mg^{2+} were depleted from the adhesion assays, RAW117-P and -H10 cell adhesion to endothelial cell monolayers was reduced by ~80% and ~70%, respectively (Tressler and Nicolson, 1992).

After first solubilizing the RAW117 cell membrane components with CHAPS detergent, we found that detergent-solubilized components could significantly inhibit the cation-dependent adhesion of RAW117-P and -H10 cells to fixed endothelial cell monolayers (Tressler and Nicolson, 1992). Fixed HSE cell monolayers were then incubated with ^{125}I-surface-labeled, CHAPS-solubilized RAW117-P or -H10 cell membrane components, the endothelial cell monolayers rinsed with buffer and eluted with detergent, and the eluted components analyzed. Alternatively, the cell monolayers were eluted three times with Ca^{2+}-Mg^{2+}-free buffer or once with Ca^{2+}-Mg^{2+}-free buffer plus 2 mM EDTA prior to detergent elution, and the various eluted components were analyzed by polyacrylamide gel electrophoresis. Divalent cation-dependent tumor cell components of ~32, ~35, and ~70 kDa bound to the HSE cell monolayers and were eluted in the Ca^{2+}-Mg^{2+}-free buffer (Tressler et al., 1992). The eluted components in Ca^{2+}-Mg^{2+}-free buffer containing EDTA were then reconstituted with excess $CaCl_2$ and $MgCl_2$ in buffer and reincubated with HSE cell monolayers, and the endothelial cell-bound components were analyzed by polyacrylamide gel electrophoresis. We found that the isolated ~32 kDa and ~35 kDa components bound again to the endothelial cells, whereas the ~70 kDa component did not (Tressler et al., 1992).

Next, the ~35 kDa molecule was selectively eluted along with a component of ~14 kDa by washing the endothelial monolayers with 0.5 M NaCl in buffer. When the salt-eluted material was desalted, reconstituted, and reincubated on fixed HSE cell monolayers, the ~35 kDa component was able to bind again to the endothelial cell monolayers, whereas the other low molecular weight proteins did not. The endothelial cell-bound, surface-labeled components remaining after 0.5 M salt elution were then released from the endothelial cell monolayers with EDTA, reconstituted with Ca^{2+} and Mg^{2+}, and reincubated with HSE monolayers. The ~35 kDa component eluted from HSE cells was able to bind again to the endothelial cells. The ~35 kDa high-salt-sensitive molecule also could be eluted by EDTA, suggesting that the ~35 kDa and ~32 kDa molecules bind independently to endothelial cells and are not noncovalently associated in a heterodimeric complex (Tressler et al., 1992).

When detergent lysates of ^{125}I-surface-labeled RAW117 cells were preincubated with various types of fixed endothelial monolayers for various times ranging from 30 min to overnight and then eluted from the endothelial cells, analysis indicated that the ~35 kDa component was bound by several endothelial cell types, such as lung microvessel, brain microvessel, and bovine aortic endothelial cells. We found that the ~35 kDa component bound to murine hepatocytes but not to murine or human erythrocytes, indicating that the ~35 kDa component is probably a common endothelial cell-binding protein. Preincubating the RAW117 detergent cell lysates on albumin-coated Petri dishes did not result in removal of the ~35 kDa molecule, indicating that binding of the ~35 kDa component was not due to nonspecific protein interactions. When crude membrane fractions of RAW117-H10 cells were extracted with 3 M KCl and 8 mM EDTA, the ~35 kDa component remained with the membrane fraction, whereas the ~32 kDa molecule was released from the membranes, suggesting either that the ~35 kDa but not the ~32 kDa molecule is an integral membrane component or that its membrane association is divalent cation-independent (Tressler et al., 1992).

IDENTIFICATION OF THE ~35 kDa TUMOR CELL SURFACE COMPONENT AS AN ANNEXIN INVOLVED IN TUMOR CELL-ENDOTHELIAL CELL ADHESION

The ~35 kDa component was partially purified by cell-affinity binding and EDTA elution, and then assessed for its ability to inhibit the adhesion of viable RAW117 cells to unfixed or fixed endothelial cell monolayers. The purified ~35 kDa component inhibited RAW117-H10 adhesion to HSE cells by ~34% at 45 min, reducing the percent cell adhesion of RAW117-H10 to HSE cell monolayers from $24.1 \pm 1.0\%$ to $16.0 \pm 1.7\%$ (Tressler et al., 1992).

The endothelial cell-adherent ~35 kDa component from RAW117-H10 cells was purified to homogeneity by elution from the fixed HSE cell monolayers with NaCl, separation by electrophoresis, and transblotting onto nitrocellulose paper. Then the ~35 kDa band was removed from the blot, and the amino acid sequence of several tryptic peptides determined. Sequence analysis of peptide fragments constituting ~40% of the protein indicated identity with murine but not bovine annexin II (Figure 1). This indicated that the cell-surface-associated annexin II was not derived from the fetal bovine serum present in the growth medium. When RAW117 cells were incubated with antibodies specific for annexin II prior to adhesion assays with fixed or viable endothelial cell monolayers, adhesion was inhibited by 40-50% (Figure 2). The inhibition was specific for anti-annexin II. Excess EDTA did not remove annexin II molecules from the surfaces of RAW117 cells, and killing RAW117 cells by heat shock did not increase the amounts of cell-surface-associated annexin II, indicating that the cell surface-bound annexin II was not released from dead cells and bound non-specifically to adjacent cells (Tressler et al., 1992).

Annexin II is important in a variety of intracellular events (Burgoyne and Geisow, 1989) and probably functions in the divalent cation-dependent adhesion of RAW117 cells to microvessel endothelial cells. In addition, we have found that it is probably one of the common RAW117-endothelial cell adhesion molecules that allows adhesion to different endothelial cells. Lim et al. (1990) reported that a high molecular weight (~67 kDa) annexin-related molecule associated with CAM105 is expressed at the cell surface of rat hepatocytes, and this molecule may function in cell-adhesive interactions. Another annexin-related molecule, anchorin II, is present on the surface of chondrocytes and has been shown to bind to collagen (Pfuttle et al., 1988). We have identified an annexin associated with the adhesion of RAW17 tumor cells to endothelial cells, and this and related molecules may play a role in the cell-cell adhesion of various cell types.

The mechanism of outer plasma membrane association of annexin II is presently not known. Annexin II does not contain a hydrophobic sequence that could serve as a transmembrane domain, nor does it have an N-terminal hydrophobic sequence for secretion or endoplasmic reticulum translocation. Recently, however, Christmas et al. (1991) reported that annexin I was secreted by human prostate gland epithelial cells using a novel pathway that did not involve targeting to the endoplasmic reticulum. The mechanism of annexin membrane attachment to RAW117 cell membranes is also not known. A calcium-independent, membrane-bound form of annexin I that behaves as an integral protein has been described by Sheets et al. (1987), who found that sonicating high-salt- and EDTA-treated membranes did not release annexin in a soluble form. They concluded that annexin I must behave as an integral membrane component. There is a cysteine residue in the carboxy termini of murine annexin II that could serve as a site for myristoylation. Support for this comes from a study by Wice and Gordon (1992), who described a myristoylated annexin of ~35 kDa that is expressed in a membrane-associated form in the gut epithelium.

We found that RAW117 cells express several cation-dependent endothelial cell-binding components. It is likely that some of these are also annexins, such as the ~32 kDa and ~70 kDa components. We do not at this time, however, have evidence as to whether

```
murine-annexin II  M S T V H E I L C K L S L E G D H S T P P P S A Y G S V K P Y T N F D A E R D A L N I E T A I K T K G V D E V T I V N I L T N R S N V Q
human-annexin II   M S T V H E I L C K L S L E G D H S T P P P S A Y G S V K A Y T N F D A E R D A L N I E T A I K T K G V D E V T I V N I L T N R S N A Q
bovine-annexin II  - S T V H E I L C K L S L E G D H S T P P P S A Y G S V K A Y T N F D A E R D A L N I E T A I K T K G V D E V T I V N I L T N R S N E Q
p35 peptides                   L S L E G D H S T P P P S A Y G S V K P Y                       K T K G V D E V T I V N I L T N R S N V

murine-annexin II  R Q D I A F A Y Q R R T K K E L P S A L K S A L S G H L E T V I L G L L K T P A Q Y D A S E L K A S M K G L G T D E D S L I E I I C S R T N
human-annexin II   R Q D I A F A Y Q R R T K K E L A S A L K S A L S G H L E T V I L G L L K T P A Q Y D A S E L K A S M K G L G T D E D S L I E I I C S R T N
bovine-annexin II  R Q D I A F A Y Q R R T K K E L A S A L K S A L S G H L E T V I L G L L K T P A Q Y D A S E L K A S M K G L G T D E D S L I E I I C S R T N
p35 peptides                                           T P A Q Y D A S E L         G L G T D E D S L I E I I C S R

murine-annexin II  Q E L Q E I N R V Y K E M Y K T D L E K D I I S D T S G D F R K L M V A L A K G R R A E D G S V I D Y E L I D Q D A R E L Y D A G V
human-annexin II   Q E L Q E I N R V Y K E M Y K T D L E K D I I S D T S G D F R K L M V A L A K G R R A E D G S V I D Y E L I D Q D A R D L Y D A G V
bovine-annexin II  Q E L Q E I N R V Y K E M Y K T D L E K D I V S D T S G D F R K L M V A L A K G R R A E D G S V I D Y E L I D Q D A R D L Y D A G V
p35 peptides                     T D L E K D I I S D T S G       L M V A L A

murine-annexin II  K R K G T D V P K W I S I M T E R S V C H L Q K V F E R Y K S Y S P Y D M L E S I K K E V K G D L E N A F L N L V Q C I Q N K P L Y F A D R
human-annexin II   K R K G T D V P K W I S I M T E R S V P H L Q K V F D R Y K S Y S P Y D M L E S I R K E V K G D L E N A F L N L V Q C I Q N K P L Y F A D R
bovine-annexin II  K R K G T D V P K W I S I M T E R S V C H L Q K V F E R Y K S Y S P Y D M L E S I K K E V K G D L E N A F L N L V Q C I Q N K P L Y F A D R
p35 peptides       K R K G T D V P K W I S I M T           S Y S P Y D M L E S I K

murine-annexin II  L Y D S M K G K G T R D K V L I R I M V S R S E V D M L K I R S E F K R K Y G K S L Y Y Y I Q Q D T K G D Y Q K A L L Y L C G G D D
human-annexin II   L Y D S M K G K G T R D K V L I R I M V S R S E V D M L K I R S E F K R K Y G K S L Y Y Y I Q Q D T K G D Y Q K A L L Y L C G G D D
bovine-annexin II  L Y D S M K G K G T R D K V L I R I M V S R S E V D M L K I R S E F K K K Y G K S L Y Y Y I Q Q D T K G D Y Q K A L L Y L C G G D D
p35 peptides       K G K G T R D K V L I R I M V             L Y Y Y I Q Q D T K G
```

Figure 1. Comparison of p35 Peptides with Human, Murine, and Bovine Annexin II. Amino Acid Sequence of Murine (Mu), Human (Hu), and Bovine (Bov) Annexin II. The ~35 kDa cell surface-labeled RAW117 component isolated by adherence to HSE cell monolayers was first digested with trypsin and separated by HPLC, then selected peptides were sequenced. The sequence data of peptides obtained from the ~35 kDa component are shown below the annexin sequences (from Tressler et al., 1992).

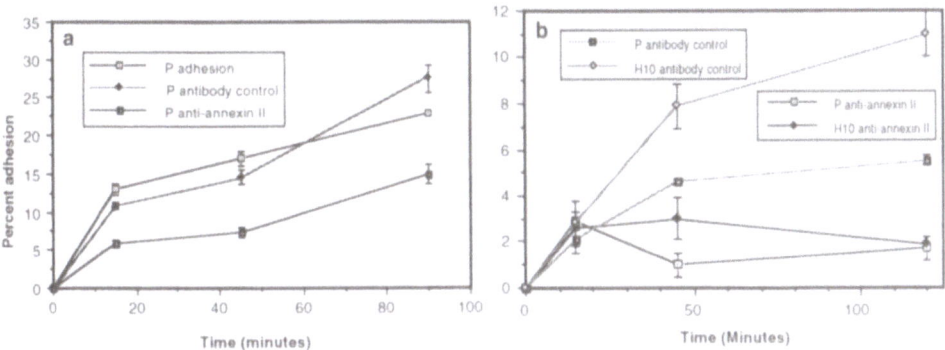

Figure 2. Inhibition of RAW117 Cell Adhesion to Unfixed or Fixed HSE Cell Monolayers by Anti-Annexin II.

a. Adhesion of RAW117-P (**P**) cells to unfixed HSE cell monolayers and its inhibition by anti-annexin II (P-anti-Annexin II); NS, normal serum control.

b. Adhesion of RAW117-P (**P**) or RAW117-H10 (**H10**) cells to fixed HSE cell monolayers and inhibition by anti-annexin II; NS, normal serum control. The data indicate quadruplicate samples ± S.D (from Tressler et al., 1992).

these components are actually expressed at the cell surface of RAW117 cells. There are at least two cell surface molecules of ~70 kDa on RAW117 cells. One is the RNA tumor virus envelope glycoprotein gp70 (Reading et al., 1980), and the other is an antiproliferative molecule that inhibits the mitogen-induced proliferation of normal spleen cells (Joshi et al., 1991). Previously we found that gp70 was expressed in much lower amounts on the highly metastatic H10 subline. Since the ~70 kDa component that is adherent to microvessel endothelial cells appears to be equally expressed in poorly and highly metastatic RAW117 cells and the highly metastatic H10 subline adheres at higher rates to HSE cells, gp70 is probably not involved in tumor cell-endothelial adhesion. Öbrink and collaborators (Obrink et al., 1976; Ocklund et al., 1984) have described a cation-dependent ~70 kDa molecule that is involved in rat hepatocyte cell-cell adhesion which could be similar to the ~70 kDa molecule on RAW117 cells. Recently we have found that antibodies against annexin VI (~70 kDa) partially block RAW117 cell adhesion to HSE cell monolayers (R. Tressler and G.L. Nicolson, unpublished data). Therefore, in light of such data, it is likely that at least some of the major cation-dependent HSE cell-binding components are annexins.

CELL SURFACE ANNEXIN EXPRESSION ON VARIOUS RODENT AND HUMAN TUMOR CELLS

Upon examining several rodent and human tumor cell lines for the presence of various annexins, we found that some but not all tumor cells express annexins on their cell surfaces (Yeatman et al., 1992). Cytofluorographic analysis using monoclonal antibodies directed against annexins I, II, III, IV, V, and VI demonstrated that rat, murine, and human neoplastic cells express variable amounts of annexins on their cell surfaces (Figure 3 and Table 1) (Yeatman et al., 1992). We also examined the expression of cell surface annexin on tumor cells from four histological types, including murine large-cell lymphoma (RAW117) lines (see section 3, above), murine (B16) and human (A375 and MeWo) melanoma lines, rat (MTLn3, see example in Figure 3) and human (MDA-231) mammary carcinoma lines, and human colon carcinoma (KM12, HT29) cell lines. Annexins I to VI were differentially expressed on the various cell lines, whereas the melanoma cell lines expressed little, if any, cell surface annexins. Similarly, cell surface annexin expression was not demonstrable with the MDA-MB-231 breast line (Table 1).

Some differences in cell surface annexin expression amongst cell lines and clones categorized as poorly or highly metastatic in the syngeneic host or nude mouse was seen with RAW117 and the KM12 cell lines but not with the HT29 cell lines. In both the RAW117 and the KM12 cell lines, there was strong cell-surface annexin II expression on cells from the highly metastatic lines (RAW117-H10 and KM12-SM), whereas expression tended to be less intense for the poorly metastatic clones (RAW117-P and KM12-C) (Table 1). This observation was noteworthy for the expression of the annexin II p11 subunit on the KM12 cell line. Cell-surface expression was high for the KM12-SM line, whereas it was low for the KM12-C line (Table 1). The highly metastatic RAW117-H10 line expressed high amounts of cell-surface annexin I, whereas its low metastatic counterpart (RAW117-P)

Table 1. Expression of Various Annexins on Human and Rodent Tumor Cell Lines [a]

Cell Lines	Degree of Cell Surface Expression of Indicated Annexin[b]						
	I	II p11	II	III	IV	V	VI
Mouse Lines							
RAW117-H10	3+	3+	3+	-			
RAW117-P	2+	3+	3+	-			
B16-F1	1+			-			
B16-F1Agg10	1+			-			
Rat Lines							
MTLn3	3+	-	-	2+	3+	2+	3+
Human Lines							
KM12-C	1+	1+	2+	-	1+	-	-
KM12-SM	1+	3+	3+	-	3+	-	1+
KM12-HX	-	-		-	-	-	-
KM12-LX	-	-		-	-	-	-
HT29-P	1+		3+	2+	3+	-	2+
HT29-LM	1+		3+	2+	3+	-	2+
A375-P	-	-			-	-	
A375-LM	-	-			-	-	
MDA-MB-231	-	-	-	-	-	-	

[a] From Yeatman et al., 1992.
[b] Quantitative cell-surface expression is indicated as the degree of labeling with reference to controls using a cytofluorographic method: (-) = none; (1+) = < 2 times controls; (2+) = 2-10 times controls; (3+) = > 10 times controls.

expressed lower levels (Table 1). By comparison, cell surface annexin expression was not detectable on cells from the KM12 lines KM12-LX and KM12-HX, which were derived from the parental cell line (KM12-C) based on their expression of low (LX) or high (HX) levels of sialyl-dimeric Le[x] antigen (Matsushita et al., 1991). Annexin V was not expressed on the surfaces of the majority of the cells tested, but annexin VI was found on the surfaces of cells from both the colon and breast cancer lines (Table 1).

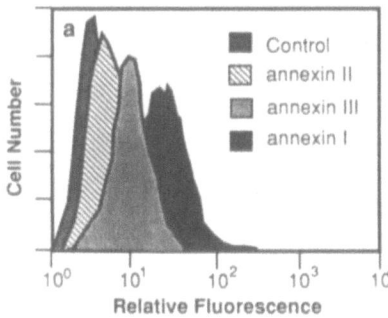 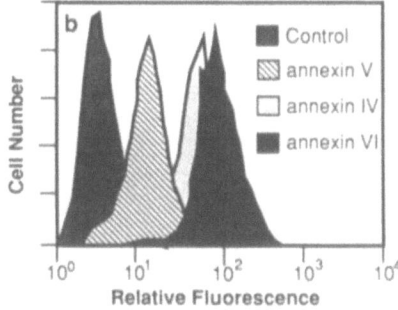

Figure 3. Flow-Cytofluorographic Demonstration of the Expression of Annexins on the Surface of a MTLN3 Rat Mammary Adenocarcinoma Cell Line Selected for Lung Meta-stasizing Ability.

a. Cells alone, as well as an irrelevant antibody, demonstrate low relative fluorescence indicative of a lack of antibody binding. By comparison, annexin I expression is high (10 times greater than controls), whereas annexin II expression is low or undetectable (-). Annexin III expression was intermediate (2+).

b. Annexin V shows moderate (2+) levels of surface binding, whereas annexins IV and VI bind very avidly (3+). Annexin VI expression is even greater than annexin IV and annexin I expression (from Yeatman et al., 1992).

Our data indicate the presence of various annexins on the surfaces of many cultured neoplastic cells. That annexins are present on the cell surface due to cell death and artifactual release of intracellular annexins is refuted by the findings that (a) annexins are differentially expressed on the surfaces of cells from various cell sublines, (b) annexins are expressed on cell surfaces independent of their concentration inside cells, (c) EDTA does not remove or inhibit the detection of cell-surface annexins, (d) our cell preparations are > 95% viable, and (e) the presence of released, soluble annexins could not be detected in the supernatants of our cell cultures. Furthermore, using fluorescent microscopy and fluorescent anti-annexins to examine live cells we found ring fluorescence on the tumor cell surface (Figure 4).

ORGAN-SPECIFIC MOTILITY FACTORS FOR METASTATIC CELLS

Another important property of organ-preferring metastatic cells is their ability to selectively invade secondary-site tissues that are targets for metastatic colonization (Nicolson, 1988a; Nicolson et al., 1985; Nicolson, 1989). Since tumor cell motility is an essential element of metastatic cell invasion (Nicolson, 1988a; Strauli and Haemmerli, 1984; Liotta et al., 1991a), the ability of tumor cells to respond to organ-specific or organ-associated chemoattractants may be an important determinant in the selective migration of tumor cells into specific tissues (Varani, 1982; Nicolson, 1988a; Nicolson et al., 1989).

Organ-derived chemotactic factors for tumor cells have been identified and in some cases partially purified. Orr et al. (1980) partially purified a bone-derived chemotactic factor for cultured Walker carcinosarcoma cells, and Hujanen and Terranova (1985) demonstrated that tumor cells that metastasize to brain or liver migrate toward a concentration gradient of soluble components extracted from brain or liver, respectively. Using a murine colon carcinoma metastatic system, Bresalier et al. (1987) observed that highly liver-metastasizing cells preferentially migrated toward a concentration gradient of liver extract rather than toward gradients made from extracts of lung or brain. In addition to soluble factors, factors present in lung extracellular matrix can also stimulate the migration of highly lung-colonizing melanoma cells at higher rates than poorly lung-colonizing or liver-colonizing cells (Cerra and Nathanson, 1991). Thus, organ-derived chemotactic factors present in secondary sites may play an important role in organ-selective tumor cell invasion.

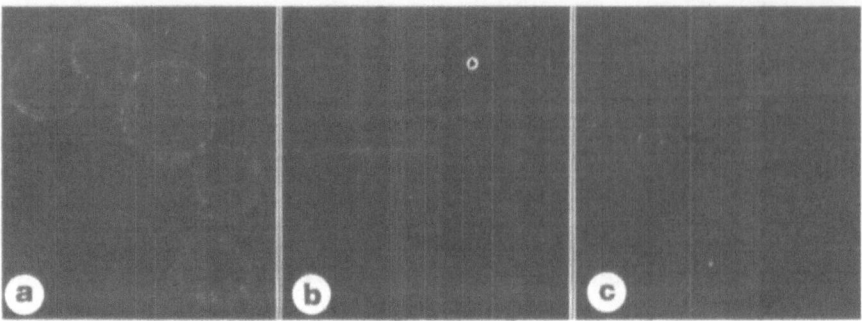

Figure 4. Photomicrograph of RAW117-H10 Cells Treated with Antibody against Annexin II (from Oncogene Science, Inc.).

a. Ring fluorescence visible on RAW117-H10 cells treated with biotinylated anti-annexin II antibody and visualized with streptavidin-R-phycoerythrin.

b. RAW117-H10 cells treated with control antibody and visualized with streptavidin-R-phycoerythrin.

c. RAW117-H10 cells treated with streptavidin-R-phycoerythrin only. Magnification X 600 (from Yeatman et al., 1992).

We have found that conditioned medium from hepatic sinusoidal endothelial cells stimulated the migration of highly liver-metastatic (H10) and highly liver- and lung-metastatic (L17) RAW117 cells at higher rates than the poorly metastatic parental (P) cell line (Figure 5). The results paralleled the liver-metastatic potentials (H10 > L17 > P) of these cell lines (Nicolson et al., 1989). We purified the motility factor present in HSE-conditioned medium by a five-step purification process involving hydroxylapatite affinity chromatography, DEAE Sephacel anion exchange chromatography, Sephacryl S-200 gel filtration, and preparative native gel electrophoresis (Hamada et al., 1992). When the HSE motility factor was analyzed by polyacrylamide electrophoresis under nonreducing conditions, it migrated as a single silver-stained component of >200 kDa. After reduction with 2-mercaptoethanol, the HSE motility factor migrated as two components of ~110 kDa and ~67 kDa, respectively (Hamada et al., 1992). Since efforts at separating these two components were unsuccessful, we concluded that purified HSE motility factor probably contains two subunits of ~110 kDa and ~67 kDa. Subsequent sequence analysis of the N-termini of the two subunits of the purified HSE motility factor indicated that the factor was a fragment of complement component C3 (Figure 6).

ORGAN-DERIVED GROWTH FACTORS FOR METASTATIC CELLS

One of the most important characteristics of metastatic cells that might explain their organ preference of metastasis is their ability to preferentially respond to growth factors produced by or situated in the preferred organ for metastasis (Nicolson, 1988a; Nicolson, 1991; Cavanaugh and Nicolson, 1991a). The ability of organ-preferring tumor cells to respond to growth factors from various organs has been studied by examining the growth of various metastatic cell lines to target and nontarget organ-derived factors prepared from extracts of organs, culture medium conditioned by organs, fragments of organ tissue in culture, or by viable cells cultured from particular organs (Nicolson 1987; Nicolson, 1988b; Nicolson and Dulski, 1986; Horak et al., 1986; Cavanaugh and Nicolson, 1989; Cavanaugh and Nicolson, 1991a; Sargent et al., 1988; Yamori et al., 1988). In our laboratory, we have concentrated on experiments that identify organ (paracrine) growth factors that differentially stimulate the growth of highly metastatic cells, and we have begun to isolate and purify such factors (Cavanaugh and Nicolson, 1989; Cavanaugh and Nicolson, 1990; Cavanaugh and Nicolson, 1991b).

In addition to soluble growth factors released from organ tissues in culture, cell contact-dependent mitogens in organ tissues could be important in stimulating growth of organ-metastasizing tumor cells. These mitogens probably include extracellular matrix- and

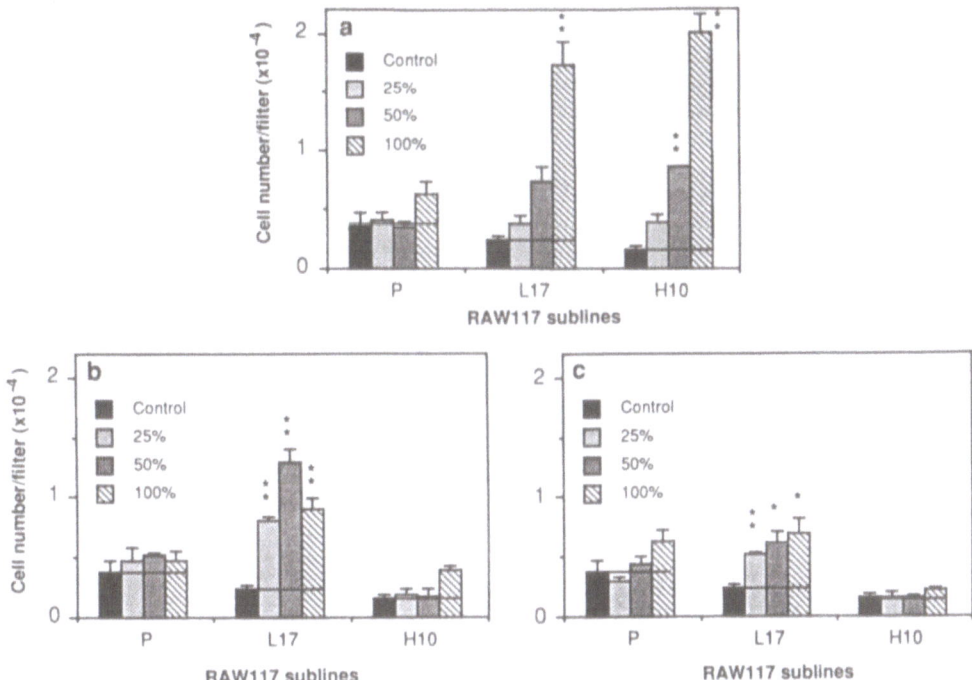

Figure 5. Chemotactic Activity of Endothelial Cell-conditioned Medium for RAW117 Sublines P, L17, and H10.

a. medium conditioned with hepatic sinusoidal endothelial cells.
b. medium conditioned with lung endothelial cells.
c. medium conditioned with brain endothelial cells.

The cell motility assay was performed as described in Hamada et al., (1992a). Columns and error bars show means and standard deviations of triplicate samples, respectively. P-values were calculated according to Student's t-test: *,$P<0.01$ compared with controls; **,$P<0.001$ compared with controls.

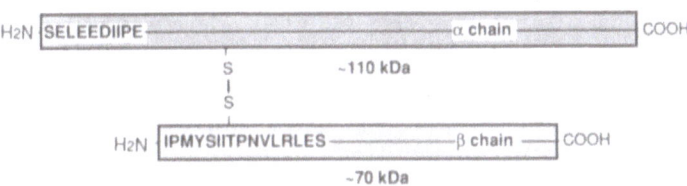

Figure 6. Subunit Structure of C3b-related Motility Factor Released by Hepatic Sinusoidal Endothelial Cells.

plasma membrane-bound growth factors, matrix components themselves, and other relatively insoluble growth factors or integral plasma membrane components (Sargent et al., 1988; Doerr et al., 1989; Mollenhauer et al., 1987). For example, extracellular matrix components can modify the growth properties of some tumor cells (Vlodavsky et al., 1980), and growth factors can be found complexed to integral membrane proteoglycans (Sakela and Rifkin, 1990). Doerr et al. (1989) not only found that extracellular matrix prepared from various organs differentially stimulated metastatic cell growth at low cell densities but also that organ matrix prepared from the target organ of metastatic colonization stimulated growth of highly metastatic rat mammary carcinoma and human hepatoma cells better than poorly metastatic cells. Fractionation of extracellular matrix from the organ matrix preparations revealed that a glycosaminoglycan fraction contained the most active growth promoter. Although highly metastatic cells responded differently to extracellular matrix heparan sulfate, Redini et al. (1990) did not find an obvious relationship between the growth response of various rhabdomyosarcoma tumor and myoblast cell lines to heparan sulfate and their metastatic properties. In general, they found that the most metastatic lines were the cell lines least growth inhibited by heparan sulfate.

The differential stimulation of metastatic cell growth by insoluble growth factors displayed by organ parenchymal cells has been studied by Sargent et al. (1988), who found that a B16 melanoma subline sequentially selected to colonize liver proliferated more rapidly when cocultured with normal murine hepatocytes than did the original parental B16 line. The same parental line was also selected to colonize the lung, but the lung-colonizing variant did not possess the increased growth response when cocultured with hepatocytes. The growth-stimulating activity was not attributed to soluble factors released from the hepatocytes but was thought to be an insoluble or membrane-bound growth factor because cell-cell contact appeared to be required for growth stimulation.

PURIFICATION OF SOLUBLE ORGAN (PARACRINE) GROWTH FACTORS FOR HIGHLY METASTATIC TUMOR CELLS

Some of the mitogens secreted by organ tissues that differentially stimulate the growth of highly metastatic tumor cells have been purified or partly purified. Szanlawska et al. (1983) found that lung tissue-conditioned medium contained a growth stimulating activity for both normal and tumor cells. Separation of the activity by gel filtration demonstrated that it possessed a molecular weight of ~50-70 kDa. Using gel filtration of lung tissue-conditioned medium Yamori et al. (1988) partially purified a lung-derived tumor cell mitogen. The mitogenic activity had a molecular weight of ~90-120 kDa, and the activity was partially destroyed by heat or totally inactivated by trypsin treatment.

Using rat or porcine lung tissue-conditioned medium, we extensively purified a lung mitogen for highly lung-metastatic rat MTLn3 mammary adenocarcinoma cells (Cavanaugh and Nicolson, 1989; Cavanaugh and Nicolson, 1990; Cavanaugh and Nicolson, 1991b). Purification of the activity from the lung tissue-conditioned medium by hydroxylapatite affinity chromatography, anion exchange chromatography, chromatofocusing, gel filtration, and preparative native gel electrophoresis resulted in a pure preparation of a growth-stimulating glycoprotein for lung-metastasizing tumor cells. We named this mitogen lung-derived growth factor-1 (LDGF-1). This protein had a molecular weight of ~66 kDa in unreduced gels and ~72 kDa in reduced gels, and it had a pI of 6.9-7.0. The mitogenic activity of LDGF-1 was abolished by exposure to high temperature (95° C for 1 hour) or treatment with reducing agents.

Using different animal tumor metastatic systems where cell sublines of poorly and highly lung-metastatic potential were available we assayed the mitogenic activity of LDGF-1. This mitogen differentially stimulated the growth of highly lung-metastatic epithelial, lymphoid, and mesenchymal tumor cells. For example, LDGF-1 differently stimulated the highly metastatic sublines of murine B16 melanoma, rat 13762NF mammary adenocarcinoma, and murine RAW117 large-cell lymphoma (Figure 7). We also found that LDGF-1 was active in stimulating the growth of human MCF-7 and other breast cancer cell lines (data not shown). The mitogenic activity of LDGF-1 was not inhibited by antibodies

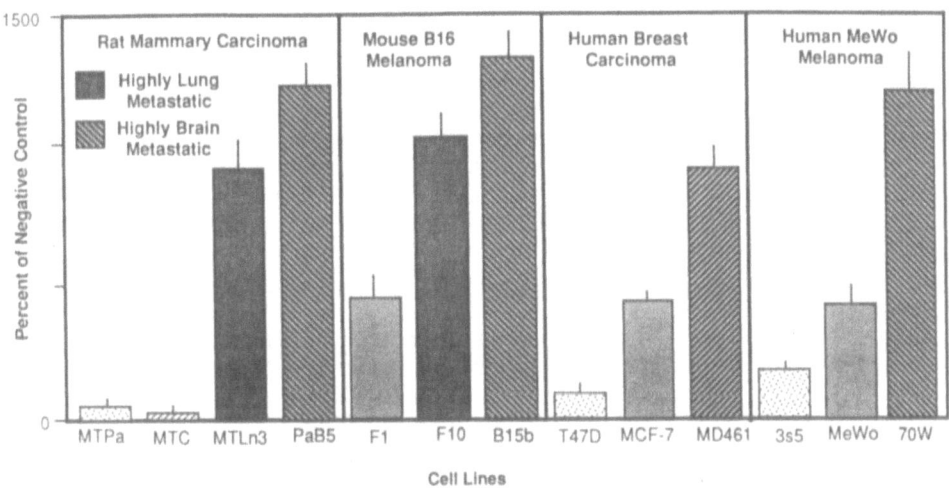

Figure 7. Growth-stimulating Properties of Lung-Derived Growth Factor-1 (LDGF-1) on Various Cell Lines. Cell growth assay was performed as described in Cavanaugh and Nicolson (1990). The values indicate the relative cell numbers at day 4 after cells were plated in 96-well plates and 5 units of growth stimulatory activity were added. Bars indicate the mean ± S.D. of four samples, respectively.

against several known growth factors, such as insulin, granulocyte-macrophage colony stimulating factor, platelet-derived growth factor, or epidermal growth factor, suggesting that LDGF-1 was not antigenically related to these known growth factors.

IDENTIFICATION OF LDGF-1 AS A TRANSFERRIN-LIKE PROTEIN

Although our initial studies did not reveal any similarity of LDGF-1 to known growth factors, further characterization of LDGF-1 indicated that it was either a member of the transferrin (Tf) family of growth factors or a Tf-like molecule (Cavanaugh and Nicolson, 1991b). The similarities of LDGF-1 to Tf were demonstrated by the nearly identical migration of rat and porcine LDGF-1 compared with human and rat Tf in SDS-polyacrylamide gels, cross-reactivity of anti-porcine LDGF-1 with human and rat Tf, and cross-reactivity of anti-human Tf with rat LDGF-1. In addition, the molecular weight of polypeptides obtained after trypsin cleavage were identical for porcine LDGF-1 and porcine Tf.

Since we had identified LDGF-1 as a Tf-like molecule, we examined the ^{125}I-Tf binding properties and growth response to Tf of tumor cell sublines of different metastatic properties. In the murine B16 melanoma system, the brain-colonizing B16-B15b subline exhibited the greatest growth response to Tf, followed in order by the ovary-colonizing B16-O13, the highly lung-colonizing B16-F10 and finally the poorly lung-colonizing B16-F1 subline (Nicolson et al., 1990). Since the B15b and O13 lines colonize the lung as well as brain or ovary, we expected these B16 sublines to respond to Tf. Cell binding of ^{125}I-Tf to the B16 cell sublines paralleled their growth responses to Tf. Although the differences in quantitative ^{125}I-Tf binding to the different B16 sublines were not as great as the differences in growth responses among the B16 sublines, the rank order of ^{125}I-Tf binding and growth responses were similar (B15b > O13 > F10 > F1) (Nicolson et al., 1990).

Using the rat mammary adenocarcinoma metastatic system we also found a relationship between the binding of ^{125}I-Tf and growth responses to Tf and spontaneous metastatic potential. The high brain- and lung-metastasizing sublines (PaB10 and PaB5) showed the best ability to bind ^{125}I-Tf; cell binding was 5 to 6 times higher than in any of the other lines. The highly lung-metastasizing MTLn3 line bound less ^{125}I-Tf than the brain-metastasizing lines, followed by the intermediate lung-metastasizing MTF7 and the poorly lung-metastasizing MTPa line (Figure 8). We calculated the numbers of Tf receptors on these cell lines by analyzing Scatchard plots obtained when increasing concentrations of ^{125}I-Tf were allowed to bind to cell monolayers. The results indicated that receptor numbers increased as spontaneous metastatic capability increased in the following order: high brain-metastasizing ability > high lung-metastasizing ability > intermediate lung-metastasizing ability > poor metastatic capability (Cavanaugh and Nicolson, 1991a).

We have examined several human breast cancer cell lines established from brain or lung metastases for the presence of Tf receptors and response to Tf. The results indicate that human tumor cells may also respond to Tf similar to animal metastatic tumor systems. Although the true biological potential of the human breast cancer cell lines in the syngeneic host is not known, the human breast cancer cell lines with the potential to metastasize in nude mice were the ones that expressed high numbers of Tf receptors. Thus, our data using animal metastatic tumor systems may be highly relevant to human cancers.

To demonstrate that a Tf-like activity in lung tissue-conditioned medium was responsible, in part, for the stimulation of mammary adenocarcinoma cell growth, we used a specific anti-Tf biotinylated antibody to remove Tf and Tf-like molecules from lung tissue-conditioned medium. When lung tissue-conditioned medium was treated with control biotinylated goat antibodies and streptavidin-agarose, growth activity was not removed; however, when the lung tissue-conditioned medium was treated with biotinylated goat anti-Tf and strepavidin-agarose, there was a significant reduction ($P < 0.001$) in the the growth potential of the lung tissue-conditioned medium.

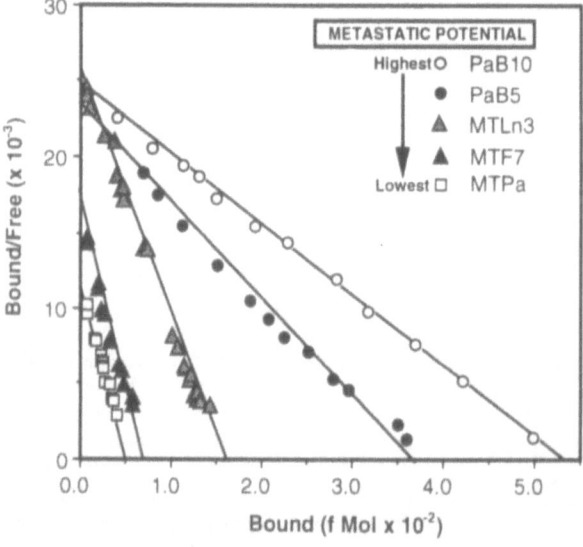

Figure 8. **Scatchard Analysis of** ^{125}**I-transferrin Binding to Various Rat 13762NF Mammary Adenocarcinoma Sublines.** Cells were grown to confluency in 12-well plates and placed in serum-free conditions 24 hours prior to assay. Increasing amounts of ^{125}I-transferrin were added to duplicate wells and incubated at 4° C for 2 hours. Monolayers were washed, solubilized and total bound counts were determined as described in Cavanaugh and Nicolson (1991). Points are mean values of two wells. In all cases, bound radioactivity could be reduced to background by the addition of excess cold transferrin.

ROLE OF ORGAN-ASSOCIATED GROWTH FACTORS AND OTHER PROPERTIES IN ORGAN PREFERENCE OF METASTASIS

The ability of malignant cells to proliferate in certain tissue compartments may be facilitated by their expression of certain receptors and response to paracrine secretion of particular growth factors. We found that one of the most potent growth factors obtained from organ-conditioned medium is a Tf or Tf-like molecule and that the most metastatic cell sublines of murine melanoma and rat mammary adenocarcinoma were the most responsive to Tf in growth factor-limiting medium. In addition, the most metastatic cells expressed the highest numbers of Tf receptors (Cavanaugh and Nicolson, 1991a; Cavanaugh and Nicolson, 1991b). Tumor cells that express high numbers of Tf receptors should be able to respond to low, limiting concentrations of Tf that exist in some tissue compartments, such as the brain where Tf is probably used as a paracrine growth factor during development (Mescher and Muniam, 1988). When malignant cells metastasize to sequestered compartments, such as the brain, it may be advantageous for them to express high numbers of Tf receptors and respond to low concentrations concentrations of Tf.

Metastatic cell response to organ-associated growth factors (and inhibitors) is probably only one of a number of tumor cell properties required for the successful formation of metastases. For instance, although highly liver-metastatic murine RAW117-H10 cells respond well to lung tissue-conditioned medium, they fail to colonize the lungs of their syngeneic host. In this case, the liver-metastatic RAW117 cells show very poor rates of adhesion to lung-derived microvessel endothelial cell and rapid rates of clearance through the lungs when injected intravenously into mice (Nicolson et al., 1989). Thus, it should be obvious that additional properties, such as adhesion to microvessel endothelial cells in the target organ, secretion of degradative enzymes, response to chemotactic and haptotactic motility factors, and other properties are collectively important in determining whether a malignant cell can successfully colonize a particular site.

REFERENCES

Belloni, P.N., Carney, D.H., and Nicolson, G.L., 1992, Organ-derived endotehlial cells exhibit differential responsiveness to thrombin and other growth factors. *Microvasc. Res.* 43:20-45.

Bresalier, R.S., Hujanen, E.S., Raper, S.E., Roll, F.J., Itzkowitz, S.H., Martin, G.R., and Kim, Y.S., 1987, An animal model for colon cancer metastasis: Establishment and characterization of murine cell lines with enhanced liver-metastasizing ability. *Cancer Res.* 47: 1398-1406.

Burgoyne, R.D. and Geisow, M.J., 1989, The annexins are a family of calcium-binding proteins. *Cell Calcium* 10:1-10.

Cavanaugh, P. G. and Nicolson, G L., 1989, Purification and some properties of a lung-derived growth factor that differentially stimulates the growth of tumor cells metastatic to the lung. *Cancer Res.* 49:3928-3933.

Cavanaugh, P.G. and Nicolson, G.L., 1990a, Purification and characterization of a Mr ~66,000 lung-derived (paracrine) growth factor that preferentially stimulates the *in vitro* proliferation of lung-metastasizing tumor cells. *J. Cell. Biochem.* 43:127-138.

Cavanaugh, P.G.and Nicolson, G.L., 1990b, Partial purification of a membrane-associated growth factor for metastatic tumor cells from lung tissue. *Proc. Am. Assoc. Cancer Res.* 31:54.

Cavanaugh, P.G. and Nicolson, G.L., 1991a, Organ preference of metastasis: Role of organ paracrine growth factors. *Cancer Bull.* 43:9-16.

Cavanaugh, P.G. and Nicolson, G.L., 1991b, Lung-derived growth factor that stimulates the growth of lung-metastasizing tumor cells: Identification as a transferrin. *J. Cell. Biochem.* 47:261-271.

Cerra, R.F., and Nathanson, S.D., 1989, Organ-specific chemotactic factors present in lung extracellular matrix. *J. Surg. Res.* 46:422-426.

Cerra, R.F. and Nathanson, S.D., 1991, Chemotactic activity present in liver extracellular matrix. *Clin. Exp. Metastasis* 9:39-49.

Christmas, P., Callaway, J., Fallon, J., Jones, J., and Haiglier, H.T.,1991, Selective secretion of annexin I, a protein without a signal sequence, by the human prostate gland. *J. Biol. Chem.* 266:2499-2507.

Doerr, R., Zvibel, I., Chiuten, D., Dolimpio, J. and Reid, L.M., 1989, Clonal growth of tumors on tissue-specific biomatrices and correlation with organ site specificity of metastasis. *Cancer Res.* 49:384-392.

Hamada, J.-I., Cavanaugh, P.G., Lotan, O., and Nicolson, G.L., 1992, Separable growth and migration factors for large-cell lymphoma cells secreted by microvascular endothelial cells derived from target organs for metastasis. *Br. J. Cancer* 66:349-354.

Horak, E., Darling, D.L., and Tarin, D., 1986, Analysis of organ specific effects on metastatic tumor formation *in vitro*. *J. Natl. Cancer Inst.* 76:913-922.

Hujanen, E.S. and Terranova, V.P., 1985, Migration of tumor cells to organ-derived chemoattractants. *Cancer Res.* 45:3517-3521.

Joshi, S.S., O'Connor, S.J., Weisenburger, D.D., Sharp, J.G., Gharpure, H.M., and Brunson, K.W. ,1991, Enhanced antiproliferative activity by metastatic RAW117 lymphoma cells. *Clin. Exp. Metastasis* 9:27-37.

Lim, Y. P., Tai, J. H., Josic, D., Callanan, H., Reutter, W. and Hixson, D. C. ,1990, Association between the rat hepatocyte cell adhesion molecule, cell-CAM105 and a 65/67 calcium-binding protein. *J. Cell Biol. 111:355a (abstract)*.

Liotta, L.A., Stracke, M.L., Aznavoorian, S.A., Beckner, M.E., and Schiffmann, E., 1991, Tumor cell motility. *Semin. Cancer Biol.* 2:111-114.

Liotta, L.A., Steeg, P.S., and Stetler-Stevenson, W.G., 1991b, Cancer metastatis and angiogenesis: An imbalance of positive and negative regulation. *Cell* 64:327-336.

Matsushita, Y., Hoff, S.D., Nudelman, E.D., Otaka, M., Hakomori, S., Ota, D.M., Cleary, K.R., and Irimura, T., 1991, Metastatic behavior and cell surface properties of HT-29 human colon carcinoma variant cells selected for their differential expression of sialyl-dimeric LE antigen. *Clin. Exp. Metastasis* 9:283-299.

Menter, D.M., Patton, J.T., Updyke, T.V., McIntire, L.V., and Nicolson, G.L., 1992, Transglutaminase stabilizes melanoma adhesion under laminar flow. *Cell Biophys.* 18:123-143.

Mescher, A.L. and Muniam, S.I., 1988, Transferrin and the growth promoting effect of nerves. *Int. Rev. Cytol.* 110:1-26.

Mollenhauer, J., Roether, I., and Dern, H.F., 1987, Distribution of extracellular matrix proteins in pancreatic ductal adenocarcinoma and its influence on tumor cell proliferation *in vitro*. *Pancreas* 2:14-24.

Nicolson, G.L., 1982, Metastatic tumor cell attachment and invasion assay utilizing vascular endothelial cell monolayers. *J. Histochem. Cytochem.* 30:214-220.

Nicolson, G.L., 1987, Differential growth properties of metastatic large cell lymphoma cells in target organ-conditioned media. *Exp. Cell Res.* 168:572-577.

Nicolson, G.L., 1988a, Cancer metastasis: Tumor and host organ properties important in colonization of specific secondary sites. *Biochim. Biophys. Acta* 948:175-224.

Nicolson, G.L., 1988b, Differential organ tissue adhesion, invasion, and growth properties of metastatic rat mammary adenocarcinoma cells. *Breast Cancer Res. Treat.* 12:167-176.

Nicolson, G.L., 1989, Metastataic cell interactions with endothelium, basement membrane and tissue. *Curr. Opin. Cell Biol.* 1:1009-1019.

Nicolson, G.L., 1991, Molecular mechanisms of cancer metastasis: Tumor and host properties and the role of oncogenes and suppressor genes. *Curr. Opin. Oncol.* 3:75-92.

Nicolson, G.L., 1992, Tumor cell adhesion and growth properties and organ-specific metastasis, in: "Homing Mechanisms and Cellular Targeting," B.R. Zetter, ed., Marcel Dekker Publishers (*in press*).

Nicolson, G.L. and Dulski, K.M., 1986, Organ specificity of metastatic tumor colonization is related to organ-selective growth properties of malignant cells. *Int. J. Cancer* 38:289-294.

Nicolson, G.L., Dulski, K., Basson, C., and Welch, D.R., 1985, Preferential organ attachment and invasion *in vitro* by B16 melanoma cells selected for differing metastatic colonization and invasive properties. *Invasion Metastasis* 5:144-158.

Nicolson, G.L., Inoue, T., Van Pelt, C.S., and Cavanaugh, P.G., 1990, Differential expression of a Mr 90,000 cell surface transferrin receptor-related glycoprotein on murine B16 metastatic melanoma sublines selected for enhanced brain or ovary colonization. *Cancer Res.* 50:515-520.

Nicolson, G.L., Belloni, P.N., Tressler, R.J., Dulski, K., Inoue, T., and Cavanaugh, P.G., 1989, Adhesive, invasive, and growth properties of selected metastatic variants of a murine large-cell lymphoma. *Invasion Metastasis* 9:102-116.

Öbrink, B. Lindsröm, H., and Svennung, N. ,1976, Calcium requirement for a reversible binding of membrane proteins to rat liver plasma membranes. *FEBS Lett.* 70:28-32.

Ocklund, C., Odin, P., and Öbrink, B. ,1984, Two different cell adhesion molecules--cell-CAM105 and a calcium-dependent protein--occur on the surface of rat hepatocytes. *Exp. Cell Res.* 151:29-45.

Orr, F.W., Varani, J., Gondek, M.D., Ward, P.A., and Mundy, G.R., 1980, Partial characterization of a bone-derived chemotactic factors for tumor cells. *Am. J. Pathol.* 99:43-52.

Pauli, B.U., Augustin-Voss, H.G., El-Sabban, M.E., Johnson, R.C., and Hammer, D.A., 1990, Organ-preference of metastasis: The role of endothelial cell adhesion molecules. *Cancer Metastasis Rev.* 9:175-189.

Pfuttle, M., Ruggiero, F., Hofmann, H., Fernandez, M.P., Sclmin, O., Yamada, Y., Garrone, R., and von der Mark, K.,1988, Biosynthesis, secretion and extracellular localization of anchorin C11, a collagen-binding protein of the calpactin family. *EMBO J.* 7:2335-2342.

Reading, C.R., Brunson, K.W., Torianni, M., and Nicolson, G.L.,1980, Malignancies of metastatic murine lymphosarcoma cell lines and clones correlate with decreased cell surface display of RNA tumor virus envelope glycoprotein gp70. *Proc. Natl. Acad. Sci. USA* 77:5943-5947.

Redini, F., Moczar, E., and Poupon, M.-F., 1990, Effects of glycosaminoglycans and extracellular matrix on metastatic rhabdomyosarcoma tumor and myoblast cell proliferation. *Clin. Exp. Metastasis* 8:491-502.

Sakela, O. and Rifkin, D.B., 1990, Release of basic fibroblast growth factor-heparan sulfate complexes from endothelial cells by plasminogen activator-mediated proteolytic activity. *J. Cell Biol.* 110:765-7775.

Sargent, N.S.E., Oestreicher, M., Haidvogl, H., Madnick, H.M., and Burger, M.M., 1988, Growth regulation of cancer metastases by their host organ. *Proc. Natl. Acad. Sci. USA* 85:7251-7255.

Sheets E.E., Giugni, T.D., Coates, G.G., Schaepfer, D.D., and Haigler, H.T.,1987, Epidermal growth factor-dependent phosphorylation of a 35-kilodalton protein in placental membranes. *Biochemistry* 26:1164-1172.

Strauli, P. and Haemmerli, G., 1984, The role of cancer cell motility in invasion. *Cancer Metastasis Rev.* 3:127-141.

Sugerbaker, E.V., 1983, Patterns of metastasis in human malignancies. *Cancer Biol. Rev.* 2:235-278.

Szanlawska, B., Majewski, S., Kaminski, M.J., Noremberg, K., Swierz, M., and Janik, P., 1983, Stimulatory and inhibitory activities of lung conditioned medium on the growth of normal and neoplastic cells *in vitro. J. Natl. Cancer Inst.* 75:303-306.

Tressler, R.J., Updyke, T.V., Yeatman, T.J., and Nicolson, G.L., 1992, Extracellular annexin II is associated with divalent cation-dependent tumor cell-endothelial cell adhesion of liver-metastatic large-cell lymphoma cells. *Proc. Natl. Acad. Sci. USA.*.

Tressler, R.J. and Nicolson, G.L., 1992, Butanol-extractable and detergent-solubilized cell surface components from murine large cell lymphoma cells associated with adhesion to organ microvessel endothelial cells. *J. Cell. Biochem.* 148:162-171.

Varani, J., 1982, Chemotaxis of metastatic tumor cells. *Cancer Metastasis Rev.* 117-28.

Vlodavsky, I., Lui, G.M., and Gospadarowicz, D., 1980, Morphological appearance, growth behavior and migratory activity of human tumor cells maintained on extracellular matrix versus plastic. *Cell* 19:607-616.

Wice, B.M. and Gordon, J.I.,1992, A strategy for isolation of cDNAs encoding proteins affecting uman intestinal epithelial cell growth and differentiation: characterization of a novel gut-specific N-myristoylated annexin. *J. Cell Biol.* 116:405-422.

Yamori, T., Iida, H., Tsukagoshi, S., and Tsuruo, T., 1988, Growth stimulating activity of lung extract on lung-colonizing colon 26 clones and its partial characterization. *Clin. Exp. Metastasis* 6:131-139.

Yamori, T., Iida, H., Tsukagoshi, S., and Tsuruo, T., 1988, Growth stimulating activity of lung extract on lung-colonizing colon 26 clones and its partial characterization. *Clin Exp. Metastasis* 6:131-139.

Yeatman, T.J., Updyke, T.V., Kaetzel, M.A., Dedman, J.R., and Nicolson, G.L., 1992, Expression of annexins on the the surfaces of nonmetastatic and metastatic human and rodent tumor cells. *Clin. Exp. Metastasis* (*in press*).

Zetter, B.R., 1990, The cellular basis of site-specific tumor metastasis. *New Engl. J. Med.* 322:605-612.

DISCUSSION

R. JULIANO

I was very interested in your observation on transglutaminases. As I am sure you know, Pete Davies transfected transglutaminase into fibroblastic cells and saw some very interesting changes in terms of increased cell spreading. Another observation along those lines is that transglutaminase is a calcium-regulated enzyme, and in this may be of interest in view of some of the stories that have been emerging about calcium transients that are associated with cell adhesion, I think that is an interesting set of observations.

G. NICOLSON

Yes, I did notice that.

N. HOGG

I was very intrigued by the results you had with the annexins that apparently were behaving as adhesion molecules under sheer force conditions.

G. NICOLSON

These experiments were actually performed under static conditions without shear. We have not examined them under shear force conditions.

N. HOGG

But the annexins are usually found inside the cell. They are bound to the plasma membrane, but to the lipid in the sub-membranous layer and then the additional factor is that they do not have transmembrane sequences.

G. NICOLSON

That is correct.

N. HOGG

So how do you reconcile all of that with a role in adhesion?

G. NICOLSON

Well, we do not know how they get outside cells, if that is what you are alluding to. But there are a number of molecules that get outside cells that do not have signal sequences and we have no idea of how they do it.

N. HOGG

Like IL1 for example.

G. NICOLSON

There are a number of secretory molecules, for example that do not have signal sequences and we do not know how they are secreted. It is a mystery because not all tumor cells express annexins on their extracellular surfaces. Actually, we have just begun to look at clinical samples, and we find, for example, in colorectal cancer tissues and on both human and murine colorectal carcinoma cells expression of extracellular annexins, but not of every annexin type. This is interesting, in that it is not a non-specific sort of process, because there are certain annexins which apparently have a capacity to be expressed on the exterior of the cell, and others that do not, and we do not have any idea of what signals control this process. There is a lot of unknown territory here. We approached this very carefully and very conservatively because we realized that some people would just conclude "Well, everyone knows that annexins are not expressed on the cell surface", and write it off. But if you surface label certain cells, you will find it. You can functionally show their involvement in adhesion using antibodies or excess added annexin molecules. You can do a number of experiments that indicate that, in fact, these are cell surface annexins. We also think they have a lipid tail, and that is how they are cell surface-adherent. They are not adherent through calcium/magnesium-dependent mechanisms to the cell surface, because we can put in EDTA with cells and that does not knock off the cell surface annexins, although that prevents the annexins from interacting in a calcium/magnesium-dependent way with other cells or other components.

M. HEMLER

I just want to comment on your statements regarding shear forces. I do not know if you are aware that at least one or two integrins have been shown, by Tim Springer and others, to withstand normal shear stresses. So perhaps upon triggering, the strength achieved through integrins might be sufficient. But more interesting to me, I did not catch the details about how this novel integrin was isolated. Did that come from the melanoma or from the endothelial cell?

G. NICOLSON

No, that came from the RAW117 large-cell lymphoma, a very metastatic line. We harvested cell membranes, isolated it from membranes in detergent, ran it through the RGD-polymer-affinity column. But first we pre-cleared with $\alpha_5\beta_1$ antibodies, because there is a fibronectin receptor on these cells that is present in fairly high amounts. It has to be pre-cleared, otherwise you do not have clean gels. And then we just simply ran it on SDS-PAGE gels. That is not good, definitive evidence for its role in cell adhesion.

M. HEMLER

So this does not necessarily have any involvement in the cell-cell interaction.

G. NICOLSON

We have made an antibody against it, and that antibody blocks cell adhesion, and so in that respect, like other people have shown throughout this conference, it is probably involved in adhesion.

M. HEMLER

Was the antibody against both the α and the β subunit?

G. NICOLSON

I can only tell you that it immunoprecipitates both of the subunits together. That does not tell you what subunits it is against, but the antibody will not Western blot, or at least it does not blot well. It does immunoprecipitate, so we feel it is against the molecule we isolated but we do not know which subunit, or both, or the complex, contains the antigenic site.

M. HEMLER

Have you tested to see if it would have α_V in it?

G. NICOLSON

We are leaning in that direction. But again we have not done the definitive experiments. This is all fairly new and unpublished data, so we just have not had time to work it out.

E. BROWN

I now want to ask you about another part, which is the C3 part of the story. There actually are a couple of situations in which C3 fragments have been alleged to cause cell motility. One of them was a mistake, namely, C5 was isolated as a minor contaminant of C3 preparation, so when people made C3 what they thought was C3-A actually contained C5-A. Have you checked the effect of C5 on your cells? Second, there is a peptide isolated by Toni Hugli, some years ago. He showed that injection of that peptide caused leukocyte migration from the bone marrow to the blood. It is a fragment of the c3 β_i molecule, and your gels look like a c3 β_i.

G. NICOLSON

Yes, that is exactly the direction we are leaning towards that it is a c3β fragment. We have done the sequencing, so we know that it is not C5, and we have very clean sequencing data on it, and we have no secondary signal.

E. BROWN

But C5-A is a thousand times more potent that C3-A in some chemotactic assays so that will not be evidence.

G. NICOLSON

We purified the C3β using a method that separates the C5. The final and other separations would not lead to isolation of C5.

J. CALVETE

If I have understood correctly, the only data you have to implicate the D4 molecule in the integrin superfamily is that it is an heterodimer which binds to RGD column in a divalent-dependent manner and is eluted by RGD polymer.

G. NICOLSON

Correct. It is preliminary data.

J. CALVETE

But you do not have any sequence data?

G. NICOLSON

No, we do not have any sequence data.

J. CALVETE

Then it could be from another family, in principle.

G. NICOLSON

No, as I kind of mentioned along the way, I really do not like to make a story unless we have sequence data. I think you can make a mistake much too easily in this business without the actual sequence. I agree with you completely. It needs to be sequenced.

T. CAREY

Dick Bankert has interesting results that I think are most applicable to this talk. He has been looking at the homing behavior of human tumors and by isolating clones from these found that they differ in their integrin expression.

R. BANKERT

We reported that human non-small cell lung cancer is associated with an increased expression of VLA-2 (*J. Exp. Med. 173:1111-1119, 1991*). This increased expression of the VLA-2 molecule was detected on the cell surface and there was at least a 20-fold increase in the α_2-chain message in the lung tumor compared to the normal adult human lung tissue. The increased expression of the VLA-2 was coincident with the magnesium-dependent binding of collagen by the tumor cells. These data raised the possibility that the over-production of VLA-2 may be involved in the pathogenesis of human lung tumors by modulating the invasive and/or potential of the tumor. To address this possibility we established primary human lung tumor cell lines from non-small cell lung tumor biopsies. Clones were established from the primary cultures and these clones were characterized with respect to the expression of six different β_1 integrins and for their magnesium-dependent binding to collagen, laminin, and fibronectin. Each clone exhibited a distinct and characteristic phenotype with respect to integrin expression and adhesive properties. These results demonstrate that primary human lung tumors are heterogeneous with respect to integrin expression and their adhesive potential. One clone (2E9) exhibited a marked increase in adhesion to collagen and laminin and a coincidental increase in the expression of the α_1 and $\alpha_2\beta_1$ integrins. This clone exhibited a marked increase in the experimental metastasis in a severe combined immunodeficient (SCID) mouse xenograft model. Following the intravenous inoculation of these cells into SCID mice we observed disseminated tumor growth in multiple organ site including the lung, ovary, adrenal glands and bone marrow. In contrast another clone derived from the same primary tumor expressed very low levels of α_1 and α_2 and bound poorly or not at all to collagen and laminin. This tumor clone failed to produce tumor engraftment in the majority of SCID mice inoculated with tumor cells intravenously. These data establish that specific integrins (VLA-1 and VLA-2), and the associated increase in adhesion to collagen and laminin correlate with a substantial increase in the experimental metastasis. The results are consistent with the notion that the up-regulation of specific integrins modulate tumor behavior and may be involved in tumor progression.

G. NICOLSON

I think it is a very interesting observation, and I think you are going to have to follow it through by examining a number of different clones, and then try and go back to the original material *in situ*. This question came up this morning in comparing tumor cells *in situ* versus tumor cells cultivated *in vitro* and cloned *in vitro*, and I think it might be an important point that we need to go back whenever we make these observations and examine directly the clinical material *in situ*. Thus, take some of the original frozen material and do the same type of study. Have you done that?

R. BANKERT

Well, we have the frozen material and we are just getting started on it. I was going to wait and comment on your talk that I think you ruled out many of the integrins but have you ruled out VLA-1 and -2? I wonder if you specifically looked at this. I mean VLA-1 and -2 binding is not blocked by RDG peptide. Have you looked at VLA-1 and -2 on your RAW117 cell lines?

G. NICOLSON

We have not been able to detect it.

S. SHAW

I have a question on this, actually for Martin. What is the evidence that α_2 functions in any physiologic way in homing via endothelial cells. For example, on resting T cells it is lacking. Activated T cells would not normally be in circulation, probably. Or at least they are very, very low frequency.

R. BANKERT

Well, the activated cells are low frequency but they would be in the circulation.

S. SHAW

Have you looked at α_2-dependent binding to endothelium?

R. BANKERT

Monocytes also express VLA-2.

S. SHAW

What is the endothelial ligand for monocyte VLA-2?

M. HEMLER

On endothelium, collagen probably.

G. NICOLSON

What is the possibility that collagen-like sequences, or some other sequences are being recognized. One of the things that bothers me about attributing matrix molecules to the apical surface of an endothelial cell, is that when you examine endothelial cells properly cultured or you look *in vivo*, you do not find matrix components on the apical surface. The subendothelium seems in a normal state, is intact. In a pathologic state it can be exposed. That is well known, and there may be a situation under pathologic conditions, where recruitment of matrix components, their arrangement, or translocation, occurs to the apical cell surface. In the normal situation is there any evidence that matrix components are expressed on the apical surface of the endothelium?

R. BANKERT

Martin, I pass to you on this.

M. HEMLER

I have seen some very surprising published results about that which suggest that there has been inhibition of cell adhesion by antibodies to a variety of things you would not expect on the apical surface. It appears as if fibronectin may be on top of the endothelium because some people are able to block cell attachment to endothelium using antibodies to 2A5 or RGD, or CS-1 peptide, in various ways, which I do not fully understand. So I think it is a tricky system.

G. NICOLSON

Well, let me respond to that because I do not fully understand that either. Because one of the situations that we found is that when endothelial cells are put into culture properly, they do not express fibronectin on their apical surface. Under conditions where you damage the monolayer, or you perturb the endothelium, let us say with proteolytic enzymes, you do get some expression on the apical surface. The cells eventually go back and form nice confluent cell monolayers and the fibronectin is then swept under the cell and is present in the subendothelium matrix, not on the apical surface. Now, this situation changes if the endothelial cells are carried for a prolonged period of time in culture. It is only the short-term passage endothelial cells that have these characteristics which is a more normal characteristic in terms of their physiology. It is true that investigators have carried endothelial cells essentially as cell lines. We never do that; we always use them as cell strains. If you try and immortalize them or passage them too long in culture, you ruin their polarity and other characteristics. If you mistreat them, you also ruin their polarity. So I do not know what those experiments mean in terms of the normal physiology.

M. HEMLER

Well, let me just comment on the experiments that I trust. They are fairly reputable and reproducible. It seems that especially for lymphomas, and also to some extent for melanomas, you can block all of the adhesion, either with antibody to VCAM or to α_4 or ELAM or some combination thereof. And there does not seem to be a lot of room left over for major other adhesion mechanisms. An enormous number of groups has been searching extremely hard, using all possible combinations of those antibodies to see if any additional adhesion still remains. This is a million dollar industry these days, and I think that every large company in the world would like to find another one of these structures that they can patent. And at this point they have not. So I am a little bit reluctant to think about lymphoma adhesion cells having another pathway that has a major involvement of these other structures that you describe.

G. NICOLSON

The types of experiment that you have both described are essentially the same types of experiments that we have performed in terms of making specific antibodies examining the antibody blocking, and in our case we used excess amount of the molecule itself to see if they would block adhesion and they do. I do not know of a comparable experiment with integrins. Whenever you use antibodies, and whenever you have a high density of an antibody-binding site on the surface of the cell, you coat the cell essentially with high molecular weight molecules. And that in itself could block any other adhesive interactions from occurring. So, I think it will come out in the long run that there are multiple components involved in adhesion selectins, annexins, integrins and other molecules. Where would you put the selections in your model? I might add that we use unstimulated endothelial cells, whereas most of the integrin studies use cytokine-activated endothelial cells. Once cytokine-activated, the endothelial cells may express a dominant ligand such as an integrin, in response to an injury-like signal.

M. HEMLER

ELAM E-selectin is critically involved as several have shown for tumor cell interaction with endothelium.

G. NICOLSON

I have certainly made the point, in my own presentation, that different types of tumor cells express different ranges of these different molecules, and some seem to have extracellular annexins, for example, and others do not. So, it is obvious that there are a number of differences, such that there is heterogeneity among different tumor systems in the expression of these various components. And probably other activities, as well.

D. LIVINGSTON

This is a general question for anybody in the session. If in one of these assays, you inject a number of tumor cells, wait some time, and sacrifice the animal and look for apparent metastasis. If they exist, they score positive, if they don no exist they score negative. And the assumption, if I got it correctly, is that the cells that end up colonizing some distant organ, in such an assay, originated from a primary tumor of a certain size. Is that an incorrect assumption?

R. BANKERT

Well, that is an incorrect assumption in my case and what we are doing initially is to inject the tumors intravenously, and then we look and score whether they are positive or negative. We do not know whether they home directly to the lung, or whether they went from the lung to the bone marrow. We are addressing that now. We have now marked these with bacterial enzymes, LAK-Z and ADH, and we are looking at immediate homing. I really think intravenous inoculations alone is an inappropriate way to go. We are now going into the left ventricle, and we are seeing obvious differences in where the tumor goes. I want to emphasize that VLA-2 clearly is correlated with a difference in the pattern of tumor growth, following intravenous inoculation. Whether that has to do with the binding to the endothelium or whether something else causes it to bind to the endothelium, which is also different in these two cell lines. But VLA has a role to play in its subsequent growth or spread from the tumor in the lung to the tumor in the bone marrow, for instance.

D. LIVINGSTON

Is there an experiment which says that, in such a metastasis assay, cells actually depart from a detectable primary tumor? Cells that end up as metastases actually depart from a detectable primary tumor as opposed to having disseminated from the site of inoculation long before the time, for example, that there is a primary tumor in existence? The implications of the two I think are potentially different.

R. BANKERT

We talk about experimental metastasis which includes injecting the tumor intravenously and then scoring the metastasis. Spontaneous metastasis is clearly a different phenomenon. There you start with a primary lesion, let us say subcutaneously, and then you look for tumors that are distant from that, say in the lung or the liver.

G. NICOLSON

Those comments are correct. There are a number of ways in which the metastatic assays have been performed. I think the assays which are the most realistic are the assays in which the tumor cells are put back in the appropriate primary site. I presented a few different metastatic models and one of the models we use is a breast cancer model. The mammary carcinoma cells were put into the mammary fat pad, and they were allowed to grow in the mammary fat pad. If we excise the primary tumor within six days, we can cure the animal. If we allow those cells to stay for a further three days in the mammary fat pad, they will spread to the regional lymph nodes, become blood borne, and eventually kill the animal with lung metastasis, or in the case of the brain metastatic line, both lung and brain metastases. We have a very good idea of the sequence of metastasis in that model which follows closely the normal pathogenesis of the process in women. There are a number of different models, of course and one of the questions that you indirectly raised, which is a valid one, is how do we know that the exact cells that are present in metastasis were the ones which we started with in the primary tumor. That is a very difficult question to answer since most of these events are very inefficient. Now, in the RAW117 model we have selected highly efficient metastatic variants, so essentially every input cell, if we inject it directly in the circulation, will end up as a metastatic lesion that we can see. But that is a rare system for the metastasis field.

D. LIVINGSTON

What about the inocula experiments? How big are the inocula?

G. NICOLSON

The inoculum go down to ten cells.

D. LIVINGSTON

Could you do it with a single cell?

G. NICOLSON

I do not know of anywhere where investigators have reduced the innoculum down to a single cell. We have been down to below ten cells but that is about as far as we can get. You have to understand that these cells once they get into circulation are subjected to the shear forces of the circulation. Many of the cells would be spontaneously killed and only a fraction of them can survive. And you can show that when they go through the microcirculation, there are cell-cell interactions with the endothelium which can lead to cell death. And I did not discuss it, but there is another whole aspect of our work. If you co-culture tumor cells with endothelial cells, you get a slightly different response, and if you take the conditioned medium from tumor cells you can activate endothelial cells and we have published this (*Li et. al., Cancer Res. 51:245-254, 1991*). Using the very same cytokines that thers use for activation, IFNγ and TNF, for example, we have shown cytolysis of tumor cells. But if you look at highly metastatic tumor cells, in general, they are less susceptible to this and other types of cytostasis and cytolysis. The most metastatic cells tend to escape these responses, but there are a lot of random events, some mechanical, some non-mechanical, that kill cells once they hit the circulation. So I do not know if we are ever going to achieve a one-to-one ratio of input cells to metastases.

ADHESION MECHANISMS IN LYMPHOMA AND CARCINOMA METASTASIS

Geertje La Rivière, Hans Kemperman, Mariëtte Driessens, and Ed Roos

Division of Cell Biology
The Netherlands Cancer Institute
121 Plesmanlaan
1066 CX Amsterdam, THE NETHERLANDS

SUMMARY

We have studied the role of adhesion molecules in metastasis formation by lymphoma and carcinoma cells, particularly in the liver. Certain lymphomas invade the liver extensively, and this process can be mimicked *in vitro* in hepatocyte cultures. Normal activated T-lymphocytes invade similarly, suggesting that lymphomas use normal lymphocyte invasion mechanisms. Indeed, T-cell hybridomas made by fusion of activated T-cells with a non-metastatic T-lymphoma, metastasize extensively to the liver. In addition, the hybrids disseminate to many other sites, and this correlates with their ability to invade monolayers of embryonic fibroblasts. It is noteworthy that cytotoxic T-cell (CTL) clones are highly invasive in these models, even in the resting state, and CTL hybridomas metastasize widely, suggesting that resting memory CTL can move freely into many tissues. Analysis of known T-cell adhesion molecules in panels of hybrids showed that CD2, VLA4-α and L-selectin are not expressed, and are therefore not indispensable. LFA-1, VLA-6 and CD44 are present on all invasive hybrids.

We have generated LFA-1-deficient mutants of a T-cell hybridoma that showed greatly reduced invasive and metastatic potential. Revertants with LFA-1 expression above a threshold level reacquired invasiveness, further supporting the notion that LFA-1 is required. Transfection of LFA-1α- and β-cDNA into α- and β-deficient mutants, respectively, reversed invasive and metastatic capacity only to a limited extent, suggesting that additional involved molecules were also repressed in the mutants.

Invasion and metastasis of the hybrids is strongly reduced by pretreatment with pertussis toxin, implying that an extracellular ligand that binds to a G-protein-coupled receptor, is required for efficient invasion. We present evidence that this ligand is neither a serum factor nor an autocrine motility factor, but is probably derived from the fibroblasts and hepatocytes. Furthermore, we show that one of the effects of the ligand is the functional activation of LFA-1, because pertussis toxin inhibition can be partially reversed by artificial activation of LFA-1 by Mn^{2+}.

Adhesion of TA3 mammary carcinoma cells to hepatocytes is one of the steps in liver metastasis formation. This adhesion was inhibited by Fab fragments made from polyclonal anti-TA3 IgG. We found that a relevant antigen is a 195 kD molecule that we identified as the β_4-integrin subunit. Monospecific anti-$\alpha_6\beta_4$ antibodies purified from the anti-TA3 serum inhibited TA3-hepatocyte adhesion. This shows that $\alpha_6\beta_4$, which so far has only been implicated in cell-matrix adhesion, can also mediate intercellular interactions, and suggests that the high levels of $\alpha_6\beta_4$ often expressed by metastatic carcinoma cells contributes to liver metastasis formation.

Cell Adhesion Molecules, Edited by M.E. Hemler
and E. Mihich, Plenum Press, New York, 1993

INTRODUCTION

MB6A lymphoma and TA3 mammary carcinoma cells that reach the liver via the portal vein, invade this organ in a similar fashion (Roos, E. et al., 1977; Roos, E. et al., 1978). Many small protrusions appear to induce gaps within the sinusoidal endothelial cells through which the protrusions are extended. There is no basement membrane around hepatic sinusoids, so the tumor cells have free access to the sinusoidal surface of the underlying hepatocytes. The main difference between the two tumor cell types is the next step: the invasion of the hepatocyte layer. As much as 40% of MB6A cells are in an extravascular location within 1-2 hours, giving rise to diffuse infiltration of the liver parenchyma. For the TA3 mammary carcinomas, this invasion occurs very infrequently, so that only approximately 0.01% of the cells succeed in establishing nodular metastatic lesions.

These interactions with hepatocytes can be mimicked *in vitro* (Middelkoop, O.P. et al., 1982; Roos, E. et al., 1981). MB6A cells invade monolayers of hepatocytes rapidly and massively, whereas TA3 cells adhere to hepatocytes and a minority of the cells subsequently invades the monolayer in a comparatively slow process. The hepatocyte cultures are also invaded by normal activated T-lymphocytes, in a similar fashion as the lymphoma cells (Roos, E. and Van de Pavert, I.V., 1983), suggesting that the lymphoma cells use the inherent invasive machinery of lymphocytes. To provide evidence for this notion, we fused activated normal T-lymphocytes with the non-invasive and non-metastatic BW5147 T-cell lymphoma. The resulting hybridomas exhibited high invasive and metastatic potential (Roos, E. et al., 1985), indicating that induction of invasiveness is sufficient for a benign lymphoma to become a highly malignant tumor. The hybrids were not only invasive in hepatocyte cultures, but also in monolayers of embryonic fibroblasts (Collard, J.G. et al., 1987; La Rivière, G. et al., 1988; Verschueren, H. et al., 1987). This invasion correlates with widespread metastasis: the MB6A lymphoma, which metastasizes only to the liver, invades hepatocyte but not fibroblast monolayers (La Rivière, G. et al., 1988). As will be described, analysis of expression and function of known adhesion molecules on panels of highly invasive hybridomas has provided extensive evidence for an essential role of LFA-1. Furthermore, VLA-6 and CD44 may be involved, but we have no evidence for their role as yet.

Pertussis toxin (PT) strongly inhibits invasion and metastasis of T-lymphoma and T-cell hybridoma cells (La Rivière, G. et al., 1988; La Rivière, G. et al., 1992; Roos, E. and Van de Pavert, I.V., 1987). PT ADP-ribosylates certain G-proteins, and thus uncouples them from associated cell surface receptors, thereby inhibiting signal transduction of an extracellular ligand binding to the receptor (Simon, M.I. et al., 1991). This implies that an extracellular signal is required for efficient invasion, but neither the nature of the ligand nor its effect on the invasion machinery is known. One of the possible effects of the PT-affected signal is the induction of the high avidity state of LFA-1. If so, artificial induction of this state by Mn^{2+} might reverse the inhibition by PT, and that was indeed the case.

As to the nature of the ligand, three types of factors may be considered: a serum component, an autocrine factor produced by the hybridoma cells, or a factor produced by the monolayer to be invaded. In particular, we have searched for an autocrine motility factor (AMF), similar to those produced by other types of metastatic tumor cells (Evans, C.P. et al., 1991; Liotta, L.A. et al., 1986; Silletti, S. et al., 1991; Watanabe, H. et al., 1991), also because the effect of a 55 kD melanoma-derived AMF is susceptible to pertussis toxin (Nabi, I.R. et al., 1990; Stracke, M.L. et al., 1987). We provide evidence that the ligand is neither a serum factor nor an AMF, but rather a factor produced by hepatocytes and fibroblasts.

The studies with LFA-1-deficient T-cell hybridomas indicated that adhesion to cells in metastasis target tissues is a major determinant of metastasis formation. The TA3 mammary carcinoma cells we used, grow in suspension, do not adhere to the extracellular matrix components fibronectin, vitronectin, laminin and collagen, and not to freshly isolated hepatic sinusoidal endothelial cells (Roos et al., unpublished), suggesting that adhesion to hepatocytes is of major importance for liver metastasis formation by these TA3 cells. We have attempted to generate rat and hamster monoclonal antibodies against TA3 cells that inhibit this adhesion, but failed so far. However, Fab fragments made from a polyclonal anti-TA3 serum did inhibit invasion. We describe here the identification of the relevant antigen as the integrin $\alpha_6\beta_4$.

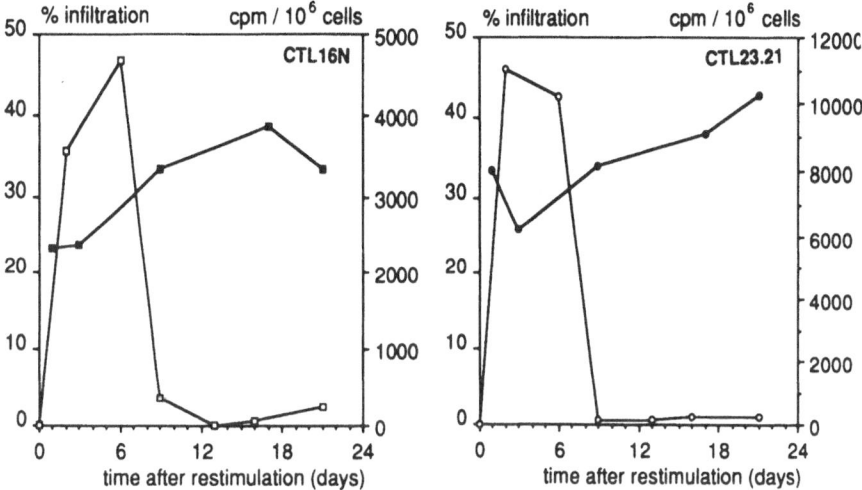

Figure 1. Invasion and Proliferation of CTL Clones. Invasion into fibroblast monolayers (*closed symbols*), and proliferation (*open symbols*) of two CTL clones (CTL23.21 and CTL16N), measured at the indicated time points after restimulation with irradiated B10BR spleen cells as antigen. Invasion is expressed as the percentage of infiltrated cells 4h after addition of cells to the monolayer, as counted using phase-contrast microscopy. Proliferation is expressed as the incorporated ^{3}H-thymidine cpm per 10^6 cells in 24h. The experiment shown is a representative one out of three.

RESULTS AND DISCUSSION

Invasiveness of Cytotoxic T-cells and Dissemination of CTL Hybridomas

To generate T-cell hybridomas, we have previously used T-cells from the spleen or from a mixed lymphocyte culture. The subset of the T-cell fusion partner was unknown and could not be traced because all relevant subset markers were switched off after fusion. We have therefore repeated our experiments with more defined T-cells, in particular two murine CTL clones, directed against minor histocompatibility antigens of B10BR mice (Di Pauli, R. and Opalka, B., 1982). The clones were restimulated regularly with irradiated B10BR spleen cells. Proliferation, as measured by ^{3}H-thymidine uptake during 24h, was high between 2 and 6 days after restimulation and then declined rapidly to basal levels (Figure 1). Furthermore, the expression of the T-cell activation markers MHC class II, CD25 (IL-2 receptor), CD54 (ICAM-1), Ly6C and asialo-GM1 was transiently increased, whereas CD2, CD11a (LFA-1 α-chain) and CD45 (T200) expression remained constant. CD44 expression also was transiently upregulated, but a substantial level of CD44 was maintained in the resting state.

Invasiveness of CTL and CTL Hybridomas. The CTL clones were highly invasive in rat embryo fibroblast monolayers, not only in the activated state but, remarkably, also after return to the resting state (Figure 1). CTL hybridomas were generated by fusion of one of the CTL clones with non-invasive BW5147 T-lymphoma cells. All CTL hybridomas studied were very highly invasive in hepatocyte and fibroblast monolayers. Invasiveness of CTL hybridomas was substantially higher than that of T-cell hybridomas previously generated from activated spleen T-cells (La Rivière, G. et al., 1988; Roos, E. et al., 1985): 30-70% of added cells infiltrated the monolayer, as compared to 8-27%. Also, invasion into hepatocyte cultures was quite extensive. Thus, CTL clones are very highly invasive, irrespective of their activation state, and the invasive potential of the CTL is dominantly expressed in the hybrids.

Dissemination of CTL Hybridomas *in vivo*. Consistent with the very high invasive potential of CTL hybrids, metastasis formation was very rapid: 10^6 i.v. injected cells were lethal in 100% of animals within 7-14 days, compared to 13-36 days for spleen T-cell hybridomas (Roos, E. et al., 1985). Histological examination revealed accumulation in liver, kidneys, lung, spleen, thymus, bone marrow and lymph nodes, ovaries, tubae and uterus, and in fat and mesenchymal tissues. The liver parenchyma was diffusely infiltrated, especially periportally. In the kidney, hybridoma cells were found exclusively in the cortex between glomeruli and tubules Large numbers of cells were also seen in the renal capsule. In the lung, hybridoma cells accumulated mainly around large blood vessels and bronchi.

Since the BW5147 fusion partner does not metastasize, this capacity to disseminate *in vivo* must be derived from the CTL. Therefore, these observations suggest that the CTL are able to migrate into the non-inflamed tissues that are susceptible to CTL hybridoma metastasis. It is noteworthy, that among the non-affected tissues are the skin and the intestine, shown to contain specific subsets of T-cells (Mackay, C.R., et al., 1992). Given the correlation that we have repeatedly found between invasiveness in fibroblast cultures and wide-spread dissemination (Collard, J.G. et al., 1987; La Rivière, G. et al., 1988; Roos, E. et al., 1985), these results suggest that resting CTL also may be able to migrate into many tissues. This notion is in agreement with the finding that memory T-cells move to the lymph nodes through tissues rather than through lymph node endothelium (Mackay, C.R et al., 1990).

Adhesion Molecules on CTL and CTL Hybridomas. The expression of several T-cell molecules able to mediate cell adhesion, was determined by FACS analysis (Figure 2). CD8 and ICAM-1 were expressed on the CTL, but not on the hybrids. CD2 was present on the CTL, but only on three of the five CTL hybrids. CD49d, the α_4 subunit of the mucosal homing receptor LPAM-1 (Holzmann, B. et al., 1989; Holzmann, B. and Weissman, I.L, 1989) and of the β_1-integrin VLA-4 (Humphries, M.J., 1990) was not detected on either CTL or CTL hybridomas. Also the 'peripheral lymph node homing receptor, L-selectin, was not present on either CTL or CTL hybrids. Thus, CD2, CD8, CD54, L-selectin, VLA4 and LPAM-1 are not required for invasion and dissemination, including migration into lymph nodes. In contrast, LFA-1, CD44 and VLA-6 were expressed on all cell lines (Figure 2) and are thus potential candidates for a role in invasion and metastasis.

Effect of Adhesion-inhibiting Antibodies on CTL Hybridoma Invasion. We have tested the effect of monoclonal antibodies (mAb) that inhibit at least one of the functions of the three consistently expressed adhesion molecules. Anti-LFA-1 mAb inhibited invasion of CTL hybrids, in agreement with our previous results with other hybrids (Roossien, F.F. et al., 1989). The anti-CD44 mAb KM201 that interferes with CD44/hyaluronate interaction and thereby with B-cell/fibroblast adhesion (Miyake, K. et al., 1990), had no effect. Also the anti-VLA-6 mAb GoH3, that inhibits VLA-6/laminin binding (Sonnenberg, A. et al., 1988), did not inhibit. However, this does not exclude a role for CD44 and VLA-6 *in vivo*.

Invasive and Metastatic Potential of LFA-1-deficient Mutants, Revertants and Transfectants

Invasive and Metastatic Potential of Mutants and Revertants. To study the role of LFA-1 *in vivo*, we have generated LFA-1-deficient mutants of the TAM2D2 T-cell hybridoma (Roossien, F.F. et al., 1989). We obtained three independent mutants, two that did not synthesize the α-chain, and one that did not produce the β-chain. All three mutants showed greatly diminished invasive and metastatic potential, indicating that LFA-1 is indispensable for efficient metastasis formation by these cells. This was further confirmed by revertants of these mutants obtained after *in vivo* passage: those expressing LFA-1 at more than 25% of the parental level had regained invasive potential (Roossien, F.F. et al., 1990). However, it was not yet clear whether LFA-1 was the only invasion-relevant molecule affected by the mutations.

Transfection of LFA-1 α- and β-chain cDNA. We have transfected the human LFA-1 α- and the mouse β-chain cDNA into, respectively, α- and β-chain-deficient mutants, using different retroviral and plasmid vectors. In all cases, the LFA-1 level obtained was rather low. To obtain clones with higher expression levels, bulk cultures of transfected cells

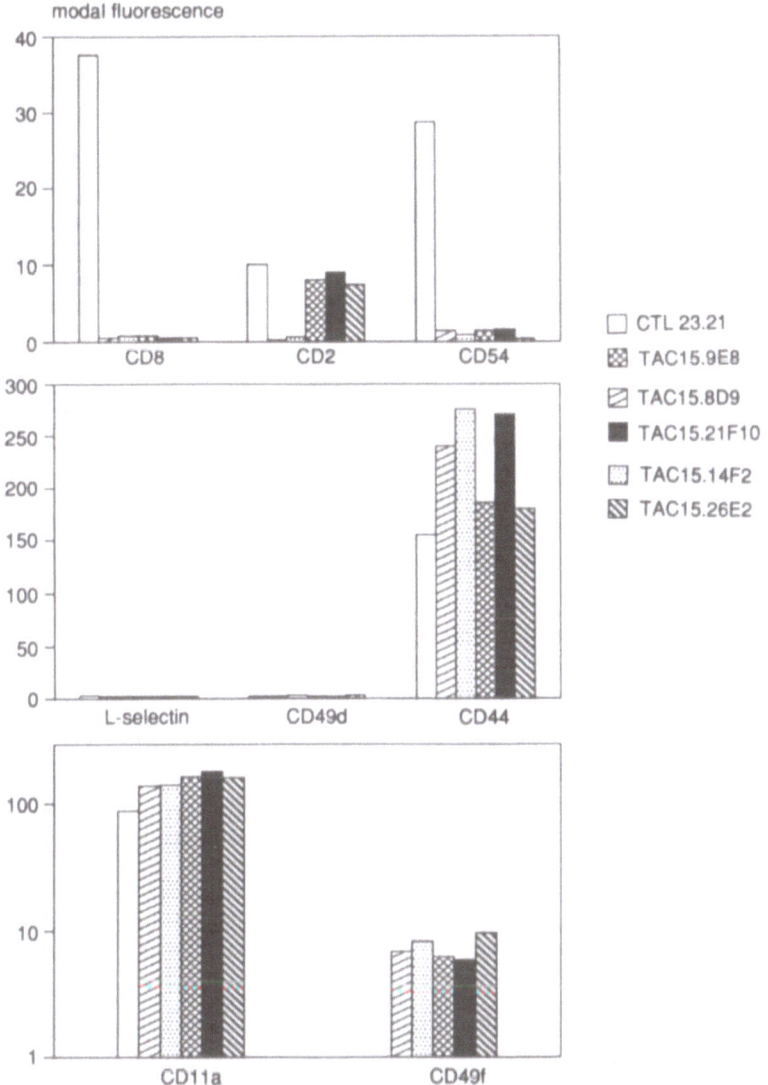

Figure 2. Adhesion Molecule Expression on CTL Clone CTL23.21 and CTL Hybridomas. Expression of CD2, CD8, ICAM-1 (CD54), L-selectin, VLA-4α (CD49d), CD44, LFA-1α (CD11a) and VLA-6α (CD49f) on a CTL clone (23.21) and five different CTL hybridomas (TAC15) was determined by FACScan analysis. Values are averages of 3 to 10 experiments and represent the modal fluorescence. Measurements on CTL23.21 cells were performed on day 3 after restimulation, except for CD8 and LFA-1, the expression of which was not dependent on the activation state. LFA-1 and CD8 values are averages of results obtained at various time points after restimululation. VLA-6α expression on CTL23.21 was not determined. Anti-VLA-4α and anti-L-selectin mAb did react with appropriate control cells.

were selected on the FACSorter. However, because of this strong selection we obtained revertants in addition to transfectants, even amongst neomycin- or hygromycin-resistant transfected cells.

In case of the β-chain mutants, we transfected the murine β-chain cDNA, and transfectants and revertants were distinguished by Northern blot analysis using probes for 3'-sequences of the vector and the endogenous β-chain, respectively. A stable transfectant thus obtained showed limited reversion of invasive and metastatic capacity. Another transfectant with a very broad expression profile could be sorted into high and low LFA-1 populations, that were tested for invasiveness immediately after sorting. The former had regained substantial invasive capacity. However, because these LFA-1 expression levels were not maintained during subsequent culture, we could not obtain sufficient cells to test their metastatic capacity.

One of the α-chain mutants was transfected with the human α-chain cDNA. Transfectants could be readily distinguished using mouse- and human-specific anti-LFA-1 monoclonal antibodies. Remarkably, selection by FACS-sorting for human LFA-1 yielded only clones that also expressed mouse LFA-1. Simultaneous selection for human LFA-1 and against murine LFA-1, followed by cloning, yielded 23 clones, 22 of which expressed murine LFA-1. The twenty-third clone expressed only human LFA-1, but at a rather low level. Its invasiveness is moderately increased, compared to the mutant. The metastatic capacity is being tested in nude and SCID mice.

Due to the low LFA-1 levels obtained, and the problems with reversion, it is difficult to draw definitive conclusions from these results. Some reversion of invasiveness can be acquired upon transfection, but not consistently. The LFA-1 β-chain transfectant with hardly increased invasiveness expresses LFA-1 at levels comparable to highly invasive revertants. Thus, it is still possible that other proteins, in addition to LFA-1, required for efficient invasion, have also been affected by the mutations. The mutants are clearly regulatory, and it is conceivable that genes encoding proteins required for proper function of the α- and β-chain have been silenced by the same mutations that affected the genes encoding either subunit.

Pertussis Toxin Inhibition of Invasion Is Reversed by Mn^{2+}

Effect of Pertussis Toxin on Invasion and Metastasis. Pertussis toxin (PT) causes lymphocytosis *in vivo*, that is an accumulation of lymphocytes in the blood, due to inhibition of lymphocyte entry into lymph nodes (Spangrude, G.J. et al., 1985) and of migration into tissues such as the skin (Spangrude, G.J. et al., 1984). In addition, PT inhibits infiltration of thymocytes into monolayers of thymic reticulo-epithelial cells *in vitro* (Sugimoto, M., et al., 1983). Recently, Chaffin et al. (1990) showed that thymocytes accumulate in the thymus of transgenic mice, in which the catalytic subunit of PT is expressed under the control of the *lck* promotor, due to a defect in migration. Thus, a PT-sensitive G-protein is required for lymphocyte migration.

This is also relevant for lymphoma invasion and metastasis, since we have demonstrated that PT strongly inhibits invasion by the MB6A lymphoma and by T-cell hybridomas, that have acquired their invasive and metastatic potential from the lymphocyte fusion partner (La Rivière, G. et al., 1988; La Rivière, G. et al., 1992; Roos, E. and Van de Pavert, I.V., 1987). Invasiveness remained inhibited for at least 5 days after treatment with 1 μg PT per ml for 4h, whereas proliferation was not affected at all, allowing the effect of PT to be tested *in vivo*. This PT-treatment resulted in a reduction of liver metastasis formation to 10-25% of controls (Roos, E. and Van de Pavert, I.V., 1987). Thus, a PT-sensitive signaling pathway is involved in invasion *in vitro* and in migration into tissues *in vivo* of both normal and malignant lymphoid cells. The inhibition of invasion by PT indicates that a PT-sensitive G-protein is involved. Because G-proteins transduce signals delivered by binding of extracellular ligands to certain cell surface receptors (Simon, M.I. et al., 1991), this implies that efficient invasion depends on an extracellular 'invasion-enhancing' ligand.

Suramin Mimics Effect of PT. In further experiments, we used the T-cell hybridomas TAM2D2 (Roos, E. et al., 1985) and TAM8C4 (La Rivière, G. et al., 1988). Pretreatment with 200 ng/ml PT for 2h inhibited invasion to 12± 6 % of controls for TAM2D2 and 20± 9 % for TAM8C4 (Figure 3). Suramin inhibited invasion to a similar extent when it was present during the invasion assay (10± 3 and 27± 11 %). This effect was dose-dependent and maximal at 0.25 to 0.5 mM, and instantaneously reversible. Neither PT nor suramin affected proliferation and viability of the cells. Suramin is known to dissociate ligands from receptors, including growth factors, low density lipoprotein and scatter factor (Adams, J.C. et al., 1991; La Rocca, R.V. et al., 1990; Schneider, W.J. et al., 1982), although clearly suramin can have other effects as well (La Rocca, R.V. et al., 1990). Moreover, suramin

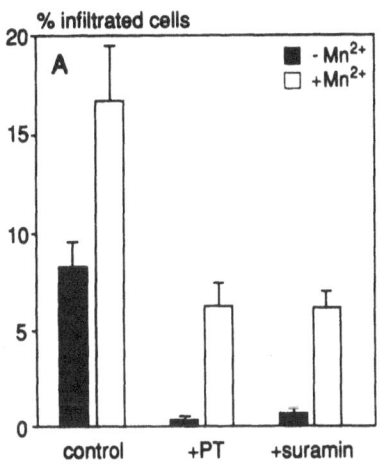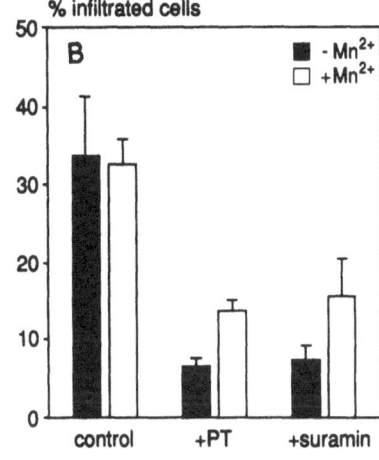

Figure 3. Pertussis Toxin and Suramin Inhibit Invasion and This Is Partially Reversed by Mn^{2+}.
Invasion by untreated (*control*), PT-pretreated (200 ng/ml for 2h), and suramin-treated (0.5 mM suramin for
15 min) TAM2D2 (**A**) and TAM8C4 cells (**B**) in the absence (black bars) or presence of 2 mM Mn^{2+} (white
bars). Suramin, but not PT, remained present during the invasion assay. Invasion is expressed as the
percentage of infiltrated cells after 1h. Standard deviations represent the variability between counted fields.
Data shown are of one representative experiment out of nine.

also dissociates small phospholipids, such as lysophosphatidic acid, from G-protein-coupled
receptors (Van Corven, E.J. et al., 1992; Van Corven, E.J. et al., 1989). Thus, the effect of
suramin supports the notion that an extracellular ligand is required for efficient invasion.

Mn^{2+} partially Reverses PT and Suramin Inhibition of Invasion. Recently, Lorant
et al. (Lorant, D.E. et al., 1991) showed that platelet activating factor (PAF), that is
produced by activated endothelium and remains associated with the membrane, induces
functional activation of the β_2-integrins Mac-1 and LFA-1 on polymorphonuclear leukocytes
(PMN), via a signal transmitted through the PMN PAF-receptor. PAF is a small
phospholipid, its receptor is coupled to a G-protein, and PAF-induced GTPase activity in
PMN membranes is inhibited by PT (Prescott, S.M. et al., 1990). We therefore hypothesized
that the effect of the postulated ligand was to functionally activate LFA-1. Since LFA-1, like
many other integrins (Altieri, D.C. 1991; Conforti, G. et al., 1990; Dransfield, I. et al., 1992;
Elices, M.J. et al., 1991; Gailit, J. and Ruoslahti, E. 1988; Kirchhofer, D. et al., 1990;
Sonnenberg, A. et al., 1988), can be artificially induced by Mn^{2+} to adopt the high avidity
state, Mn^{2+} should reverse the inhibitory effect of PT and suramin on invasion. We found
that 2 mM Mn^{2+} strongly stimulated invasion of PT- pretreated TAM2D2 cells and, to a
lesser extent, invasion by PT-pretreated TAM8C4 cells (Figure 3). Mn^{2+} also stimulated
invasion of untreated TAM2D2 cells, but much less, and did not enhance invasion by
untreated TAM8C4 cells (Figure 3). The effects of Mn^{2+} in the presence of suramin were
similar. As a result, inhibition by PT and suramin was much reduced by Mn^{2+}. Thus, Mn^{2+}
bypasses the inhibitory effect of both PT and suramin on invasion, indicating that it acts
distal of the PT-sensitive signal transduction pathway.

Mn^{2+}-induced Invasion is LFA-1-dependent. Anti-LFA-1 mAb inhibited invasion in
the presence of 2 mM Mn^{2+} to a similar extent for untreated as for PT-pretreated cells
(Figure 4). Furthermore, Mn^{2+} did not stimulate invasion of the LFA-1-deficient mutants
described above (Figure 4). Thus, Mn^{2+} appears to influence LFA-1-dependent adhesion
and not to induce independent alternative adhesion mechanisms.

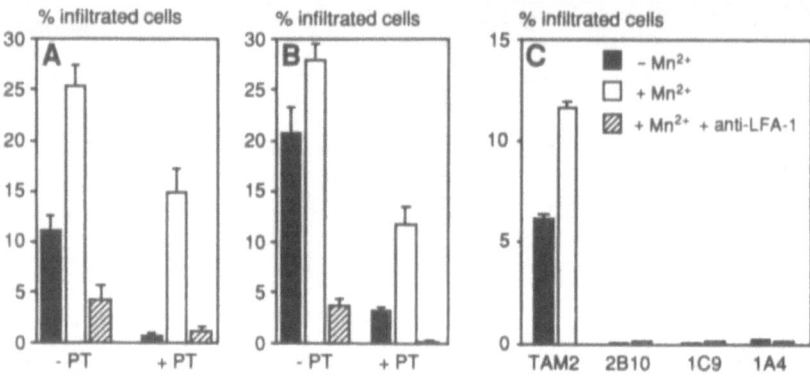

Figure 4. Contribution of LFA-1 to Mn^{2+}-stimulated Invasion. Invasion of untreated and PT-pretreated (200 ng/ml PT for 2h) TAM2D2 (*A*) and TAM8C4 (*B*) cells in the absence (black bars) and presence of 2 mM Mn^{2+} without (*white bars*) and with (*hatched bars*) purified anti-LFA-1 mAb. Cells were preincubated with a saturating concentration of anti-LFA-1 mAb (5 μg/ml) for 30 min at 21°C and then added to the fibroblast monolayer. (*C*) invasion of TAM2D2.2.6 (*indicated as TAM2*) and of three independent LFA-1-deficient mutants: 2B10, 1C9 and 1A4, in the absence (*black bars*) and presence (*white bars*) of 2 mM Mn^{2+}. Invasion is expressed as the percentage of infiltrated cells after 1h. Standard deviations represent the variability between counted fields. Data of one out of four experiments are shown.

Manganese Activates T-cell Hybridoma LFA-1. Mn^{2+} induced homotypic cell aggregation of TAM8C4 cells, and this was inhibited substantially by anti-LFA-1 Fab fragments, that were used because in the conditions of an aggregation assay, LFA-1 mAb caused some aggregation itself. Thus, at least on TAM8C4 cells, LFA-1 is constitutively not active, but higher avidity for its ligand(s) can be induced by Mn^{2+}, supporting the notion that Mn^{2+}-induced invasion is due to activation of LFA-1. Mn^{2+} did not induce aggregation of TAM2D2 cells, possibly due to lack of expression of LFA-1 counterstructures.

The Nature of the 'Invasion-enhancing' Ligand. There are three conceivable sources for the 'invasion-enhancing' ligand: serum; the T-cell hybridoma cells themselves; or the fibroblast and hepatocyte monolayers. Since we observed the same effects of PT and suramin on invasion by serum-free-cultured cells in an assay performed in serum-free medium, a serum factor can be ruled out. An alternative possibility was that the T-cell hybridoma cells produced an autocrine motility factor (AMF) to stimulate their own motility and thus invasion, similarly as other tumor types (Evans, C.P. et al., 1991; Liotta, L.A. et al., 1986; Silletti, S. et al., 1991; Watanabe, H. et al., 1991). A melanoma-derived AMF is in fact sensitive to PT (Liotta, L.A. et al., 1986). AMF stimulates spontaneous migration through Nucleopore filters (Liotta, L.A. et al., 1986). We have therefore tested the effect of PT and suramin on migration by T-cell hybridoma cells through polycarbonate filters with 8 μm pores, and observed no inhibition, which argues against involvement of an AMF. If our hypothesis that the PT-affected ligand acts to induce LFA-1 activation is correct, autocrine factors can not be involved, since continuous production of such factors by the T-cell hybridoma cells should result in constitutive LFA-1 activation. This leaves the fibroblast and hepatocyte monolayers as the only possible source. This ligand would play a similar role as the PAF produced by activated endothelium (Lorant, D.E. et al., 1991).

The Integrin α$_6$β$_4$ on TA3/Ha Mammary Carcinoma Cells Is Involved in Adhesion to Hepatocytes

The reduced metastatic capacity of LFA-1-deficient mutants, described above, argues strongly for a decisive role of adhesion molecules in metastasis formation. Direct evidence for this notion has been provided by Chan et al. (Chan, B.M.C. et al., 1991), who transfected the cDNA encoding the VLA-2 integrin α-chain into non-metastatic rhabdomyosarcoma cells. The transfectants expressed VLA-2 ($\alpha_2\beta_1$), exhibited increased adhesion to collagen and laminin, and produced lung metastases upon i.v. injection, whereas untransfected cells and control transfectants did not. For carcinomas, however, there is so far no direct evidence for involvement of particular cell adhesion molecules in colonization of metastasis target tissues. The liver is exceptional in that it contains no basement membrane around the sinusoids where we have observed invasion of tumor cells to occur (Roos, E. et al., 1978). Therefore, as discussed above, the only relevant interactions are those with sinusoidal endothelial cells and hepatocytes, and since TA3 cells do not seem to adhere to hepatic sinusoidal endothelial cells (Roos et al., unpublished), adhesion between TA3 cells and hepatocytes appears to be particularly important. To identify the adhesion molecules involved, we have attempted to generate inhibitory rat and hamster monoclonal antibodies against TA3 cells, so far without success. However, Fab fragments made from a polyclonal rabbit serum against the TA3 cells did inhibit, as described below.

Adhesion of TA3 Cells to Hepatocytes Is Inhibited by Polyclonal Anti-TA3 Antibodies. Fab fragments prepared from a polyclonal serum against TA3 mammary carcinoma cells inhibited adhesion of TA3 cells to hepatocytes. Maximal inhibition, at 0.1 mg/ml, varied between experiments from 40 to 70%. The results from one of three experiments is shown in Figure 5. Similar inhibition was seen after preincubation of TA3 cells and removal of Fab fragments before adding the TA3 cells to the hepatocyte cultures. Thus, the anti-TA3 Fab affected the TA3 cells and not the hepatocytes.

Identification of a 195 kD Protein on TA3 Cells Potentially Involved in TA3-hepatocyte Adhesion. Our strategy to identify the relevant TA3 antigen was to affinity purify Fab fragments on blotted separated membrane proteins, and test eluted Fab for effect on adhesion. As a first step, we used total membrane protein. Octylglucoside-solubilized TA3 plasma membrane proteins were immobilized on nitrocellulose. Anti-TA3 Fab fragments were affinity purified on these blots and tested in an adhesion assay, but surprisingly did not inhibit TA3-hepatocyte adhesion (Figure 5). Western blot analysis showed that the eluted Fab fragments reacted with the same TA3 plasma membrane proteins as the original anti-TA3 serum with the exception of a 195 kD protein that was not solubilized by octylglucoside. This suggested that the 195 kD was the relevant antigen. To demonstrate this, we purified Fab fragments specific for the 195 kD protein on Western-blotted 195 kD protein. The obtained fragments were highly specific for the 195 kD protein as shown by Western blot analysis, and inhibited TA3-hepatocyte adhesion to an extent varying between 30 and 60% (Figure 5). The same procedure, applied to a major 57 kD TA3 plasma membrane protein, yielded highly specific Fab fragments, that did not inhibit TA3-hepatocyte adhesion at all. This strongly suggested that the 195 kD protein was the relevant antigen.

Purification of the 195 kD Protein and Production of a Specific Antiserum. To obtain additional evidence, we purified the 195 kD protein by preparative SDS-PAGE, and rabbits were immunized with the purified denatured protein. The obtained polyclonal rabbit serum (7880) specifically stained the 195 kD protein on a Western blot of TA3 plasma membrane proteins. Immunoprecipitation using ^{125}I-surface labelled TA3 cells showed a major band of 195 kD and three minor bands of 180, 150 and 115 kD. However, antibodies isolated from this serum did not inhibit TA3-hepatocyte adhesion. FACScan analysis showed that the antibodies did not react with intact cells, indicating that the serum was directed against intracellular epitopes. This notion was confirmed by FACScan analysis of permeabilized TA3 cells.

Identification of the 195 kD Protein as the Integrin β_4 Subunit. Immunoprecipitation using ^{125}I-surface labeled TA3 cells yielded a pattern of bands that was similar to that observed with antibodies against the α_6 or β_4 subunit of the $\alpha_6\beta_4$ integrin (Figure 6): a major band of 200 kD which is the mature β_4 subunit, two minor bands at 180 and 150 kD which are proteolytic products of β_4, and a faint band of 115 kD which is the α_6 subunit (Hemler, M.E. et al., 1989). After preclearing of a TA3 cell lysate with an α_6-specific mAb, the anti-195 kD serum (7880) no longer precipitated any protein, conclusively showing that the 195 kD protein is in fact the β_4 subunit of the $\alpha_6\beta_4$ integrin. Since the serum reacts only with intracellular epitopes, it is apparently directed against the cytoplasmic domain of β_4.

Affinity Purified $\alpha_6\beta_4$-specific Antibodies from the Anti-TA3 Serum Inhibit TA3-hepatocyte Adhesion. Next, we attempted to affinity purify the native $\alpha_6\beta_4$ integrin from TA3 cell lysates. However, $\alpha_6\beta_4$ bound so strongly to the Sepharose-coupled 7880 IgG that it could not be eluted by low or high pH, high salt or ethylene glycol. This enabled us to purify $\alpha_6\beta_4$-specific Fab fragments prepared from the anti-TA3 serum on the $\alpha_6\beta_4$ bound to the column. The eluted Fab fragments were highly specific for β_4 as demonstrated by Western blot analysis (Figure 6), immunoprecipitated the $\alpha_6\beta_4$ integrin (Figure 6), and reacted with intact cells as shown by FACScan analysis. These antibodies inhibited adhesion, as quantitated with ^{51}Cr-labeled TA3 cells, to an extent varying between 30 and 50%. The result of one of three experiments is shown in Figure 7. This was confirmed by direct counting of adherent TA3 cells in sections of embedded hepatocyte cultures. Again, inhibition was approximately 40%. This result demonstrates that the integrin $\alpha_6\beta_4$ is involved in the intercellular interaction between TA3 carcinoma cells and hepatocytes.

The $\alpha_6\beta_4$ Integrin Is Often Expressed at High Levels in Metastatic Carcinoma Cells. The $\alpha_6\beta_4$ integrin was first described as the tumor associated complex 180 (TSP-180) on carcinoma cells (Falcioni, R. et al., 1986; Falcioni, R. et al., 1988; Kennel, S.J. et al.,

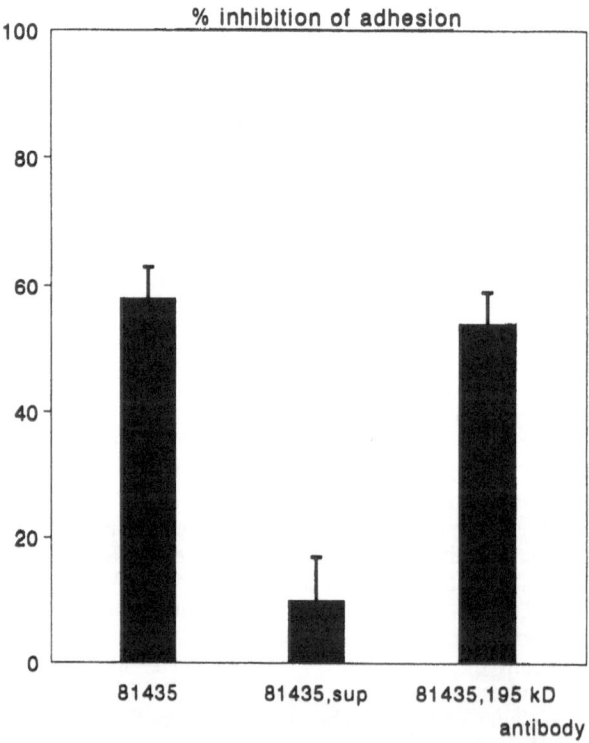

Figure 5. Effect of Antibodies on the Adhesion of TA3 cells to Hepatocyte Cultures. ^{51}Cr-labeled TA3 cells were preincubated for 30 min with Fab fragments prepared from the anti-TA3 serum (81435), Fab that had been affinity-purified on octylglucoside-solubilized TA3 plasma membrane proteins (81435, sup) and Fab affinity-purified on Western-blotted 195 kD protein (81435,195 kD). Hereafter, cells were allowed to adhere to hepatocyte cultures for 1h, in the presence of the Fab fragments, and non-adherent cells were washed away. Cultures were lysed and counted in a gamma counter. Each bar is the average of three separate wells. Shown is the maximal percentage of inhibition compared to control values (cells incubated with PBS only).

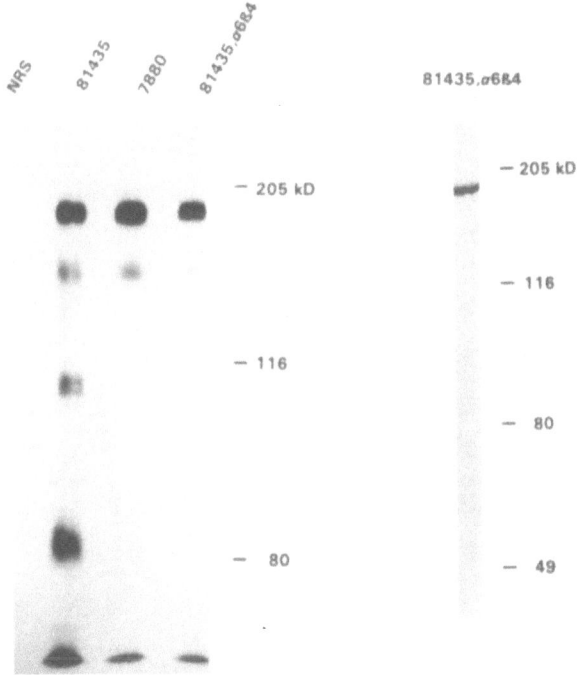

Figure 6. Characterization of Antibodies Affinity-purified on $\alpha_6\beta_4$ (81435,$\alpha_6\beta_4$).

a) Antigens were precipitated from ^{125}I-labeled TA3 cells with the anti-TA3 serum (81435), the anti-195 kD serum (7880), or antibodies affinity purified on $\alpha_6\beta_4$ (81435,$\alpha_6\beta_4$). Normal rabbit serum (NRS) was used as a negative control. Precipitates were analyzed by reduced 7% SDS-PAGE.

b) Western blot of TA3 plasma membrane proteins separated by reduced 7% SDS-PAGE, electrophoretically transferred to nitrocellulose and probed with antibodies affinity-purified on $\alpha_6\beta_4$ (81435,$\alpha_6\beta_4$).

1989). In several tumor cell types, a correlation was found between TSP-180 expression and malignant potential: The complex is expressed at low levels on non-metastatic basal cell epithelioma cells but highly expressed on metastatic squamous carcinomas, particularly at the growing edge of the tumors (Kimmel, K.A. and Carey, T.E., 1986). Furthermore, $\alpha_6\beta_4$ expression correlates with aggressiveness of human squamous carcinomas (Van Waes, C. et al., 1991), and $\alpha_6\beta_4$ is only present on highly metastasizing variants of murine Lewis lung carcinomas (Cimino, L. et al., 1991). Our present results suggest that these high levels of $\alpha_6\beta_4$ may in fact play a role in metastasis formation, at least in the liver.

The Counterstructure of $\alpha_6\beta_4$. In normal tissues, $\alpha_6\beta_4$ is expressed by a variety of epithelial cell types but most strongly at the basal region of stratified squamous epithelium, where it is located in hemidesmosomes (Carter, W.G. et al., 1990; Jones, J.C.R., et al., 1991; Sonnenberg, A. et al., 1991), suggesting interaction with a basement membrane component. De Luca et al. (De Luca, M. et al., 1990) postulated that $\alpha_6\beta_4$ binds to laminin, based on the ability of polyclonal antibodies that were affinity-purified on $\alpha_6\beta_4$, to detach keratinocytes from Matrigel or purified laminin. Also Lotz et al. (Lotz, M.M., et al., 1990) suggested this, based on the inhibition by an anti-α_6 mAb of carcinoma cell adhesion to laminin. However, this has not been widely excepted, because other groups were unable to block adhesion to laminin of cells expressing $\alpha_6\beta_4$ but not $\alpha_6\beta_1$, with monoclonal antibodies to α_6 or β_4. Furthermore, attempts to purify $\alpha_6\beta_4$ on a laminin column were unsuccessful (Carter, W.G. et al., 1991; De Luca, M. et al., 1990; Quaranta, V. and Jones, J.C.R., 1991; Sonnenberg,

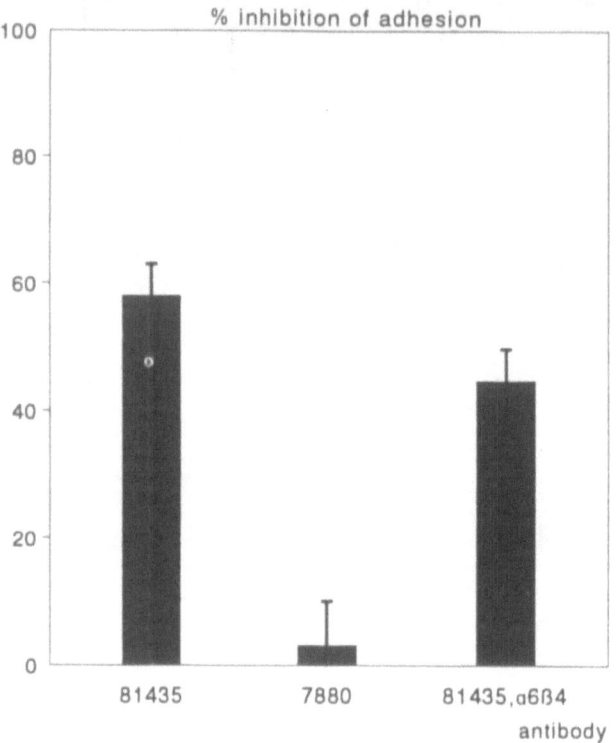

Figure 7. Effect of Antibodies Affinity-purified on $\alpha_6\beta_4$ (81435,$\alpha_6\beta_4$) on Adhesion of TA3 cells to Hepatocytes. ^{51}Cr-labeled TA3 cells were preincubated for 30 min with anti-TA3 Fab (81435), anti-195 kD Fab (7880), or Fab affinity purified on $\alpha_6\beta_4$ (81435,$\alpha_6\beta_4$). Hereafter, cells were allowed to adhere to hepatocyte cultures for 1h, in the presence of the Fab fragments, and non-adherent cells were washed away. Cultures were lysed and counted in a gamma counter. Each bar is the average of three separate wells. Shown is the maximal percentage of inhibition compared to control values (incubation with PBS only).

A., et al., 1990). Recently, however, Lee et al. (Lee, E.C. et al., 1992) showed that cells expressing α_6 only in association with β_4, bind to the laminin E8 fragment and this adhesion is blocked by both a mAb to β_4 (30%) and a mAb to α_6 (60%). Moreover, these authors did succeed in purification of $\alpha_6\beta_4$ on a laminin column, strongly suggesting that laminin is in fact a ligand of $\alpha_6\beta_4$.

It is unlikely, however, that $\alpha_6\beta_4$ on TA3 cells binds to hepatocytes via laminin, because TA3 cells do not adhere to purified laminin coated on plastic or glass, and because cultured hepatocytes produce little if any laminin (La Rivière et al., unpublished results). Moreover, the α_6-specific mAb GoH3 which completely blocks $\alpha_6\beta_1$ binding to laminin (Sonnenberg, A. et al., 1988) and partly the binding of $\alpha_6\beta_4$ to laminin (Lee, E.C. et al., 1992), did not inhibit TA3-hepatocyte adhesion (Kemperman et al., unpublished results). This suggests that hepatocytes express an alternative ligand to which $\alpha_6\beta_4$ binds at a different domain that is not affected by GoH3. This notion is not without precedent, because the adhesion of the integrin VLA-4 ($\alpha_4\beta_1$) to its matrix ligand fibronectin and its cellular ligand VCAM-1 is mediated by different binding domains that are blocked by distinct antibodies (Elices, M.J. et al., 1990). A candidate ligand on hepatocytes is an antigen detected with the OPAR mAb, directed against rat hepatocytes, that inhibits adhesion of TA3 cells (Middelkoop, O.P. et al., 1985). This will be tested in the near future, since we have recently been able to purify substantial amounts of this antigen from rat liver plasma membranes.

ACKNOWLEDGEMENTS

We thank Cor Schipper, Jacqueline W.T.M. Klein Gebbinck, Yvonne Wijnands, Diana de Rijk, Felix Rodriguez Erena and Eric Nooteboom for expert technical assistance, and Conny Van Niele-Pouw for secretarial assistance. We thank many collegues for their generous gifts of monoclonal antibodies.

This work was supported by grants NKI 88-18, 91-03, and 91-04 from the Dutch Cancer Society.

REFERENCES

Adams, J.C., Furlong, R.A., and Watt. F.M., 1991, Production of scatter factor by *ndk*, a strain of epithelial cells, and inhibition of scatter factor activity by suramin. *J. Cell Sci.* 98.385-394.

Altieri, D.C., 1991, Occupancy of CD11b/CD18 (Mac-1) divalent ion binding site(s) induces leukocyte adhesion. *J. Immunol.* 147:1891-1898.

Carter, W.G., Kaur, P., Gil, S.G., Gahr P.J., and Wayner, E.A., 1990, Distinct functions for integrins α3β1 in focal adhesions and α6β4/bullous pemphigoid antigen in a new stable anchoring contact (SAC) of keratinocytes: Relation to hemidesmosomes. *J. Cell Biol.* 111:3141-3154.

Carter, W.G., Ryan, M.C., and Gahr, P.J., 1991, Epiligrin, a new cell adhesion ligand for integrin α3β1 in epithelial basement membranes. *Cell* 65:599-610.

Chaffin, K.E., Beals, C.R., Wilkie, T.M., Forbush, K.A., Simon, M.I., and Perlmutter, R.M., 1990, Dissection of thymocyte signaling pathways by *in vivo* expression of pertussis toxin ADP-ribosyltransferase. *EMBO J.* 9:3821-3829.

Chan, B.M.C., Matsuura, N., Takada, Y., Zetter, B.R., and Hemler, M.E., 1991, *In vitro* and *in vivo* consequences of VLA-2 expression on rhabdomyosarcoma cells. *Science* 251:1600-1602.

Cimino, L., Perrotti, D., Falcioni, R., Kennel, S.J., and Sacchi, A., 1991, β4 integrin expression on *in vitro* and *in vivo* metastatic variants of Lewis Lung carcinoma. *Cytotechnol.* 5:S45-48.

Collard, J.G., Schijven, J.F., and Roos, E., 1987, Invasive and metastatic potential induced by *ras*-transfection into mouse BW5147 lymphoma cells. *Cancer Res.* 47:754-759.

Conforti, G., Zanetti, A., Pasquali-Ronchetti, I., Quaglino, D., Neyroz, P., and Dejana, E., 1990, Modulation of vitronectin receptor binding by membrane lipid composition. *J. Biol. Chem.* 265:4011-4019.

De Luca, M., Tamura, R.N., Kajiji, S., Bondanza, S., Rossino, P., Cancedda, R., Marchisio, P.C., and Quaranta, V., 1990, Polarized integrin mediates human keratinocyte adhesion to basal lamina. *Proc. Natl. Acad. Sci. USA* 87:6888-6892.

Di Pauli, R. and Opalka, B., 1982, Antigen-dependent H-2-restricted cytolytic and noncytolytic T cell lines with specificity for minor histocompatibility antigens. *Eur J Immunol.* 12:365-373.

Dransfield, I., Cabañas, C., Craig, A., and Hogg, N., 1992, Divalent cation regulation of the function of the leukocyte integrin LFA-1. *J. Cell Biol.* 116:219-226.

Elices, M.J., Osborn, L., Takada, Y., Crouse, C., Luhowskyj, S., Hemler, M.E., and Lobb, R.R., 1990, VCAM-1 on activated endothelium interacts with the leukocyte integrin VLA-4 at a site distinct from the VLA-4/fibronectin binding site. *Cell* 60:577-584.

Elices, M.J., Urry, L.A., and Hemler, M.E., 1991, Receptor functions for the integrin VLA-3. Fibronectin, collagen, and laminin binding are differentially influenced by arg-gly-asp peptide and by divalent cations. *J. Cell Biol.* 112:169-181.

Evans, C.P., Walsh, D.S., and Kohn, E.C., 1991, An autocrine motility factor secreted by the Dunning R-3327 rat prostatic adenocarcinoma cell subtype AT2.1. *Int. J. Cancer* 49:109-113.

Falcioni, R., Kennel, S.J., Giacomini, P., Zupi, G., and Sacchi, A., 1986, Expression of tumor antigen correlated with metastatic potential of Lewis Lung carcinoma and B16 melanoma clones in mice. *Cancer Res.* 46:5772-5778.

Falcioni, R., Sacchi, A., Resau, J., and Kennel, S.J., 1988, Monoclonal antibody to human carcinoma-associated protein complex: Quantitation in normal and tumor tissue. *Cancer Res.* 48:816-821.

Gailit, J. and Ruoslahti, E., 1988, Regulation of the fibronectin receptor affinity by divalent cations. *J. Biol. Chem.* 263:12927-12932.

Hemler, M.E., Crouse, C., and Sonnenberg, A., 1989, Association of the VLA α_6 subunit with a novel protein. A possible alternative to the common VLA β_1 subunit on certain cell lines. *J. Biol. Chem.* 264:6529-6535.

Holzmann, B., McIntyre, B.W., and Weissmann, I.L., 1989, Identification of a murine Peyer's patch-specific lymphocyte homing receptor as an integrin molecule with an α chain homologous to human VLA-4α. *Cell* 56:37-46.

Holzmann, B. and Weissman, I.L., 1989, Peyer's patch-specific lymphocyte homing receptors consist of a VLA-4-like α chain associated with either of two integrin β chains one of which is novel. *EMBO J.* 8:1735-1741.

Humphries, M.J., 1990, The molecular basis and specificity of integrin-ligand interactions. *J. Cell Sci.* 97:585-592.

Jones, J.C.R., Kurpakus, M.A., Cooper H.M., and Quaranta, V., 1991, A function for the integrin $\alpha_6\beta_4$ in the hemidesmosome. *Cell Regulation* 2:427-438.

Kennel, S.J., Foote, L.J., Falcioni, R., Sonnenberg, A., Stringer, C.D., Crouse, C., and Hemler, M.E., 1989, Analysis of the tumor-associated antigen TSP-180. Identity with $\alpha_6\beta_4$ in the integrin superfamily. *J. Biol. Chem.* 264:15515-15521.

Kimmel, K.A. and Carey. T.E., 1986, Altered expression in squamous carcinoma cells of an orientation restricted epithelial antigen detected by monoclonal antibody A9. *Cancer Res.* 46:3614-3623.

Kirchhofer, D., Gailit, J., Ruoslahti, E., Grzesiak, J., and Pierschbacher, M.D., 1990, Cation-dependent changes in the binding specificity of the platelet receptor GPIIb/IIIa. *J. Biol. Chem.* 265:18525-18530.

La Rivière, G., Schipper, C.A., Collard, J.G., and Roos, E., 1988, Invasiveness in hepatocyte and fibroblast monolayers and metastatic potential of T-cell hybridomas in mice. *Cancer Res.* 48:3405-3410.

La Rivière, G., Klein Gebbinck, J.W.T.M., Schipper, C.A., and Roos, E., 1992, Tumour necrosis factor-α stimulates invasiveness of T-cell hybridomas and cytotoxic T-cell clones by a pertussis toxin-insensitive mechanism. *Immunology* 75:269-274.

La Rocca, R.V., Stein, C.A., and Myers, C.E., 1990, Suramin: Prototype of a new generation of antitumor compounds. *Cancer Cells* 2:106-115.

Lee, E.C., Lotz, M.M., Steele, Jr., G.D., and Mercurio, A.M., 1992, The integrin $\alpha6\beta4$ is a laminin receptor. *J. Cell Biol.* 117:671-678.

Liotta, L.A., Mandler, R., Murano, G., Katz, D.A., Gordon, R.K., Chiang, P.K., and Schiffmann, E., 1986, Tumor cell autocrine motility factor. *Proc. Natl. Acad. Sci. USA* 83:3302-3306.

Lorant, D.E., Patel, K.D., McIntyre, T.M., McEver, R.P., Prescott, S.M., and Zimmerman, G.A., 1991, Coexpression of GMP-140 and PAF by endothelium stimulated by histamine or thrombin: A juxtacrine system for adhesion and activation of neutrophils. *J. Cell Biol.* 115:223-234.

Lotz, M.M., Korzelius, C.A., and Mercurio, A.M., 1990, Human colon carcinoma cells use multiple receptors to adhere to laminin: involvement of $\alpha_6\beta_4$ and $\alpha_2\beta_1$ integrins. *Cell Regulation* 1:249-257.

Mackay, C.R, Marston W.L., and Dudler, L., 1990, Naive and memory T cells show distinct pathways of lymphocyte recirculation. *J. Exp. Med.* 171:801-817.

Mackay, C.R., Marston, W.L., Dudler, L., Spertini, O., Tedder, T.F., and Hein, W.R., 1992, Tissue-specific migration pathways by phenotypically distinct subpopulations of memory T cells. *Eur J. Immunol.* 22:887-895.

Middelkoop, O.P., Roos, E., and Van de Pavert, I.V., 1982, Infiltration of lymphosarcoma cells into hepatocyte cultures: Inhibition by univalent antibodies against liver plasma membranes and lymphosarcoma cells. *J. Cell Sci.* 56:461-470.

Middelkoop, O.P., van Bavel, P., Calafat, J., and Roos, E., 1985, Hepatocyte surface molecule involved in the adhesion of TA3 mammary carcinoma cells to rat hepatocyte cultures. *Cancer Res.* 45:3825-3835.

Miyake, K., Underhill, C.B., Lesley J., and Kincade, P.W., 1990, Hyaluronate can function as a cell adhesion molecule and CD44 participates in hyaluronate recognition. *J. Exp. Med.* 172:69-75.

Nabi, I.R., Watanabe, H., and Raz, A., 1990, Identification of B16-F1 melanoma autocrine motility-like factor receptor. *Cancer Res.* 50:409-414.

Prescott, S.M., Zimmerman, G.A., and McIntyre, T.M., 1990, Platelet-activating factor. *J. Biol. Chem.* 265:17381-17384.

Quaranta, V. and Jones, J.C.R., 1991, The internal affairs of an integrin. *Trends Cell Biol.* 1:2-4.

Roos, E., Dingemans, K.P., Van de Pavert, I.V., and Van den Bergh-Weerman, M.A., 1977, Invasion of lymphosarcoma cells into the perfused mouse liver. *J. Natl. Cancer Inst.* 58:399-407.

Roos, E., Dingemans, K.P., Van de Pavert, I.V., and Van den Bergh-Weerman, M.A., 1978, Mammary carcinoma cells in mouse liver. Infiltration of liver tissue and interaction with Kupffer cells. *Brit. J. Cancer* 38:88-99.

Roos, E., Van de Pavert, I.V., and Middelkoop, O.P., 1981, Infiltration of tumour cells into cultures of isolated hepatocytes. *J. Cell Sci.* 47:385-397.

Roos, E. and Van de Pavert, I.V., 1983, Antigen-activated T-lymphocytes infiltrate hepatocyte cultures in a manner comparable to liver-colonizing lymphosarcoma cells. *Clin. Exp. Metastasis* 1:173-180.

Roos, E., La Rivière, G., Collard, J.G., Stukart, M.J., and De Baetselier, P., 1985, Invasiveness of T-cell hybridomas *in vitro* and their metastatic potential *in vivo*. *Cancer Res.* 45:6238-6243.

Roos, E. and Van de Pavert, I.V., 1987, Inhibition of lymphoma invasion and liver metastasis formation by pertussis toxin. *Cancer Res.* 47:5439-5444.

Roossien, F.F., Bikker, A., De Rijk, D., and Roos, E., 1989, Involvement of LFA-1 in lymphoma invasion and metastasis demonstrated with LFA-1-deficient mutants. *J. Cell Biol.* 108:1979-1985.

Roossien, F.F., De Kuiper, P.E., De Rijk, D., and Roos, E., 1990, Invasive and metastatic capacity of revertants of LFA-1-deficient mutant T-cell hybridomas. *Cancer Res.* 50:3509-3513.

Schneider, W.J., Beisiegel, U., Goldstein, J.L., and Brown, M.S., 1982, Purification of the low density lipoprotein receptor, an acidic glycoprotein of 164,000 molecular weight. *J. Biol. Chem.* 257:2664-2673.

Silletti, S., Watanabe, H., Hogan, V., Nabi, I.R., and Raz, A., 1991, Purification of B16-F1 melanoma autocrine motility factor and its receptor. *Cancer Res.* 51:3507-3511.

Simon, M.I., Strathmann, M.P., and Gautam, N., 1991, Diversity of G proteins in signal transduction. *Science* 252:802-808.

Sonnenberg, A., Modderman, P.W., and Hogervorst, F., 1988, Laminin receptor on platelets is the integrin VLA-6. *Nature* 336:487-489.

Sonnenberg, A., Linders, C.J.T., Modderman, P.W., Damsky, C.H., Aumailley, M., and Timpl, R., 1990, Integrin recognition of different cell-binding fragments of laminin (P1, E8) and evidence that α6β1 but not α6β4 functions as a major receptor for fragment E8. *J. Cell Biol.* 110:2145-2156.

Sonnenberg, A., Calafat, J., Janssen, H., Daams, H., Van der Raaij-Helmer, L.H.M., Falcioni, R., Kennel, S.J., Aplin, J.D., Baker, J., Loizidou, M., and Garrod, D., 1991, Integrin α6β4 complex is located in hemidesmosomes, suggesting a major role in epidermal cell-basement membrane adhesion. *J. Cell Biol.* 113:907-917.

Spangrude, G.J., Braaten, B.A., and Daynes, R.A., 1984, Molecular mechanisms of lymphocyte extravasation. *J. Immunol.* 132:354-362.

Spangrude, G.J., Araneo, B.A., and Daynes, R.A., 1985, Site-selective homing of antigen-primed lymphocyte populations can play a crucial role in the efferent limb of cell-mediated immune responses *in vivo*. *J. Immunol.* 134:2900-2907.

Stracke, M.L., Guirguis, R., Liotta, L.A., and Schiffmann, E., 1987, Pertussis toxin inhibits stimulated motility independently of the adenylate cyclase pathway in human melanoma cells. *Biochem. Biophys. Res. Commun.* 146:339-345.

Sugimoto, M., Nakanishi, Y., Otokawa, M., Uchida, N., Yasuda, T., Sato, H., and Sato, Y., 1983, Effect of *Bordetella pertussis* leukocytosis (lymphocytosis)-promoting factor (LPF) on the physical lymphoepithelial-cell association studied with the use of an *in vitro* model of mouse thymus. *J. Immunol.* 130:2767-2774.

Van Corven, E.J., van Rijswijk, A., Jalink, K., van der Bend, R.L., Van Blitterswijk, W.J., and Moolenaar, W.H., 1992, Mitogenic action of lysophosphatidic acid and phosphatidic acid on fibroblasts. *Biochem. J.* 281:163-169.

Van Corven, E.J., Groenink, A., Jalink, K., Eichholtz, T., and Moolenaar, W.H., 1989, Lysophospatidate-induced cell proliferation. Identification and dissection of signaling pathways mediated by G proteins. *Cell* 59:45-54.

Van Waes, C., Kozarsky, K.F., Warren, A.B., Kidd, L., Paugh, D., Liebert, M., and Carey, T.E., 1991, The A9 antigen associated with aggressive human squamous carcinoma is structurally and functionally similar to the newly defined integrin $\alpha_6\beta_4$. *Cancer Res.* 51:2395-2402.

Verschueren, H., Dekegel, D., and De Baetselier, P., 1987, Development of a monolayer invasion assay for the discrimination and isolation of metastatic lymphoma cells. *Invasion Metastasis* 7:1-15.

Watanabe, H., Carmi, P., Hogan, V., Raz, T., Silletti, S., Nabi, I.R., and Raz, A., 1991, Purification of human tumor cell autocrine motility factor and molecular cloning of its receptor. *J. Biol. Chem.* 266:13442-13448.

DISCUSSION

G. TARONE

I have missed something in the last part of your talk. What is the relationship between the antigen recognized by the monoclonal antibody OPAR and the invasiveness of the TA/3 mammary carcinoma cells? If I understood correctly, this monoclonal is recognizing a molecule on liver cells. And what is the activity of the antibody?

E. ROOS

It inhibits, like the anti-$\alpha_6\beta_4$ antibody, but more efficiently, the adhesion of the mammary carcinoma cells to the hepatocytes.

G. TARONE

OK, I have a second question. You showed that the monoclonal antibody OPAR gives a very intense staining of liver cells in immuno-electronmicroscopy, suggesting that this is a very abundant antigen. However, after affinity chromatography purification the antigen corresponds to a minor cellular component. Can you comment on this apparent discrepancy?

E. ROOS

That is one of the reasons why it took us some time to purify it. We thought that because it is such an abundant protein, we would not have to use a lot of rat liver. But that does not turn out to be the case. And really, what you see in immuno-EM depends on your antibody and a lot of other factors.

G. TARONE

The staining was done by direct or indirect immunoelectron microscopy?

E. ROOS

With some antibodies against very abundant proteins it is very hard to stain, and the other way around. So what we thought for years, that this is a very abundant plasma membrane protein is probably not true.

M. HEMLER

A comment about that 145 kD band, with the other band underneath. Are you sure that is not an integrin?

E. ROOS

I am not sure that is not an integrin, no. These are rat hepatocytes so you need anti-rat integrin antibodies, which I do not have. I do not know how many of them are around.

M. HEMLER

Well, I mean, if you ran that using reduced and non-reduced conditions, does it behave as a typical integrin?

E. ROOS

No, we have not done that. But, as I have said the ratios of the two bands can differ a lot. And now that we do this extensive purification, we end up with almost only the 115 kD, so the 145 kD band is lost. So I suspect it is a proteolytic fragment, not a different protein.

M. HEMLER

Well that happens also with α_4.

E. ROOS

Yes, but then there is no β_1.

M. HEMLER

The main question I want to ask is about this complicated hypothetical process whereby LFA-1 might get activated. I am not sure that you ever showed us that it was inactive on these CTL hybrids. Have you looked, for example, at L16 epitope and have you looked to show that LFA-1 has diminished capability to bind to ICAM or to undergo aggregation or some other LFA function?

E. ROOS

Well these are mouse cells, so L16 cannot be checked for, because that is a human epitope. At the moment we are purifying enough soluble mouse ICAM-1 from a cell line given to us by Adrian Brian, and do all the experiments on LFA-1 in the LFA-1-ICAM assay. For instance, for the transfectants this should show whether the LFA-1 is really functional on these cells.

P. HERRLICH

Does this very nice trick of cell fusion permit you to get some growth preference in response to the lymphocytes that you choose, for instance, fusions with gut lymphocytes as compared to those with skin lymphocytes?

E. ROOS

Well the only thing we have done is use these activated T cells, so I cannot comment on other kinds of hybrids. Patrick De Baetselier from Brussels, with whom we collaborated in the beginning of these studies, has also done fusions with macrophages, for instance, and got cell lines that went to the lungs. But I do not know whether I have really answered your question there.

P. HERRLICH

Whether you can confer homing specificities. Would fusion products with gut lymphocytes return with preference to the gut, and so on?

E. ROOS

I would suspect so, but we have not done that. What we think we see here is perhaps a memory T cell dissemination phenotype, and if you use more specific subsets of cells you might be able to see more specific patterns.

D. LIVINGSTON

What happens if you transfect, in parallel, several revertant clones with a mutant of LFA-1, and ask whether the same distribution of modest versus no invasiveness was present?

E. ROOS

No, we have not done that. So, I do not know what this modest invasion is. It is still kind of difficult because, for instance, in the human/mouse hybrids we find an all or no difference -- if human chromosome 7 is out, then it is not invasive, if it is in, it is invasive. That is, if it is a human chromosome 7 derived from an activated T cell, because we have put in an X-7 chromosome derived from a fibroblast and that does not work. It should be from activated T cells.

D. LIVINGSTON

Should a mutant which is dead in the ligand-binding assay fail to give you such an effect?

E. ROOS

Interesting, but we have not done that.

V. QUARANTA

Some months ago there was a report in the *Journal of Experimental Medicine* by Louis Lanier and his group. They made a monoclonal antibody to colon carcinoma cells, and this monoclonal blocks killing by natural killer cells. It turns out that monoclonal reacts with a carbohydrate determinant, which is found in these cells on 80 kDalton molecule, but it is also expressed on integrin β_4. In fact, they initially thought that it was a molecule associated with β_4. So do you know whether that antiserum that you made and purified with β_4 could be directed to such a carbohydrate determinant? There are lots of receptors for carbohydrates on the hepatocytes, so that may explain the inhibition results.

E. ROOS

Well, it is definitely possible. I do not know. I showed you that the antiserum it recognizes $\alpha_6\beta_4$ but we do not know what the epitope is.

A. SONNENBERG

Have you tested any other $\alpha_6\beta_4$-positive cells, whether they can invade the hepatocyte layers?

E. ROOS

Not yet extensively, the only invade cell we have tested so far is H29 human colon carcinoma. The obvious next step is to make monoclonal antibodies against human cells before we do this. You can imagine that using these affinity chromatography approaches, you get very, very small amounts of antibody, too little to check for a lot of carcinoma cell lines, whether the adhesion to hepatocytes is blocked by anti-β_4 antibody. So, if you ask, did you pick out a lot of carcinoma cells from the freezer and throw them on the hepatocytes and see whether they adhere? We have not tested that.

A. SONNENBERG

No, I am just interested whether when you use other cell lines, which are $\alpha_6\beta_4$-positive, the cells of all cell lines can invade the hepatocytes or whether there are exceptions.

E. ROOS

Well, that would be interesting, but I tend to shy away from correlation studies, because after a lot of work you do not get results that are conclusive. So I would rather go for a good monoclonal antibody, and then check a lot of cell lines. For other cell lines, for instance lymphoma, we know that different cell lines use other mechanisms. So, I would not be surprised to find carcinoma cells that adhere to hepatocyte layers and do not use β_4 but another molecule.

A. SONNENBERG

But of course now you can test whether cells use the $\alpha_6\beta_4$ integrin to adhere to the hepatocyte layers, because in those cases adhesion should also be blocked by antibodies against the ligand for $\alpha_6\beta_4$, recognized by your OPAR antibodies.

E. ROOS

Maybe. So I first want to know whether this is true. We will have to see whether the TA3 cells adhere to purified OPAR antigen, for instance, and whether we can block that with the $\alpha_6\beta_4$ antibodies.

S. GOODMAN

You said that your invasive carcinoma cells grew as an ascites and had no attachment to matrix. How widely did you screen? Would you just like to go through the list of what matrix molecules you checked over for the record?

E. ROOS

Well, there are just four: fibronectin, laminin, collagen type-1, and vitronectin. So there are a few other candidates, right?

R. BANKERT

I know that at the University of Utah, Ray Daines has suggested that pertussis toxin may have an effect on the cytoskeleton, particularly, spectrin and talin, and I guess you would just want to include that in the possible interpretation of what may be playing a role in your invasiveness.

E. ROOS

Yes, that is quite possible, but I do not see how manganese would revert that. But I must add that the reversion by manganese is partial, and it is clear that, for instance the effect of platelet-activating factor and IL8 on neutrophils, not only includes activation of Mac-1, but also includes polarization, extension of protrusions, etc., which are clearly also involved here. So I am more surprised to find any reversion at all than that I do not see complete reversion. The pertussis toxin affected signal has a lot of effects on the cell.

G. NICOLSON

But you used a manganese control?

E. ROOS

Only in one of the cell lines. In the other cell lines the manganese control was the same as the control, and we had a 2-3 fold enhancement of invasion by the toxin-treated cells, which in absolute numbers was a very, very large increase because that was a very highly invasive cell line.

G. TARONE

You have mentioned a cell line that does not attach to any matrix protein. Did you check whether it expressed β_1 integrins? It would be very interesting to have a mutant that does not express β_1 integrins for many different reasons, so, this cell line may be a good candidate.

E. ROOS

These are mouse cell lines and up to recently there were very few reagents. Now we have obtained a cytoplasmic domain polyclonal antiserum that can be used for that so we are doing the immunoprecipitations now with all these cell lines. So I do not know.

G. DOUGHERTY

Maybe I missed this as well. But in the LFA-1 lymphoma binding to the fibroblasts and hepatocytes, what ligand are you assuming that it is recognizing? I presume that they are ICAM-1 negative.

E. ROOS

Yes. We have worked on that and we are working on that also *in vivo*. It is a point that most of the adhesive interactions that are discussed in these kinds of meetings occur in inflamed tissues. However, tumor cells that are injected into a mouse would interact with normal non-inflamed tissues. So we have thought for a long time that it should be ICAM-2, because ICAM-1 was supposed not to be present. And then data from Tim Springer's laboratory showed, interestingly enough, that there is constitutive expression of ICAM-1 in the liver and in the kidney, which are two tissues very heavily invaded by these cell lines. So we are now testing the effects of anti LFA-1 and anti-ICAM-1 antibodies, *in vivo*, to see whether you see differences. About the *in vitro* systems -- we know that on the hepatocytes there is ICAM-1, so it can be involved, and it seems like invasion can be blocked, but we have not extensively done this yet by anti-rat ICAM-1 antibody. In the case of the fibroblasts that is not clear. If you enhance ICAM-1 expression, which is low on the fibroblasts, and you only get a 2-fold increase by TNF etc., you do not see any effect on invasion. So either it is not really the limiting factor, or ICAM-1 is not really very important, and it is adhesion to ICAM-2. I do have to mention that there is an enormous amount of LFA-1 on the surface of the hybrids, so it is not clear whether you really need a high affinity ligand. But we have no data at the moment, because there is no anti-mouse ICAM-2 antibody.

CD44 AND SPLICE VARIANTS OF CD44 IN NORMAL DIFFERENTIATION AND TUMOR PROGRESSION

Peter Herrlich[1], Wolfgang Rudy[1], Martin Hofmann[1], Robert Arch[2], Margot Zöller[2], Volker Zawadzki[1], Cornelia Tölg[1], Armin Hekele[1], Gerrit Koopman[3], Steven Pals[3], Karl-Heinz Heider[1], Jonathan Sleeman[1], and Helmut Ponta[1]

1 Kernforschungszentrum Karlsruhe, Institut für Genetik und Toxikologie
   P.O. Box 3640, D-7500 Karlsruhe 1, GERMANY

2 Deutsches Krebsforschungszentrum Heidelberg, Institut für Radiologie und
   Pathophysiologie, Im Neuenheimer Feld 280, D-6900 Heidelberg 1
   GERMANY

3 Academic Medical Center, Department of Pathology, Meibergdreef 9
   NL-1105 AZ Amsterdam, THE NETHERLANDS

ABBREVIATIONS USED

CD44 = Cluster Determinant-44; **PCR** = Polymerase Chain Reaction; **AP-1** = Activating Protein 1 (the prototype is a heterodimer of *fos* and *jun*); **CMV** = Cytomegalovirus; **CTL** = Cytotoxic T-lymphocytes; **TNP** = Trinitrophenol; **TPA** = 12-0-Tetradecanoylphorbol-13-acetate; **DCC** = "Defective in Colon Carcinoma" (surface glycoprotein described by Fearon et al., 1990); **mAb** = Monoclonal Antibody

INTRODUCTION

CD44, originally defined by antibodies as a leukocyte surface protein (reviewed in Haynes et al., 1989), has become a polymorphic family of proteins expressed in various cells and conditions. The polymorphism is due to differential modifications and to splice variation (Hughes et al., 1983; Omary et al., 1988; Kansas et al., 1989; Picker et al., 1989; Goldstein et al., 1989; Goldstein and Butcher, 1990; Stamenkovic et al., 1991, Dougherty et al., 1991; Brown et al., 1991; Günthert et al., 1991; Shtivelman and Bishop, 1991; Jackson et al., 1992). Because of the polymorphism a multitude of functions can be expected. Yet these need to be defined. All or part of the CD44 glycoproteins have affinity for hyaluronic acid, some forms are linked to chondroitin sulfate and bind fibronectin and collagen (Aruffo et al., 1990; Miyake et al., 1990; Stamenkovic et al., 1989; Goldstein et al., 1989; Wolffe et al., 1990; Carter and Wayner, 1988; Jalkanen and Jalkanen, 1992). These affinities may have functions in association with different primary structure domains introduced by splice variation.

The smallest form of membrane-anchored CD44 (CD44s) in the rat is 342 amino acids long. 71 amino acids in the C terminus reach into the cytoplasm. A single transmembrane domain connects this cytoplasmic domain to the large extracellular N terminal part. CD44s is expressed in many cells, particularly mesenchymal cells, throughout the body of mouse, man and rat, e.g. hemopoietic cells carry 10^4 to 10^5 molecules on their surface. Also dermal fibroblasts are strongly positive. Certainly CD44s is the most abundant isoform expressed in normal tissues. CD44s is also found on many cancer cells. The larger splice variants are by far more restricted in their expression. Some are even extremely rare in the adult organism but do occur at selected sites in the embryo.

Our own entry into the field of CD44 originates from the generation of a monoclonal antibody that later turned out to recognize an epitope encoded by a novel exon of CD44 (within the domain designated v_6 in Figure 1). The antibody was generated by injection into mice of membrane proteins of the metastasizing pancreatic rat tumor BSp73ASML and by screening for antibodies that would stain the metastatic tumor cells and not the presumably isogenic variant cell line BSp73AS that does not metastasize (Matzku et al., 1983; Matzku et al., 1989). One of the antibodies was particularly clear in this distinction. Using this antibody, we cloned the epitope encoding cDNA (pMeta-1) from a bacterial expression library and found, by sequencing, homology to CD44 that had also just been cloned (Stamenkovic et al., 1989; Goldstein et al., 1989; Idzerda et al., 1989; Nottenburg et al., 1989; Zhou et al., 1989; Günthert et al., 1991). The "metastasis-specific" epitope was located in an "extra" region "inserted" into a juxta-membrane position of the extracellular portion of CD44 (Günthert et al., 1991; see Figure 1). The interest in this variant has been tremendously stimulated by the observation that the variant confers metastatic behavior. We will summarize here what is presently known about the gene, its expression and function.

SPLICE VARIATION AND GENOMIC MAP

The "extra" sequences are the result of alternative splicing. The cloned splice variant of CD44 represented by cDNA clone pMeta-1, carries 162 amino acids of extra sequences. By Northern and PCR analyses it has become apparent that the cloned variant is just one of many variants expressed in the metastatic tumor cells. If primer oligonucleotides were chosen from either side of the "insertion" site in CD44s, a whole range of splice variants could be identified (Figure 2). BSp73AS cells express only CD44s. The indicative PCR product is 420 bp as calculated from the region between the oligos used. BSp73ASML cells produce CD44s plus many different larger variants, with PCR products between 630 and more than 859 bp in length.

The amplification product of 859 bp corresponds to pMeta-1. One additional splice variant, the smallest prominent one (pMeta-2), was also cloned from BSp73ASML cells. Other splice variants were cloned by PCR from other tumor cell lines and used to obtain a complete list of variant exons expressed. These were then compared with the genomic structure as derived from genomic clones (not shown). Mouse, rat or human cells can choose from a total of at least 10 variant exons (Figure 3).

One of the largest splice variants was found in the HPV-transformed keratinocyte cell line HPKII. Its protein core is 521 amino acids long and includes information from all exons between v3 and v10. A still larger form was isolated from the murine mammary carcinoma line GR.

The synthesis of such a diversity of splice variants in BSp73ASML suggests that this tumor cell had acquired a splice disorder in the CD44 gene. A polymorphism was indeed found in one of the alleles which could be compatible with a cis-acting splice mutation (Figure 4). Possibly residual CD44s is synthesized from the other allele.

THE CD44 PROMOTER AND THE REGULATION OF ALTERNATIVE SPLICING

From the rather ubiquitous expression of CD44s we would expect that the promoter would obey one of the standard signalling networks present in most or all cells. The rather rare expression of CD44 variants should be the result of strict splice regulation. In order to study promoter regulation, the promoters of the human and mouse genes were cloned and sequenced.

Computer comparison indicates a member of putative cis-acting elements that, in part, are differently spaced in the human and murine promoter (Figure 5). Both promoters

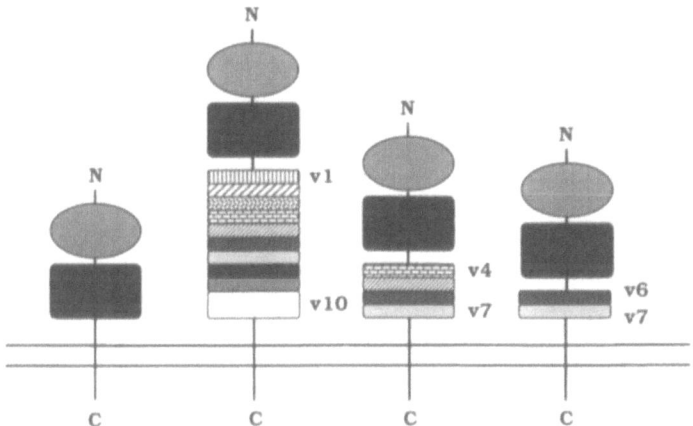

Figure 1. CD44 variants. The four examples shown here represent, from left to right, DC44s, the largest variant found in GR mammary tumor cells, pMeta-1 and pMeta-2. See description in text.

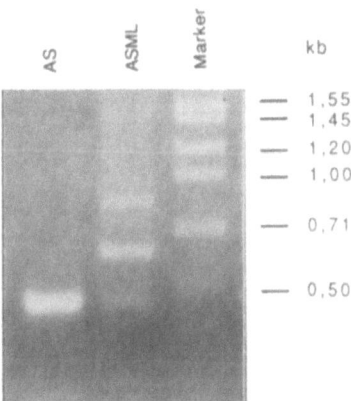

Figure 2. Reverse PCRs of RNAs from the Pancreatic Tumor Cell Lines BSp73AS and BSp73ASML. cDNA synthesis followed by PCR amplification was performed from poly A$^+$ RNA prepared from the non-metastasizing rat pancreas carcinoma line BSp73AS (AS) and the metastasizing isolate BSp73ASML (ASML) (Matzku et al., 1983). The two oligonucleotides comprise the sequence from positions 554 to 581 and 1383 to 1413, respectively. Positions as decribed by Günthert et al. (1991).

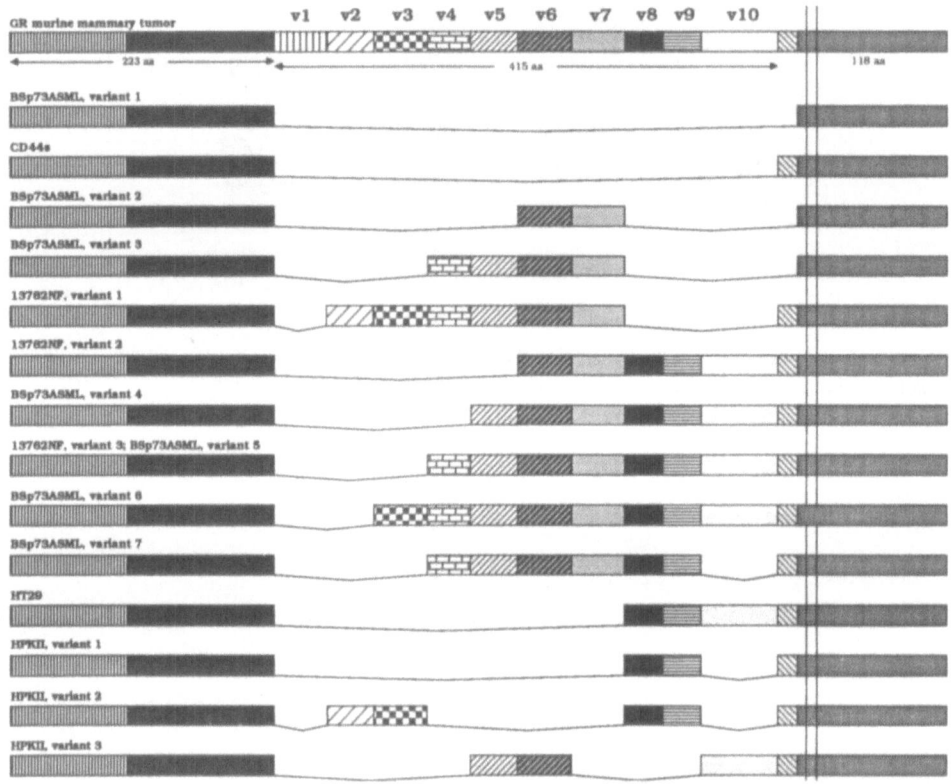

Figure 3. Splice Variants of CD44 Isolated to Date. The largest splice variant was isolated from a murine mammary tumor cell line (GR). It encodes sequences of all ten exons present in the genome from which splice variant sequences can be derived (v1 to v10). The sources for the identification of the other splice variants are indicated and have been explained in detail in Günthert et al. (1991) and Hofmann et al. (1991). In brief, BSp73ASML refers to the rat pancreas adenocarcinoma line described in Matzku et al. (1983), BSp73ASML variant 3 is identical with pMetal (Günthert et al., 1991), BSp73ASML variant 2 is identical with pMeta2 (*see text*). 13762NF refers to cell lines that have been established from a rat mammary carcinoma (Neri et al., 1982). HT29 is a colon carcinoma cell line that was kindly provided by Dr. Wolfgang Dippold and HPKII is a keratinocyte cell line, obtained from Drs. Matthias Durst and Lutz Gissmann.

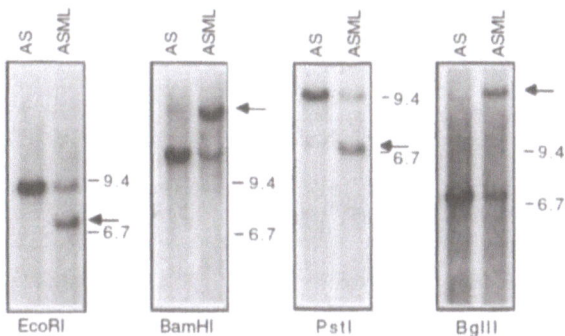

Figure 4. Genomic Rearrangement in the Epitope-encoding Region of BSp73ASML. 10 μg each of BSp73ASML DNA or BSp73AS DNA were subjected to digestions with either EcoRI, BamHI, PstI or BglIII as indicated, and, after gel electrophoresis and blotting, hybridized with a variant specific probe (position 941-1108 of pMeta-1. Günthert et al., 1991). This probe covers roughly exons v_5 and v_6. The arrows indicate the altered restriction fragments in BSp73ASML DNA.

carry a *bona fide* TATA box and an AP-1 site at almost identical position. The human promoter was linked to the CAT gene and transiently transfected into several types of cells: human fibroblasts, GM637; mouse teratocarcinoma, F9; and rat embryonal fibroblasts, CREF. This promoter fragment exerted activity comparable in magnitude to the collagenase promoter. Apparently basal expression was governed by similar abundant transcription factors. The similarity to the collagenase promoter suggests regulation by *jun*. *jun* expression indeed could enhance the promoter activity seven to fifteen fold. So did *ras* overexpression upon induction from a stably transfected chimeric construct that is driven by the mouse mammary tumor virus promoter. The promoter was also responsive to phorbol ester. Upon mutating the AP-1 site as indicated in Figure 5, the basal level and inducibility were severely reduced (not shown).

Thus, the CD44 promoter resembles those of housekeeping genes active in proliferating cells. This promoter property matches the observation that CD44s is expressed in many tumor cells. *ras* activation is an early event in many "naturally" occuring tumors and may turn on CD44 expression.

The expression of v_6 exon splice variants does not seem to be turned on "automatically" with increased promoter activity. Rather specific splice regulation permits recruitment of variant exons. We have come across one cell system, CREF, in which *ras* transformation led to some low level appearance of the v_6 epitope. In these cells hormone dependent overexpression of *ras* causes a moderate increase in the ratio of CD44v/CD44s suggesting that *ras*, in the genetic background of CREF cells, can possibly influence the splicing of CD44.

CD44 VARIANT EXPRESSION CONFERS METASTATIC BEHAVIOR

In view of the presumably very complex nature of the metastatic process some of us considered it unlikely that CD44v would be more than one of many properties required to metastasize. Nevertheless we tested whether transfection would produce a phenotype. Against expectation, by expressing defined CD44 variants, we have been able to convert several non-metastatic cell lines to tumor cells that potently spread and subsequently colonize the lung and other tissues. All our metastasis assays involve injection of tumor cells into the footpad or subcutaneously into the flank and testing for lymph node and lung metastases. BSp73ASML in isogenic rats spread lymphogenically. Two variants have been transfected as constructs under the control of the SV40 or CMV promoters. They carry the variant exons v_4 through v_7 (pMeta-1) or v_6,v_7 (pMeta-2). Both confer similar metastatic properties. CD44s was totally negative for metastasis, while pMeta-1 transfected into BSp73AS cells converted these cells to highly metastatic clones, metastasizing in 60/60 animals injected subcutaneously with $2x10^5$ cells. pMeta-2 was similarly effective (11/12 animals). These variants suffice for metastasis formation and no additional step seems to be

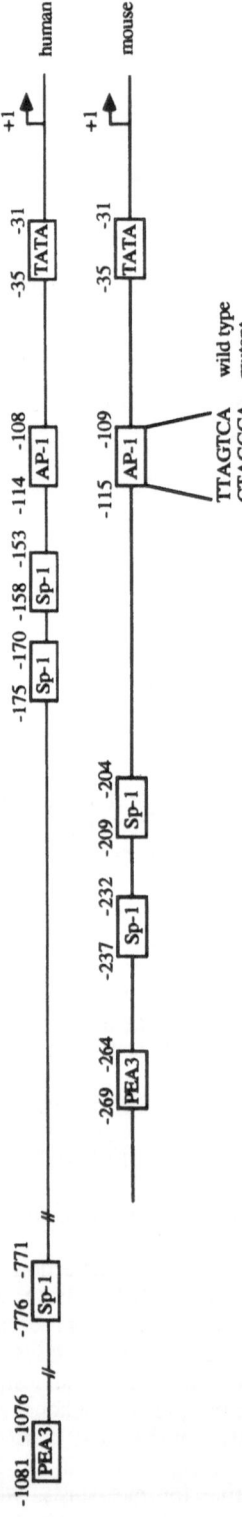

Figure 5. AP-1-dependent Regulation of the CD44 Promoter. Schematic representation of the human and murine CD44 promoters. The sites indicated have not been proven to be functional except for the AP-1 site. Its response to *jun*, *ras*, and TPA is obliterated by point mutation.

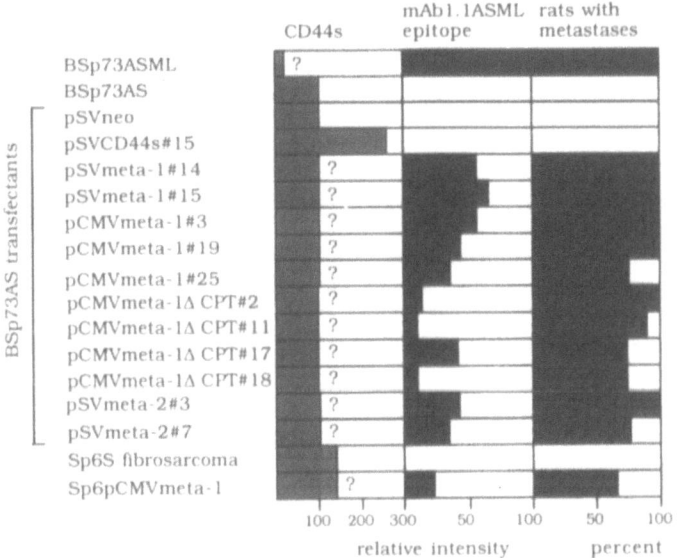

Figure 6. CD44 Variants Confer Metastatic Behavior. Tumor cells and transfectants were examined for CD44s and CD44v (1.1 ASML epitope) expression. CD44s was measured using the mAb Ox49 which stains both CD44s and CD44v. Therefore in the CD44v transfectants, it was assumed that the CD44s expression was identical to the recipient cell lines (*indicated by question marks*). In BSp73ASML, CD44s levels were estimated from PCR data (*see Figure 2*). The metastatic potential of cells was determined in isogenic rats by the spontaneous metastasis assay (Günthert et al., 1991). Sp6S is a spontaneous fibrosarcoma. SV and CMV indicate viral promoters used for cDNA expression. ΔCPT = truncation of the cytoplasmic tail. The numbers refer to individual transfected cell clones.

required since lung colonies reisolated and then reinjected were homogeneously positive for v_6 exon sequences and metastasized with identical rapid kinetics (not faster) and efficiency as compared with the original transfectant.

Two other types of non-metastatic tumor cells could also be converted to metastatic clones (one is shown in Figure 6). This suggests that either the tumor cells used so far carry all properties required in addition to CD44v except for CD44v, or CD44v acts as a pleiotropic gene influencing several properties relevant to metastatic behavior. Interestingly, however, the cytoplasmic tail of CD44v seems not to be required (Figure 6).

From the comparison of v_6 epitope expression and metastatic properties of individual cell clones, no dose dependence is apparent, as if the levels were above threshold and metastatic efficiencies were perhaps influenced only by other parameters. Interestingly, however, the *ras*-transformed CREF clones mentioned above express such low CD44v levels that the efficiency of spontaneous metastasis formation was only 40% (Hofmann et al., in preparation).

In the transfectants, CD44s and CD44v are co-expressed at comparable levels. The metastasis function of CD44v is obviously dominant (CD44v could thus be called a dominant metastogene).

The fulminant effect on metastatic properties raises the point whether the metastasis assay used mimics at all the natural course of tumor progression. Strictly speaking it does of course not. Monoclonal tumor cells are injected while in natural tumor progression presumably only a small minority gains metastatic properties. Further, most cells injected never manage to organize in a vascularized manner at the site of injection. Also subcutaneous tissue is an infrequent site of tumor formation. Still we feel that this is currently the only assay available with such dissecting ability.

The successful transfer of metastatic functions with CD44v encouraged us to screen human cancer cells for expression of corresponding CD44 splice variants. The rat specific monoclonal antibody did not unfortunately recognize the human variant. By cross-hybridization we could, however, find human tumor cell lines that express exon v_6 (Hofmann et al., 1991). The variant sequences were cloned and used, as bacterial fusion peptides, to raise antibodies. With extensively purified v_6-specific antibodies, various human tumor samples were screened (Table 1).

The preliminary message from these data is that normal tissue does not express CD44v (not shown) and that many carcinoma cells carry CD44v on their surface at the site of the primary tumor. Almost all metastasis samples were strongly positive for v_6. Those few that did not stain were derived from an equally negative primary tumor (not shown).

In some examples of human cancer, stages of tumor progression are well characterized on the basis of morphological criteria. One is colon cancer. We have examined CD44v expression at different stages in human colon cancer. Figure 7 shows the result. CD44v, defined by a v_6 exon specific antibody, is expressed. Expression starts in the polyp stage. Particularly dysplastic polyps are frequently positive. 100% positive tumors are reached in carcinomas and in metastases.

Table 1. Expression of CD44s and CD44v in Human Tumors

Carcinoma	Origin of Sample	Number of Immune-positive / Total number	
		CD44s	CD44v
Breast	primary tumor	30/41	30/41
	metastasis	11/16	11/16
Stomach	primary tumor	20/21	19/21
	metastasis	17/17	17/17
Colon	primary tumor	17/17	17/17
	metastasis	8/8	8/8
Ovarian	primary tumor	0/3	0/3
	metastasis	0/3	0/3
Cervix	primary tumor	5/5	5/5
	metastasis	5/5	5/5
Bladder	primary tumor	5/5	5/5
	metastasis	n.d.	n.d.

Cryosection of surgically removed tumor tissues were stained with either antibodies recognizing the CD44s sequences or with polyclonal antibodies reacting with the v_6 region of CD44v. Since the variant molecules carry CD44s sequences, reactivity for CD44s epitopes does not indicate CD44s expression. However, data from PolyA$^+$ RNA samples examined by PCR suggest overexpression of both CD44s and CD44v in tumors. N.D. = not done.

The following features are particularly interesting: a variant of CD44 is expressed in human cancer. Its onset may be early, at a still benign state of the development of cancer. The variant may therefore be unrelated to the metastatic phenotype in humans, or the metastatic phenotype is generated by several genetic events all of which need to be assembled in order to confer to cells full metastatic potential. The complement of events may be more complete in the rat tumor cells that served as recipients for the transfections. We note that rat2 cells transfected with pMeta-1 become only metastatic upon *ras*

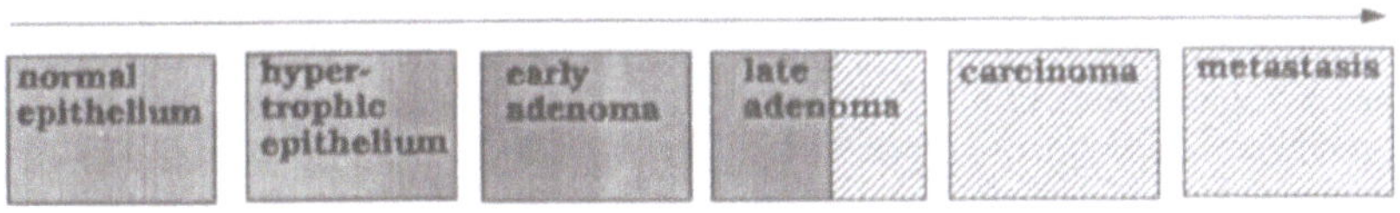

	normal epithelium	hyper-trophic epithelium	early adenoma	late adenoma	carcinoma	metastasis
CD44 (v6,v7)				82	100	100 %
Ki-ras			7	61	48	
DCC					78	
p53			3	30	75	

Figure 7. Colorectal Carcinogenesis. CD44v surface expression is negative in normal colon epithelium. In the polyp stage, cells become positive. Expression was observed particularly with dysplastic cell morphology and appeared to be clonal. Frequencies for *ras* mutation, deficiency of the DCC gene and mutation of p53 were taken from Vogelstein et al. (1988) and Fearon and Vogelstein (1990).

transformation (not shown). By inference, the colon carcinoma stage must be reached before a metastasis gene (if it is a metastasis gene in humans) can work. The clonal expression suggests, however, that the CD44 variant confers a growth advantage. Whether this is a separate property or part of the function detected in the transfection experiments for metastatic behavior, cannot be decided yet. It is further interesting that expression is maintained in cells that had already formed a distant colony. Only a small minority of all metastases examined were epitope-negative. Of course, these cases may have been positive transiently or may have never expressed a CD44v.

MOLECULAR PROPERTIES OF CD44 VARIANTS

For the remainder of this overview we will discuss experiments that may advance our understanding with respect to the mode of action of CD44v.

Obviously CD44 and its variants carry multiple domains. These could modulate each other such that the presence of one domain and possibly its interaction with intra- or intermolecular ligands could influence other domains of CD44. In this frame, the observation would apply that the so-called epithelial variant (found by Stamenkovic et al., 1991; carrying v8 through v10) and our pMeta-1 appear not to bind hyaluronate while CD44s does. Perhaps an "insertion" alters a function of the N-terminus (hyaluronate binding is a function of the N-terminal domain).

New domains could add new surfaces that interact with additional partners. Unfortunately no new ligand has yet been found. Interestingly, the v_6/v_7 region is not or barely modified while the extracellular portions shared by CD44s and CD44v, are heavily N- and O-glycosylated.

A puzzling series of results is derived from experiments that address the function of cytoplasmic tail of CD44v. The tail exists in a phosphorylated form and appears to participate in covalent homodimerization. This result has been derived from gel resolutions under reducing and non-reducing conditions. As seen in Figure 8 only a minority of CD44v molecules run in a "dimer position" in the non-reducing gel. The dimer is not visible if the gel was run with dithiotreitol. The dimer position has been cut out and rerun under reducing conditions. All material then runs at the position of the monomer (not shown). No other molecular species appears to be linked, neither CD44s nor a foreign protein. A deletion mutant of the cytoplasmic tail does not dimerize (not shown).

The puzzling aspect is that the cytoplasmic tail can be removed without loss of metastatic properties (Figure 6). Several independent transfectants are fully metastatic. One could argue in several directions. Either the cytoplasmic tail is irrelevant for metastasis formation, or it is partly redundant in that signal transduction is still made possible by a bridging protein (not yet found) -- a so-called co-transducer. Alternatively, the cytoplasmic tail has a negative function. Its deletion could be equivalent to e.g. the phosphorylation, both relieving function from repression.

EFFECTIVE ANTIMETASTATIC THERAPY BY ANTIBODIES TO CD44v

An important lead into the putative site of action of CD44v was generated by attempts to rescue animals from metastatic disease. The most effective treatment so far was by the monoclonal antibody 1.1ASML (v_6 epitope). When given i.v. at high dose (200 µg/rat twice per week) the antibody could rescue from any lymphogenic spread 75% of isogenic rats injected with a fully metastatic dose of the pMeta-1 BSp73AS transfectant (Figure 9). The presence of antibody prevented colonization of the draining lymph nodes and any further metastasis formation. We interpret this to indicate:

1. very effective accessibility of the tumor cells to antibodies from the blood stream. The monoclonal tumor cell line may, because of its high migratory potential, behave like a "single cell suspension" and meet the antibody during passage to the lymphatic tissue.

2. prevention of a ligand interaction by the antibody *in vivo*. This is deduced mainly by the finding that the antibody does not kill tumor cells *in vitro* nor inhibit *in vitro* growth. If this interaction cannot occur, tumor cells are subject to passive cell death or to apoptosis (e.g. by lack of a growth signal).

3. a critical action of CD44v prior to outgrowth in the lymph node.

The monoclonal antibody can also inhibit lung colonization by the pMeta-1 transfectant upon injection into the tail vein (unpublished). This could mean that either the variant can also promote lung colony growth or that the antibody passively interferes with adhesion to the endothelial cells of the lung. Interestingly other laboratories have reported that CD44s expression promotes lung colonization (Sy et al., 1991). Our transfectants carry both CD44s and pMeta-1 in about equimolar ratio. It is not yet clear how the CD44v specific antibody inhibits lung colonization.

To obtain access to a broader range of therapeutic molecules, single-chain antibodies were produced carrying v_6 specificity. These were combined with a number of different effector functions. Binding studies revealed specificity and avidity of similar magnitude as mAb1.1ASML. Therapy experiments are in progress.

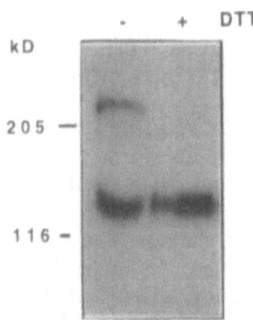

Figure 8.   Dimerization of Meta-1 Protein.   Western Blot Analysis of pMeta-1 Protein from Transfected BSp73AS Cells. The polyacrylmide gel was run in the absence of dithiotreitol (DTT). The sample was divided and parts were treated with and without dithiotreitol. The dimerization also has been made visible by metabolic labeling and subsequent antibody precipitation. Dimer size varied correspondingly with the CD44v expressed. Cytoplasmic tail deletions could not dimerize (data not shown).

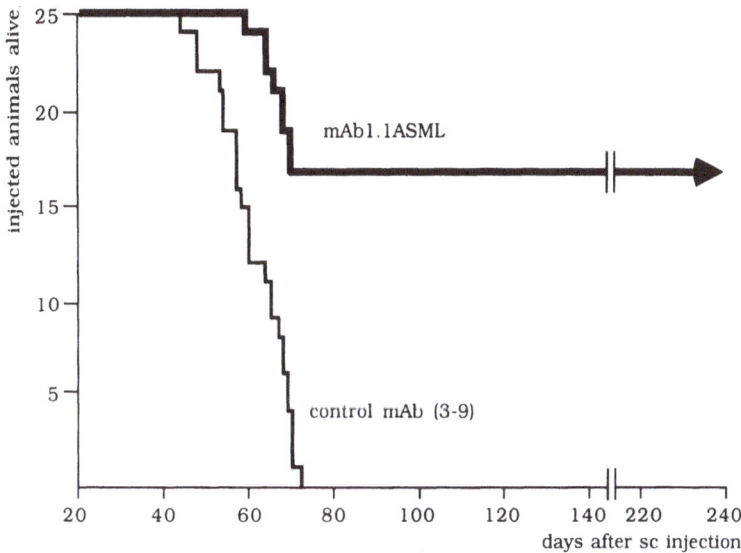

Figure 9. Rescue of Tumor-bearing Animals from Metastatic Disease by Anti-CD44v. BDX rats were injected subcutaneously (s.c.) with a lethal dose of a BSp73AS pMeta-1 transfectant. Twice per week each injected animal received 200 μg of purified antibody -- either an isotype-matched unrelated antibody, mAb (3-9), or the mAb 1.1ASML that recognizes the v_6 portion of CD44v. For details of the protocol, see Seiter et al. (1992).

PHYSIOLOGIC FUNCTIONS OF CD44 VARIANTS MAY GUIDE TOWARDS UNDERSTANDING THE MECHANISM OF METASTASIS FORMATION

Presumably, a gene with such splicing diversity must have telling physiologic functions. While CD44s is ubiquitous, v_6 exon sequences as measured by in situ hybridization or surface staining of a v_6 epitope, are rare. We found RNA carrying v_6 sequences in stages of the mouse embryo (together with Peter Gruss, Göttingen, Germany), e.g. fetal liver, the epithelium above Zahnanlagen, early pancreatic duct epithelium and bone marrow. In rats (and presumably in mouse as well) we found transient expression on cells from various lymphoid and hemopoietic organs around the time of birth (Arch et al., 1992). The neonatal expression suggested the hypothesis that the expression may be in spontaneously activated cells and that antigenic stimulation in the adult would also generate positive cells. This was tested. Indeed injection of allogeneic BDX lymphocytes into DA rats led to increased numbers of epitope-positive cells (Arch et al., 1992). Also the intensity of staining per cell was increased. Sorting revealed that small and large cells and cells of several lineages, T and B lymphocytes, and monocytes, transiently express CD44v for 3-5 days after the antigenic stimulus. The fact that small lymphocytes become positive, speaks for an early event in the activation process.

The expression can also be triggered *in vitro* by polyclonal stimulation of either rat or human peripheral lymphocytes. Onset of epitope expression in T cells is very early, clearly preceding DNA synthesis (Figure 10). By PCR we determined the major variants expressed in lymphocytes. *In vivo*-activated lymphocytes sorted for CD44v epitope expression, contained RNA for a variant that carries only v_6 exon sequences (Arch et al., 1992). Anti-CD3-stimulated T cells synthesized the same variant plus a second larger variant.

Is this expression related to a discernable physiologic function? To this end we took advantage of the conditions worked out to effectively inhibit metastasis formation by antibodies and asked whether antibodies to CD44v could interfere with the immune response *in vivo*. TNP-hapten-specific B cell responses and the allogeneic CTL response were tested. Monoclonal 1.1ASML antibody inhibited both responses when present i.v. over 4-7 days after the antigenic stimulation *in vivo* (Arch et al., 1992).

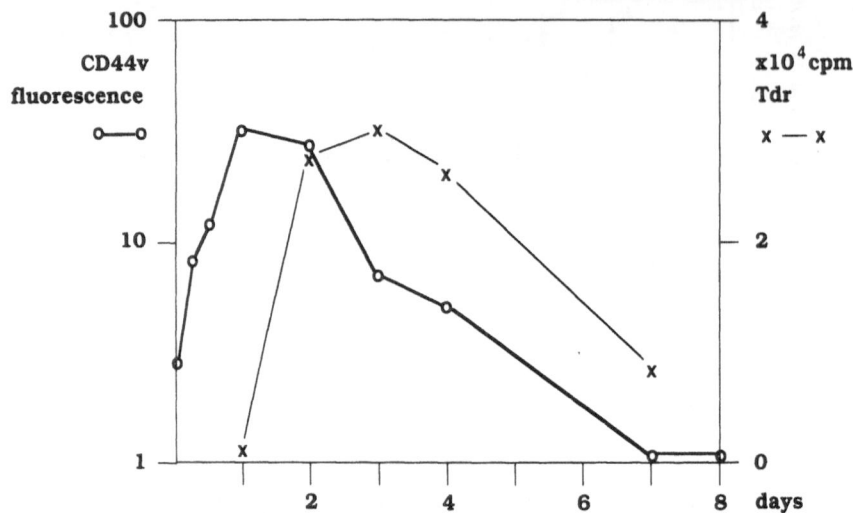

Figure 10. **Time Course of CD44v Upregulation in Polyclonally-stimulated Peripheral Human Lymphocytes.** The appearance of variant epitopes was determined after stimulation by anti-CD3 of density gradient purified peripheral blood T lymphocytes. CD44v expression precedes the onset of DNA synthesis as measured by ^{3}H Tdr incorporation.

Our working hypothesis is that immune cells during activation and metastatic cells need the v_6 exon sequences on their surface for a similar step. The presence of these may generate specific signals to the nucleus, promote migratory ability or mediate adhesion and growth in the lymphoid tissue. If we draw the parallel, CD44v seems to act in the same compartment during lymphocyte activation and tumor cell spread: activation of lymphocytes occurs in the periphery prior to or within the draining lymphoid tissue. The tumor cells are also generated in the periphery and need to pass through the lymphoid tissue as the first growth station. This hypothesis raises of course a number of questions whose investigation will keep our laboratories busy for some time.

ACKNOWLEDGEMENTS

This work was supported by a Boehringer Ingelheim Fonds fellowship to A.H., by a fellowship of the Deutsche Forschungsgemeinschaft to K.-H.H. and an EMBO postdoctoral fellowship to J.S. Further support was received through DFG grant He 551/7-1, and Boehringer Ingelheim KG.

REFERENCES

Arch, R., Wirth, K., Hofmann, M., Ponta, H., Matzku, S., Herrlich, P., and Zöller, M., 1992, Participation in normal immune responses of a splice variant of CD44 that encodes a metastasis-inducing domain. *Science* 257:682-685.

Aruffo, A., Stamenkovic, I., Melnick, M., Underhill, C.B., and Seed, B., 1990, CD44 is the principal cell surface receptor for hyaluronate. *Cell* 61:1303-1313.

Brown, T.A., Bouchard, T., St. John, T., Wayner, E., and Carter, W.G., 1991, Human keratinocytes express a new CD44 core protein (CD44E) as a heparan-sulfate intrinsic membrane proteoglycan with additional exons. *J. Cell Biol.* 113:207-221.

Carter, W.G. and Wayner, E.A., 1988, Characterization of the class III collagen receptor, a phosphorylated, transmembrane glycoprotein expressed in nucleated human cells. *J. Biol. Chem.* 263:4193-4201.

Dougherty, G.J., Lansdorp, P.M., Cooper, D.L., and Humphries, R.K., 1991, Molecular cloning of CD44R1 and CD44R2, two novel isoforms of the human CD44 lymphocyte"homing" receptor expressed by hemopoietic cells. *J. Exp. Med.* 174:1-5.

Fearon, E.R. and Vogelstein, B., 1990, A genetic model for colorectal tumorigenesis. *Cell* 61:759-767.

Fearon, E.R., Cho, K.R., Nigro, J.M., Kern, S.E., Simons, J.W., Ruppert, J.M., Hamilton, S.R., Preisinger, A.C., Thomas, G., Kinzler, K.W., and Vogelstein, B., 1990, Identification of a chromosome 18q gene that is altered in colorectal cancers. *Science* 247:49-56.

Goldstein, L.A., Zhou, D.F.H., Picker, L.J., Minty, C.N., Bargatze, R.F., Ding, J.F., and Butcher, E.C., 1989, A human lymphocyte homing receptor, the hermes antigen, is related to cartilage proteoglycan core and link proteins. *Cell* 56:1063-1072.

Goldstein, L.A. and Butcher, E.C., 1990, Identification of mRNA that encodes an alternative form of H-CAM(CD44) in lymphoid and nonlymphoid tissues. *Immunogenetics* 32:389-397.

Günthert, U., Hofmann, M., Rudy, W., Reber, S., Zöller, M., Haußmann, I., Matzku, S., Wenzel, A., Ponta, H., and Herrlich, P., 1991, A new variant of glycoprotein CD44 confers metastatic potential to rat carcinoma cells. *Cell* 65:13-24.

Haynes, B.F., Telen, M.J., Hale, L.P., and Denning, S.M., 1989) CD44 - a molecule involved in leukocyte adherence and T-cell activation. *Immunol. Today* 10:423-428.

Hofmann, M., Rudy, W., Zöller, M., Tölg, C., Ponta, H., Herrlich, P., and Günthert, U., 1991, CD44 splice variants confer metastatic behavior in rats: Homologous sequences are expressed in human tumor cell lines. *Cancer Res.* 51:5292-5297.

Hughes, E.N., Colombatti, A., and August, J.T., 1983, Murine cell surface glycoproteins. *J. Biol. Chem.* 258:1014-1021.

Idzerda, R.L., Carter, W.G., Nottenburg, C., Wayner, E.A., Gallatin, W.M., and St. John, T., 1989, Isolation and DNA sequence of a cDNA clone encoding a lymphocyte adhesion receptor for high endothelium. *Proc. Natl. Acad. Sci. USA* 86:4659-4663.

Jackson, D.G., Buckley, J., and Bell, J.I., 1992, Multiple variants of the human lymphocyte homing receptor CD44 generated by insertions at a single site in the extracellular domain. *J. Biol. Chem.* 267:4732-4739.

Jalkanen, S. and Jalkanen, M., 1992, Lymphocyte CD44 binds the COOH-terminal heparin-binding domain of fibronectin. *J. Cell Biol.* 116:817-825.

Kansas, G.S., Wood, G.S., and Dailey, M.O., 1989, A family of cell-surface glycoproteins defined by a putative anti-endothelial cell receptor antibody in man. *J. Immunol.* 142:3050-3057.

Matzku, S., Komitowski, D., Mildenberger, M., and Zöller, M., 1983, Characterization of Bsp 73, a spontaneous rat tumor and its *in vivo* selected variants showing different metastasizing capacities. *Invasion Metastasis* 3:109-123.

Matzku, S., Wenzel, A., Liu, S., and Zöller, M., 1989, Antigenic differences between metastatic and nonmetastatic BSp73 rat tumor variants characterized by monoclonal antibodies. *Cancer Res.* 49:1294-1299.

Miyake, K., Underhill, C.B., Lesley, J., and Kincade, P.W., 1990, Hyaluronate can function as a cell adhesion molecule and CD44 participates in hyaluronate recognition. *J. Exp. Med.* 172:69-75.

Neri, A., Welch, D., Kawaguchi, T. and Nicolson, G.L., 1982, Development and biologic properties of malignant cell sublines and clones of spontaneously metastasizing rat mammary adenocarcinoma. *J. Natl. Cancer Inst.* 68:507-517.

Nottenburg, C., Rees, G., and St. John, T., 1989, Isolation of mouse CD44 cDNA: structural features are distinct from the primate cDNA. *Proc. Natl. Acad. Sci. USA* 86:8521-8525.

Omary, M.B., Trowbridge, I.S., Letarte, M., Kagnoff, M.F., and Isacke, C.M., 1988, Structural heterogeneity of human Pgp-1 and its relationship with p85. *Immunogenetics* 27:460-464.

Picker, L.J., Nakache, M., and Butcher, E.C., 1989, Monoclonal antibodies to human lymphocyte homing receptors define a novel class of adhesion molecules in diverse cell types. *J. Cell Biol.* 109:927-937.

Seiter, S., Wirth, K., Hofmann, M., Ponta, H., Herrlich, P., Matzku, S., and Zöller, M., 1992, Prevention of tumor metastasis formation by anti-variant CD 44. *J. Exp. Med. (in press)*.

Shtivelman, E. and Bishop, M., 1991, Expression of CD44 is repressed in neuroblastoma cells. *Mol. Cell. Biol.* 11:5446-5453.

Stamenkovic, I., Amiot, M., Pesando, J.M., and Seed, B., 1989, A lymphocyte molecule implicated in lymph node homing is a member of the cartilage link protein family. *Cell* 56:1057-1062.

Stamenkovic, I., Aruffo, A., Amiot, M., and Seed, B., 1991, The hematopoietic and epithelial forms of CD44 are distinct polypeptides with different adhesion potentials for hyaluronate-bearing cells. *EMBO J.* 10:343-348.

Sy, M.S., Guo, Y.-J., and Stamenkovic, I., 1991, Distinct effects of two CD44 isoforms on tumor growth *in vivo*. *J. Exp. Med.* 174:859-866.

Vogelstein, B., Fearon, E.R., Hamilton, S.R., Kern, S.E., Preisinger, A.C., Leppert, M., Nakamura, Y., White, R., Smits, A.M.M., and Bos, J.L., 1988, Genetic alterations during colorectal tumor development. *New Engl. J. Med.* 319:525-532.

Wolffe, E.J., Gause, W.C., Pelfrey, C.M., Holland, S.M., Steinberg, A.D., and August, J.T., 1990, The cDNA sequence of mouse Pgp-1 and homology to human CD44 cell surface antigen and proteoglycan core/link proteins. *J. Biol. Chem.* 265:341-347.

Zhou, D.F.H., Ding, J.F., Picker, L.F., Bargatze, R.F., Butcher, E.C., and Goeddel, D.V., 1989, Molecular cloning and expression of Pgp-1 - The mouse homolog of the human H-CAM (Hermes) lymphocyte homing receptor. *J. Immunol.* 143:3390-3395.

DISCUSSION

G. NICOLSON

Have you ever tried to do a kinetic distribution study of cells after the anti-CD44 to see if implantation or clearance by an organ after i.v. injection is affected, or have you taken the transfected cells and examined their ability to be retained (organ retention) in a similar type of experiment?

P. HERRLICH

These transfected clones are extremely aggressive so we cannot really wait for any other organ colonization.

G. NICOLSON

Let me explain. If you take radiolabelled cells and you follow the cell distribution at various times, from minutes up to days, you can see if there is any difference in cell clearance rates. That can give you an indication of whether the cells are actually being retained in the microcirculation, or not. If there is no difference in clearance rate ± antibody then it is probably unlikely to be an implantation event. Thus it might be an event subsequent to tumor cell implantation.

P. HERRLICH

This experiment has not been done. It should be done perhaps. We have concentrated on this early step. The use of labelled cells was a complete failure, because most of the label disappeared indicating very rapid decay. Most of the cells never make it to the lymph nodes. So I think this type of experiment, at least for the lymphatic spread is not helpful. For the distribution in organs that may be a worthwhile experiment.

G. NICOLSON

The results also suggest that perhaps something in the microenvironment of the lymph node might be stimulating these cells, or modifying them in some way to change their properties. This could be growth properties, it could be invasive properties, or whatever. Have you tried to take lymphatic materials, say from a thoracic duct fistula or something similar, to see if the lymphatic molecules had any differential effects on the cells and whether transfection or addition of antibody could block those effects? The effects might be growth, invasion, or they might be a number of different tumor cell properties.

P. HERRLICH

For *in vitro* growth we could not find any difference +/- antibody, or +/- lymphocyte populations. So there was no difference, that is in serum. We have tried to find a ligand, some partner of the CD44 variant in lymphoid tissue, and so far we have not been successful. We speculate, however, that an interaction between the microenvironment and CD44 variants *in vivo* triggers several functions. In particular, we have not yet found any specific binding that is related to our epitope sequence.

D. LIVINGSTON

What happens, Peter, if you take v_6 and introduce it alone on a strong promoter into a variety of cells? Is it an oncogene in any assay?

P. HERRLICH

That has not been done yet, it is in the planning stage.

D. LIVINGSTON

It is a weird business. You have this tiny little exon that is expressed normally, right?

P. HERRLICH

We did not think of doing it as a single exon, rather hook it onto other surface molecules or secreted molecules.

D. LIVINGSTON

Does it have to be on the surface?

P. HERRLICH

I do not know.

D. LIVINGSTON

I mean it is a great epitope. It allows you to kill the cells, right? But the question is, where is the dirty work getting done?

P. HERRLICH

If the CD44 variant is not expressed on the surface, we do not get metastatic behaviour. This is an experiment that came from an unfortunate experience. We tried to express certain fusion constructs and could not get them expressed on the surface. They were not metastatic. But that is again in the context of a different molecule. That is not what you asked for.

D. LIVINGSTON

Or, for example, if you did something, as you would say, outrageous, like fusing v_6 to a class-1 molecule that was autocthonous for that particular animal, do you get a similar response?

P. HERRLICH

We hope to be able to tell you this experiment soon, using the EGF receptor.

D. LIVINGSTON

Yes, that is a dandy one.

N. HOGG

Do the sequences cleated by splicing look like individual protein domains? And, particularly, does the v_6 splice look like a domain?

P. HERRLICH

These exons look like individual domains. They are hydrophilic in nature. The variant sequences are not, or little, glycosylated, amazingly, while the whole rest of the molecule is full of glycosylations, that is why you saw such a tremendous diversity of bands. But this area in the middle, v_6/v_7, is not glycosylated.

D. LIVINGSTON

If you put one of those PCR products into a transcription translation system can you make something?

P. HERRLICH

It has not been done, thus I do not know, whether it has a translational start signal in front of v_6.

D. LIVINGSTON

Well, you would kind of hope it would, no?

P. HERRLICH

Not as a single exon. It is always expressed in context of the large molecule.

D. LIVINGSTON

And v_6 alone?

P. HERRLICH

Yes, it has not been specifically asked for. I do not think it is. It is just within the reading frame of the large molecule.

D. LIVINGSTON

Well, it would be interesting.

E. ROOS

First of all, this was an absolutely beautiful talk Peter. What I would like to know is your thoughts about lymphoid metastasis. You noted that the expression was only seen in newborn mice, was that right? And also, was the activation that you see only observed in lymph nodes of new born mice, right?

P. HERRLICH

Yes. Activation could then be done with adult rats. Meanwhile that has also been done with human peripheral T cells, you can activate them *in vitro* with anti-CD3 and interleukin-2 and you get also high expression.

E. ROOS

But does this mean that activated T cells have the variants?

P. HERRLICH

That is right. Activated T cells have one specific variant type.

E. ROOS

And that is v_6 alone.

P. HERRLICH

Yes.

G. DOUGHERTY

I would like to ask a couple of quite specific questions. Firstly, I assume from your talk that you believe that antibodies directed against the v_6 domain inhibit the generation of immune responses *in vivo* by altering patterns of lymphocyte recirculation. Do you have any information on the effect of these antibodies on the generation of immune responses *in vitro*? Is it possible that they could also interfere with the interaction between accessory cells and T cells, or have an inhibitory effect on some later stage in the T cell activation process?

P. HERRLICH

This experiment is still lacking. We know that we can effectively turn CD44 variant expression on, and we have not yet looked for specific T cell functions *in vitro*.

G. DOUGHERTY

A second minor point -- when you were talking about dimerization, it was unclear whether the 90kD hemopoietic form of CD44 also dimerized or whether this effect was restricted to isoforms containing the v_6 domain.

P. HERRLICH

It could not be seen in that experiment because there was no antibody that would specifically see the small form. However, we could not find any heterodimers, which I find strange. There was an excess of standard type CD44 together with the variant, thus one would expect heterodimers.

G. DOUGHERTY

So the assumption is that you need exon 6 to get the dimerization.

P. HERRLICH

Somehow the interaction between the molecules determines whether they are dimerized. But you also should remember that it is a minority which is dimerizing, the large majority is not.

G. DOUGHERTY

It seems strange that both the hemopoietic and the larger form are both with the same cytoplasmic domain which you were showing as important, so that you are implying that there is some sort of action extracellular by which is then signalling into cells.

P. HERRLICH

That is the assumption, yes.

A. ANICHINI

If I understood correctly, your experiments have shown that antibody anti-CD44 can block CTL-mediated lysis, right?

P. HERRLICH

It can block the generation of CTL *in vivo*. But the effector function is not changed.

G. TARONE

You said that the keratinocytes express a variants containing the v_6 exon plus other exons too, right? But those cells do not, of course, diffuse in the circle. Does this imply that other exons have a sort of suppressor effect on this v_6 exon, or not?

P. HERRLICH

That is a suggestion. Take this please as a very preliminary suggestion. We have tried to see whether transfectants with a large keratinocyte-type clone metastasize. To this date, they have not.

G. TARONE

So in a metastatic model actually the other exons suppress the functions of the v_6, is that correct?

P. HERRLICH

Yes.

E. BROWN

I am fascinated by the observation that *ras* turns on the alternative splicing. So two questions about that, and one is just general -- can you tell us more? The second is, is there a specificity to the alternative splicing? Do you find specific forms or is it just a whole mixture of different variants which are found in your *ras*-transformed cells?

P. HERRLICH

That happens to be a specific form. The variants were checked by Northern analysis. There are four typical classes of RNA that have different polyadenylation sites. So the sizes are defined in these *ras*-transformants. But *ras* does not induce in all cells. *Ras*-2 cells do not become metastatic upon *ras*-transformation and they do not express the splice variant. It just happens to be in these CREF cells, and it is not very efficient. So I wonder what else do we need in addition to the *ras* condition.

E. BROWN

But presumably this could be turned into an assay to find that out.

P. HERRLICH

That is right. The most efficient place to turn on is in these T cell clones. You can get turn-on of the exon in close to 100% of the cells.

J. McCARTHY

How large is the protein, with v_6, v_6, or just v_6?

P. HERRLICH

Simply by adding up the amino acids, a v_{6ev} only CD44 variant should be 40kD. Each molecule can occur in several modified forms and the largest one is about 200kD in apparent molecular weight.

J. McCARTHY

That is with its glycosylations?

P. HERRLICH

With the glycosylations.

D. LIVINGSTON

What about the core protein?

P. HERRLICH

The largest core protein is about 80kD.

J. McCARTHY

Does it carry chondroitin sulphate or heparin sulphate?

P. HERRLICH

No, we could not detect chondroitin sulfate. We saw no digestion with chondroitinase. So all of the modifications that we could see were in N- and O-linked carbohydrates.

J. McCARTHY

Was it sulphated or you would need to know if it is modified as a proteoglycan.

P. HERRLICH

Yes, it has been tried. There was no sulphate incorporation.

R. BANKERT

When you inhibit the B cell response, you are looking at inhibition of antibody secreting cell. You do not know if you have inhibited or interrupted the activation of that cell or simply inhibited the secretion of the antibody. Also, was this a primary response that you inhibited? Have you looked at a subsequent immunization after you have blocked this primary response? In other words are you establishing a long-term tolerance? And have you looked at the ability to block a secondary response, which is almost impossible to do?

P. HERRLICH

The secondary response was not blocked, in my recollection.

R. BANKERT

Have you looked at a subsequent immunization?

P. HERRLICH

Subsequent immunization with the same antigen has been done for the CTL response, but has not been blocked.

R. BANKERT

In the B cell response where you are looking at plaques to TNP, and LPS, have you looked at a secondary response?

P. HERRLICH

No, I do not think we have done a secondary with that.

A. SANTONI

I did not catch when you inject the antibody with respect to the antigen, to see block of CTL *in vivo*. You have to inject the antibody and then the antigen, or you inject the antibody with the antigen, the yellow antigen.

P. HERRLICH

All these experiments, both for inhibition of metastasis and for interference with the immune reactions, were done by i.v. injection just subsequent to antigen or metastasizing cells respectively. We would then keep up the level of antibody for some period of time, in the metastasizing experiment for three weeks, in the antigen-stimulated case for a few days - four days.

A. SANTONI

And then you measured the response of the lymph node levels or the spleen? I mean what were the effector cells? Do you prevent lymph node enlargement with this antibody?

P. HERRLICH

Yes.

S. GOODMAN

Did I get the impression, correctly, that the only place where the 6-variant is expressed is in keratinocytes normally?

P. HERRLICH

Exon 6-carrying variants are expressed in activated cells, but other than that, one of the dominant tissues is keratinocytes.

S. GOODMAN

And where else?

P. HERRLICH

In the embryo in several places. In adults it is very hard to find. There are some, bone marrow, spleen, tissues where you find individual cells that look like monocytes that are positive.

S. GOODMAN

OK, but no other?

P. HERRLICH

No other dominant tissue. I forgot to tell you, recently we checked through brain that had been exempted from earlier experiments, and there is expression on astrocytes.

S. GOODMAN

Melanocytes?

P. HERRLICH

No, melanocytes are negative.

S. GOODMAN

In the gut epithelium, is there a transition, does the cell start dropping of the end of the gut villus?

P. HERRLICH

Yes, Steven Pals has done that, and I think he did not see much.

T. CAREY

I have a question for you Peter. I have seen that staining pattern before, with a different antibody, which recognizes the H type-2 blood group in keratinocytes. You could see staining of every little finger-like projection of the surface microvilli. It is a rather characteristic ladder-like pattern between the keratinocytes, and it would be curious if the carbohydrate that is put on this molecule in the skin is the H type-2 blood group.

G. DOUGHERTY

This is more of a comment to Simon Goodman's question, we have found that normal epithelial cells express largely the hemopoietic form of CD44 but that when these cells are activated or transformed various other isoforms come to predominate particularly an isoform containing exons V_8, V_9, and V_{10} that has been designated CD44E or CD44R1. The functional importance of the inclusion of these additional exons remains, at the moment, quite unclear.

T. CAREY

The reason I mentioned the H type 2 antigen is that this molecule is also very strongly over-expressed in squamous cell carcinomas. And the final point that I was going to make is that clearly your molecule is lateral and apical only, in the keratinocytes. There is no basal stain that I could see at all. Which is another nice marker for keratinocyte people. And I wonder if you have looked at this on keratinocytes *in vitro*?

P. HERRLICH

Keratinocyte cell lines express. That is where we cloned it from, that is a large variant. The cell lines are HPV-transformed keratinocytes, and the HaCat cell line of Dr. Fusenig.

T. CAREY

That is a skin-derived cell line.

P. HERRLICH

Yes, skin. Both have a large variant, essentially the same large variant.

V. QUARANTA

Is this present in the whole thickness of the epithelium?

P. HERRLICH

Yes, throughout all the layers. We tried to find the epitope that is present on the standard type CD44 in keratinocytes, and could not find it. So it seems that the structure of the molecule is quite different. It shields the epitope that we usually use for the standard type.

V. QUARANTA

OK. So according to what Tom said, it would be absent from the basal surface of the germinal layer.

P. HERRLICH

That is what Dr. Carey derived from these slides, and they are very nice ones (made by Dr. Pals).

T. CAREY

That was pretty clear in your slides.

D. LIVINGSTON

Peter, have you had a chance to do an assay for dominance or recessiveness of the variant guys? Who wins? If you express the variant or over-express the wild-type in the presence of the variant, do you see any suppression or vice versa?

P. HERRLICH

The small standard type version is recessive.

S. SHAW

Have you looked at the cells which bear the v_6 form of CD44 in terms of whether the anti-v_6 mAb affects some of the functions inferred to involved CD44? For example, homotypic aggregation or E-rosetting?

P. HERRLICH

We have some preliminary work with Fiona Watt. Fiona received our transfected clones, and then tried to see specific binding to keratinocytes, because they express the same v_6 exon. And there is some homophilic interreaction. It was not easily blockable by antibodies. So we do not know what it means. Grossly, in culture, the transfectants do not aggregate, so it is not that these cells have an obvious cell-cell interaction.

S. SHAW

Did I understand that, as an aside in the previous answer, you implied that CD44 sticks to a variety of things?

P. HERRLICH

We do not really know what it interreacts with. It has many domains so it may interact with hyaluronic acid or it has been suggested that the chondroitin sulphated form is a collagen binder. I do not know. We have only tested hyaluronic acid and the variants do not do that, while the standard form that we have also in a soluble form, does bind. But the results are at variance with those in Graeme's laboratory, I think.

G. DOUGHERTY

We have not been looking at exactly the same isoform as Peter, but it does seem from our studies that CD44Rl/CD44E, an isoform that contains exons 8, 9 and 10, is able to bind very efficiently to hyaluronan when transfected into simian or murine fibroblasts. These findings contradict studies by Brian Seeds' group which demonstrated that Namalwa cells transfected with CD44E are unable to bind hyaluronan. One possibility is that the adhesive function of the various CD44 isoforms may be differentially regulated in different cell types.

G. NICOLSON

The suggestion that cells that are either transfected or contain, say the v6 domain, one with perhaps some other domains, might have a phenotype similar to activated T cells, suggests that there may be a number of gene products that might be turned on in their expression either by the interaction of this molecule with ligands in the environment, or by cell-cell or cell-matrix interactions. Some of these molecules are not that difficult to measure, for example, degradative enzymes. Activated extravasating white cells express a variety of different degradative enzymes that allow them to penetrate the basement membrane as well as a number of other adhesion components. Some of these are cloned and available. Is there any evidence that the expression of a v_6-containing CD44 leads to a turning on of genes such as those that encode degradative enzymes such as collagenases, heparinase, others.

P. HERRLICH

There is a set of experiments where we compared various clones that carry the v_6 CD44 on the surface and those that do not, for other molecules. We found no difference in NM23 expression; all had high levels. We found no difference in cadherin expression, but we found that all the variant expressors, so far, have higher levels of uPA and stromelysine. It is not modulated however by antibody, and so we cannot really relate it to the phenotype *in vivo*, because we would expect at the end that one of these critical functions is knocked out by the antibody.

G. NICOLSON

It may be that one of the critical functions is knocked out. We feel that metastasis is really a cascade involving a number of different properties. If you block one, then that might be sufficient to stop the whole process dead in its tracks.

N. HOGG

Have you searched these protein data banks looking to see whether any of these domains contain motifs that are seen in other sorts of proteins? That may give some ideas about what sorts of functions they might perform.

P. HERRLICH

That was our wishful thinking as well. My eye did not see anything, though. We went over to Chris Sander at the EMBL, and he had a look through the data bases, could not find any motifs that were sticking out to his eye. This does not preclude us or preclude anybody else from finding something interesting in time. I would be happy.

N. HOGG

Secondly, there have been two other studies suggesting that just over-expression of the basic hematopoietic form confers metastatic status on tumor cells. So I just wonder how you interpret those studies?

P. HERRLICH

That is Ian Hart's CD44 expression, and that done by Ivan Stamencovic. These were expressions in melanoma or in lymphoma of the standard type. The standard type CD44 promoted lung colonization in the tail vein assay. You can also sort melanoma cells for low expression. And that is what Ian Hart did, for low expression of a standard CD44, and the lung colonization goes down as well. So there are two types of evidence that suggest that lung colonization is promoted by the small form of CD44.

C. FIGDOR

You said that the expression of the v_6 form on leukocytes is transient when they are activated. Can you say something about this transiency? How long is it expressed? And, have you also looked to granulocytes when they are activated, for instance, by chemotactic factors or PMA?

P. HERRLICH

That has not been tested. The transient nature was seen on this slide. We did not wait for a zero level, but it was in the order of five days, then it was way down. The half decay was about two days.

C. FIDGOR

But, for instance, if you stimulate a T cell *in vitro* with CD3, do you see that expression is going up and down again and in what time scale?

P. HERRLICH

I do not think that we have this kinetic.

C. FIDGOR

And you did not look at the granulocytes?

P. HERRLICH

Not yet.